Finite Element Analysis Applications

Finite Element Analysis Applications
A Systematic and Practical Approach

Zhuming Bi
Purdue University Fort Wayne,
Indiana, United States

ACADEMIC PRESS
An imprint of Elsevier

Academic Press is an imprint of Elsevier
125 London Wall, London EC2Y 5AS, United Kingdom
525 B Street, Suite 1800, San Diego, CA 92101-4495, United States
50 Hampshire Street, 5th Floor, Cambridge, MA 02139, United States
The Boulevard, Langford Lane, Kidlington, Oxford OX5 1GB, United Kingdom

Library of Congress Cataloging-in-Publication Data
A catalog record for this book is available from the Library of Congress

British Library Cataloguing-in-Publication Data
A catalogue record for this book is available from the British Library

ISBN: 978-0-12-809952-0

For information on all Academic Press publications visit our
website at https://www.elsevier.com/books-and-journals

Working together
to grow libraries in
developing countries

www.elsevier.com • www.bookaid.org

Publisher: Katey Birtcher
Acquisition Editor: Steve Merken
Editorial Project Manager: Peter Jardim
Production Project Manager: Mohanapriyan Rajendran
Designer: Christian J. Bilbow

Typeset by TNQ Books and Journals

Contents

Preface

Modern products and systems are becoming more and more complicated: these include designs of complex products and systems involved in the analysis of physical behaviors in multiple disciplines. Although a design engineer cannot have refined knowledge in solving engineering problems in relation to all design aspects, it is desirable to have some generic analysis tools, which are capable of coping with the variety and complexity of engineering problems. Since its birth in the1960s, finite element analysis (FEA) has proven itself as a versatile tool for solving various design problems. The FEA theory and its applications have been introduced universally as a core course in the majority of engineering curricula. Students are required to understand FEA fundamentals and master the skills of using commercial software tools to obtain design solutions appropriately. However, course designs and resources related to FEA are so diversified, many educators face the difficulty in choosing suitable textbooks and tools to serve their teaching or self-learning purposes. The author has not found suitable ones which are concise but comprehensive enough to cover all of these aspects for beginners. This book is proposed to fill in this blank.

FEA is an advanced design tool that requires users to formulate their design problems into appropriate computer models; the commonalities of FEA teaching should be *FEA fundamental*, *computer implementation*, and *skills of using commercial tools*. For most engineers and college students who treat FEA as a design tool rather than research subject, ideal textbooks should cover all three subjects so that they can achieve their learning outcomes in a short period of time. This book is written to meet this need with four distinguishing features. (1) The axiomatic design theory is introduced as the systematic approach in dealing with the variety and complexity throughout the book. It aligns with the philosophy of *divide and conquer* of FEA seamlessly, and it fills the theoretical gap between the FEA theory and computation implementation. (2) The object-oriented approach is adopted in the computer implementation. FEA code can be written in such a way that its functional modules can be naturally linked to two critical procedures in FEA, i.e., top-down decomposition and bottom-up system modeling. (3) The selection of course materials is application oriented, and many examples and case studies are included as the practical guidance for readers to define legitimate finite element models using commercial software tools. (4) The FEA fundamentals, computer implementation, and skills for using commercial tools are covered; students can learn both the theory and skills in one course setting; in particular, two advanced topics on multiphysics applications and verifications and validations are covered. To assist educators in adopting this book as teaching materials, it is packaged with the solution manual, lecture slides, and object-oriented code for computer implementation in Matlab.

This author believes that the presented book is aligned well with the underlying philosophy of FEA methods. FEA copes with the variety and complexity of design problems by developing the solutions to partial differential equations (PDEs), and an FEA tool becomes generic and versatile because any engineering system can be formulated as mathematic models with PDEs. The book is *comprehensive* because it covers all of the critical steps systematically, especially the steps for the conversion of a design problem for an FEA model as well as verification and validation of simulation. The book is *concise* because the focus of writing is for a better understanding of FEA methods. Students are trained to provide appropriate inputs and specify simulation settings in FEA models with the minimal knowledge of computer implementation. The book is *application oriented* because the primary objective of the book is to teach students the essentials to use FEA tools rather than to advance FEA methodologies.

The book is written as an introductory course for undergraduates and engineers who need to gain basic understanding on theory and skills in using FEA tools appropriately. Minimal prerequisites in engineering, mathematics, and computer programming are needed. Students are able to learn all of the essentials of FEA in one course. However, these essentials are not sufficient for graduates and researchers who will conduct the academic research in the fields of FEA methodologies and numerical simulation.

Zhuming Bi
Professor of Mechanical Engineering
Department of Civil and Mechanical Engineering
Purdue University Fort Wayne
Fort Wayne, Indiana, United States

Acknowledgment

The author would like to express his great gratitude and appreciation to

- Editor *Steve Marken, Peter Jardim* and *Anita Mercy M. Vethankkan* at Elsevier Inc., for their guidance and supervision in the course of book writing; without their great support, this book would have never been completed.
- His colleague Professor *Donald W. Mueller* for constant encouragement at workplace and collaborative contribution to multiple projects, which have been used as design examples in the book.
- Reviewers of book proposal and chapters: Professor *Li Da Xu* at Old Domino University, Professor *Lihui Wang* at KTH Royal Institute of Technology, Professor *Jun Wang* at the University of New South Wales, Professor *Jian Wang* at Kinstong University, Professor *Yongmin Zhong* at RMIT University, Professor *Puren Ouyang* at Ryerson University, and a few of anonymous ones for invaluable feedback to improve the quality of manuscript.
- Professor *Xiaoqin Wang* from Nanjing University of Science and Technology for her constructive suggestions and corrections in finalizing this manuscript during her visit to Purdue University Fort Wayne.
- *His students* at Purdue University Fort Wayne in ME-545 Advanced Finite Element Analysis course for proofreading of draft chapters and language corrections.
- His family members, *Rongrong Wu, Chenghao Bi*, and *Chengyu Bi*, for their full understanding and cooperation in the completion of this book project.

This book proposal is supported partially by the Program of Foshan Innovation Team of Science and Technology (Grant No. 2015IT100072).

Zhuming Bi
Professor of Mechanical Engineering
Department of Civil and Mechanical Engineering
Purdue University, Fort Wayne
Fort Wayne, Indiana, United States

Overview of Finite Element Analysis

As design engineers, we have experienced the dramatic change of available techniques that were developed and used to design, model, manufacture products, and implement manufacturing processes. One of the greatest changes since the 1950s has been the broad application of finite element analysis (FEA). Nowadays, FEA becomes a default analysis tool in the design of products and systems. However, using commercial FEA tool does not guarantee that appropriate solutions can be obtained without an adequate level of the understanding on the FEA theory. FEA is a type of numerical computation to approximate the solution to a mathematic model. It is critical to model design problems adequately, so that the approximated solution can be converged (Knight, 1994). To appreciate the importance of FEA, *first*, the general engineering design process is presented to identify typical activities involved in solving an engineering problem. *Second*, the trend of product or system design is discussed to illustrate some emerging challenges in carrying out the identified activities. *Third*, FEA is introduced to elaborate its importance in helping designers to perform their design activities. *Finally*, the organization of the book is presented with the overviews of individual chapters.

1.1 ENGINEERING DESIGN PROCESS

An engineering process can be viewed as the conversion from a set of customers' needs, requirements, and intentions to a product or system with some physical existence and an intended use (Dorst, 2016). It consists of some essential steps that a designer follows to obtain the solution to a formulated problem. An engineering solution can be as simple as a gear set for the given transmission ratio or a computer program for a motor control. It can be as complex as an aircraft or spacecraft designed to transport people from one place to another. However, despite the variety of design processes, engineers follow similar steps in seeking and evaluating design candidates for engineering optimization. Fig. 1.1 illustrates a set of common steps and activities

Finite Element Analysis Applications. http://dx.doi.org/10.1016/B978-0-12-809952-0.00001-7

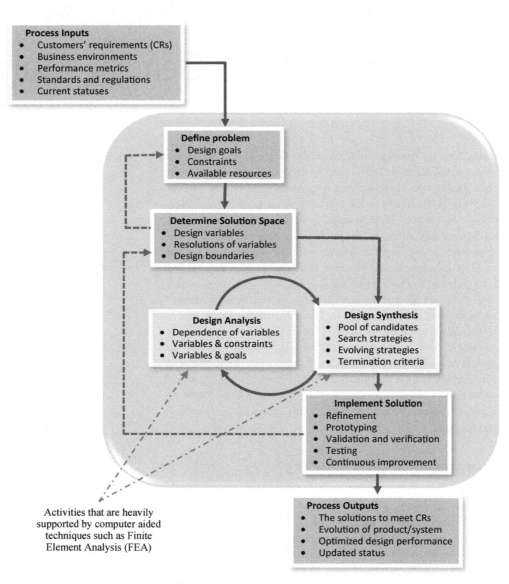

Process Inputs
- Customers' requirements (CRs)
- Business environments
- Performance metrics
- Standards and regulations
- Current statuses

Define problem
- Design goals
- Constraints
- Available resources

Determine Solution Space
- Design variables
- Resolutions of variables
- Design boundaries

Design Analysis
- Dependence of variables
- Variables & constraints
- Variables & goals

Design Synthesis
- Pool of candidates
- Search strategies
- Evolving strategies
- Termination criteria

Implement Solution
- Refinement
- Prototyping
- Validation and verification
- Testing
- Continuous improvement

Process Outputs
- The solutions to meet CRs
- Evolution of product/system
- Optimized design performance
- Updated status

Activities that are heavily
supported by computer aided
techniques such as Finite
Element Analysis (FEA)

■ **FIGURE 1.1** Steps and activities in the design process.

in the solving process of an engineering problem. Because a design process aims to find the solution with the satisfied customers' requirements (CRs), the design process is generally initialized and driven by CRs. The following typical steps are involved in a design process: "define design problems," "determine a solution space," "design analysis," "design synthesis," and "implement solution."

1.1.1 **Define design problem**

An engineering design process is to solve a design problem. One cannot solve the problem unless its design requirements are clearly defined. Therefore, one needs to initialize an engineering design process by formulating a clear, unambiguous design problem. A design problem is driven by the customer's need. Therefore, one has to gather much information relating to business goals, environments, and available resources. Once the need has been identified, the next step is to convert the need into a design problem statement. A problem can be formulated to clarify the answers to "5W" questions, i.e., Who has the need? What are the constraints of the need? When does this need occur? Where does this need occur? Why is it necessary to address the need?

Example 1.1

A railway manufacturer plans to produce new rails with a higher grade. According to relevant standards and codes, destructive bending tests must be performed on randomly selected products. Fig. 1.2 shows a test

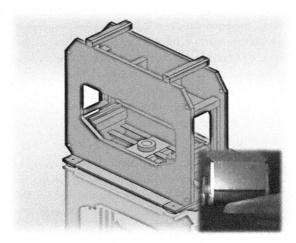

■ **FIGURE 1.2** Rails bend test machine.

Table 1.1 Example of Defining an Engineering Problem

Description	Testing of New Rails
Who has the need?	Rails manufacturer
Why is it necessary to address the need?	Destructive bend tests must be carried out to meet new product standards
When does this need occur?	Existing test machine is old-fashioned, which works only for regular rails with lower strengths. The need occurs when new rails are designed and manufactured
Where does this need occur?	The need occurs in the production line where destructive bend tests are carried out on rails randomly
What are the constraints of the need?	The company does not plan to buy a brand-new test machine with required capacities

machine where a rail sample is applied by a gradually increased load until it fails by fracture. However, the available test machine at the company is applicable only for the rails with a low grade. The price of a new test machine for the enhanced capacity is above a quart of millions, which is far beyond the company's budget. This design problem can be defined as Table 1.1, which includes the answers to "5W" questions.

1.1.2 Define solution space

A solution space is composed of all of design candidates that have maybe possible to meet expected functionalities subjected to given constraints. Once an engineering problem is formulated, designers will determine applicable techniques, components, products, and systems, which will be considered as potential solutions to meet given CRs. However, these identified solutions give the boundaries of a design space; they are conceptual and abstractive. A solution space has to be defined quantifiably by specifying the types and number of design variables, the ranges and resolutions of changes of variables, and applied constraints.

Due to the complexity of modern products or systems, a solution space at high level has to be decomposed into lower levels, so that the system structure is detailed enough to identify corresponding a design variable (DV) for each functional requirement (FR). Some system design approaches, such as Axiomatic Design Theory (Suh, 2005), can be applied to decompose FRs and DVs of an engineering problem. Fig. 1.3 gives an example that the

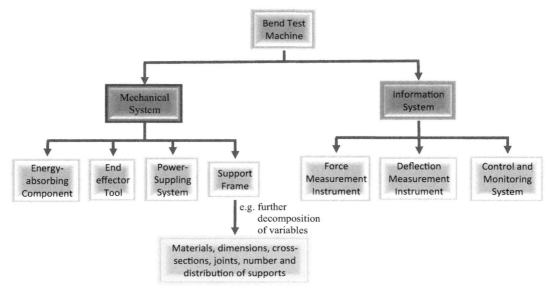

■ **FIGURE 1.3** Decomposition of design variables and solution space.

FRs of a bend test machine are decomposed in a hierarchical structure to identify the corresponding DVs in the solution space. To fulfill the destructive bend test in Example 1.1, the test machine is decomposed into *mechanical subsystem* and *information subsystem* at the highest level, and each subsystem is further decomposed at the next level until the design variables of individual components are defined clearly for their types, varying ranges, and constraints. Taking an example of the system component "support frame," it has been decomposed into a set of design variables, including "materials," "topology," "dimensions," "cross-section," and "distribution of supports." In such cases, the design subspace of each variable can be specified. Note that a design variable can be either "discrete" or "continual." For example, design variables for "dimension" are continual, and those for others are discrete.

1.1.3 **Design synthesis**

Depending on the types and number of design variables, a solution space may include a large or even infinite number of design candidates when some design variables are continual or the number of discrete design variables is large. To find the best solution for a given problem, it is impractical to evaluate and compare the performance of all of possible design candidates. Seeking the best solution from a given solution space can be

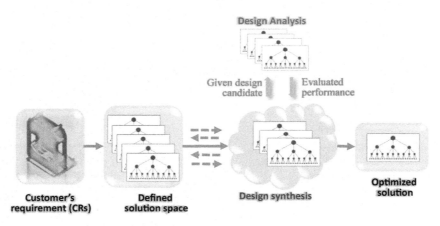

■ **FIGURE 1.4** Design analysis and synthesis in an engineering design process.

formulated as an optimization problem. As shown in Fig. 1.4, design synthesis serves as multiple functions such as to (1) select an initial pool of solutions, (2) determine searching strategies, (3) specify the ways to refine solutions, and (4) define terminating conditions.

For an uncoupled design problem where a subset of DVs can fulfill the corresponding FR independently, design synthesis over a subsolution space can be carried out. Taking an example of the design in Example 1.1, the mechanical subsystem and information subsystem are loosely coupled, and design syntheses for the solutions of these two sun systems can be carried out separately. On the other hand, all of the design variables in the mechanical subsystem contribute to the strengths and deflections of the entire test machine. To optimize the mechanical subsystem, these design variables have to be taken into consideration simultaneously.

1.1.4 **Design analysis**

As shown in Fig. 1.4, one of the critical tasks in the design synthesis is to evaluate the performance for selected design candidates. This is carried out by *design analysis*. Design analysis tells whether or not a design candidate meets all of constraints and how well it works in terms of the given design criteria. The evaluation results from design analysis are used by *design synthesis* to select better solutions. To evaluate a design candidate, the coupling of design variables, and the relational models of design variables with FRs, constraints, and design goals must be established. Taking the design of the bend test machine as an example, a simulation model must be developed to calculate the stress distribution over a frame and determine whether or not it has sufficient strength to carry the required bend load.

1.1.5 **Implementation of solution**

Because a large number of design candidates have to be analyzed and compared to determine the best solution. It is ineffective to take into account all design details at the conceptual design phase. These design details can be refined and finalized when one optimizes the design solution. One example of these types of design details is the assemblies of system components in the test machine. For example, two metal parts can be welded or fastened and different numbers of fasters can be applied. Different assemblies affect the integrity of the test machine locally. When the conceptual solution of the test machine is defined, the details of these assemblies have to be further investigated. In some extreme cases, the impact of technical details at low levels is so significant that the obtained conceptual solution has to be revised; the precedent steps have to be repeated iteratively. In addition, the step of implementing solution is a conversion from a virtual model to physical product or system. Other typical activities in the implementation include "prototyping", "verification and validation", "testing", and "continuous improvement".

1.2 **COMPUTERS IN ENGINEERING DESIGN**

Modern products are becoming more and more complicated. However, it is interesting to see that although the basic steps of engineering design processes are similar for different products, the techniques and tools that are applied to accomplish design activities have been changed significantly. Computer aided design (CAD) and computer aided engineering (CAE) have been widely applied in the practices of engineering design. In this section, engineering designs are classified in terms of the level of creativity, designers' roles are compared, and the impact of computer aided techniques on design creativities and effectiveness is discussed.

Designing a complex system or product can be decomposed into a level where each component can be designed individually to fulfill its corresponding FR. As shown in Fig. 1.5, engineering design activities can be classified into six types; Table 1.2 (DesignWorks, 2016).

The level of creativity can be characterized based on the types of *solution space* and *design variables*. Routine design, innovative design, and creative design correspond to different combination of solution space and design variables as shown in Table 1.3 (Goel, 1997).

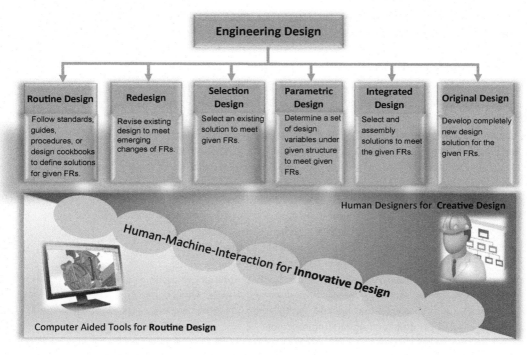

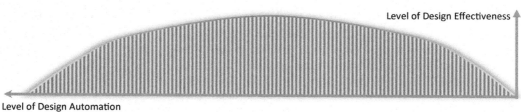

■ **FIGURE 1.5** Human designers and computers in engineering design.

The characteristics of the solution space and design variables determine if an engineering design is "routine," "innovative," or "creative." Both humans and computers can compete to accomplish some design activities; however, computers and human designers are good at different things, and it is desirable to synergize the strengths of designers and computers to achieve the effectiveness of engineering designs. Many researchers have discussed the differences of human and computers (Chakrabarti et al., 2011; Kim and Rezaei, 2008; Seo et al., 2003; Johansson, 2011; Vanderplaats, 2006). Table 1.4 summarizes the role differences of human being and computers in engineering design processes.

Table 1.2 Types of Design Activities

Type	Description	Example
Routine design	To perform design by following existing standards and codes that outline the steps and computations for certain products or systems.	Follow American Society of Mechanical Engineers (ASME) boiler and pressure vessel codes to design a pressure vessel.
Redesign	To revise an existing design when functional requirements (FRs) have been changed in dynamic environment.	Reprogram a robot when the tag points on the motion trajectory are changed.
Selection design	To find a solution by selecting appropriate components from an existing design inventory.	Select standardized fasteners to join two metal plates.
Parametric design	To determine design variables in a given conceptual structure for an optimized performance.	Minimize the materials usage of a cylindrical container subject to the given volume.
Integrated design	To design and assemble components as an integrated products or systems meet strongly coupled FRs.	Design a robotic configuration for a given task in a modular robot system.
Original design	To design a product or system from scratch to meet emerging FRs.	Design a patentable product or system.

Table 1.3 Characteristics of Solution Spaces and Design Variables

Level of Creativity		Routine Design	Innovative Design	Creative Design
Solution Space	Structure	Known	Known	Unknown
	Search procedure	Known	Unknown	Unknown
Design Variables	Types	Fixed	Fixed	Changed
	Ranges	Fixed	Changed	Changed

Table 1.4 Comparison of Human Designers and Computers

	Human Designers	Computers
Strengths	■ Identifying design needs ■ Brainstorming to think solutions "out of box" ■ Engineering intuition and big knowledge base ■ Selecting design variations ■ The flexibility to deal with changes ■ Qualitative reasoning ■ Psychologically, human decision is more trusted than artificial intelligence ■ Predict trends, patterns, or anomalies, and ■ Learn from experience	■ Fast speed, reliable, endurance, and consistent ■ Capable of exploring a large number of options ■ Carry out long, complex and laborious calculations ■ Store and efficiently search large databases ■ Provide information on design methodologies, heuristic data and stored expertise
Weaknesses	■ Easily tired and bored ■ Cannot do micromanage ■ Biased and inconsistent ■ Prone to make errors ■ Not good at quantified reasoning ■ Incapable of utilizing the data presented in awkward manner	■ Difficult to synthesize new rules ■ Limited knowledge base ■ No common sense

It is critical to take advantage of modern computing technologies to improve the effectiveness of engineering designs. The limitations of human designers can be addressed by computer programs. Design synthesis of a modular robotic system (Leger, 1999; Bi and Zhang, 2000; Bi et al., 2006, 2010) is a perfect example where computer aided tools are utilized to solve engineering design problems. A modular robotic system consists of various modules, which can be selected and used in different robot configurations. A solution space of robotic candidates can be defined by selecting different types and number of modules and assembling them in different ways for various tasks. Fig. 1.6 shows application scenarios of a modular robotic system. To configure a robot optimally for a specific task, a large number of design alternatives are evaluated and compared. It is common that a design synthesis needs to evaluate 10^4-10^5 design candidates before a feasible or optimal solution can be found. Such amount of design loads is

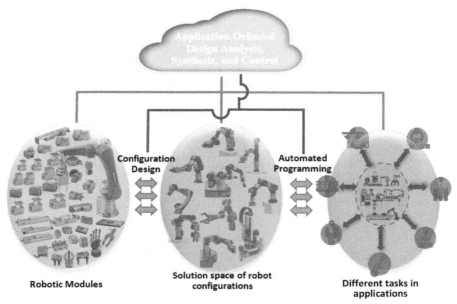

■ FIGURE 1.6 Example of computer aided synthesis (modular robotic system).

clearly impractical to human designers. Instead, CAD tool is able to search the solution space thoroughly without some aids from human designers. In addition, design analysis can be automated in evaluating a large number of design solutions without manual intervention. In such a way, human designers can focus on the creativity for the appropriate definitions of design problems and solution spaces.

1.3 COMPUTER AIDED ENGINEERING

Human designers are able to perform design analysis and synthesis only for simple objects with ideal application scenarios. The majority of textbook problems fall into this category. These problems are used by instructors to illustrate physical principles, mathematic models, and solving processes. However, the geometric shapes of real-world products or systems are generally too complex for manual solutions.

Taking an example of conventional machine design, students are taught to calculate physical quantities such as stress distribution, deflection, and fatigue life using theoretical and empirical equations when the dimensions and loading conditions of machine elements are given. Even though the

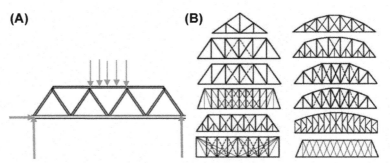

(A) **(B)**

■ **FIGURE 1.7** Computer aided engineering (CAE) tool to overcome limitations of human designers. (A) Example of textbook problem manageable by hand calculation; (B) CAE for evaluations of numerous design alternatives.

procedures of manual calculations are systematic and straightforward, it is very tedious and prone to erroneous for designers. Most importantly, it becomes impractical when the part geometry becomes complex, and/or a large number of design candidates in the solution space are evaluated and compared. Fig. 1.7 gives an example in evaluating the loading capacities of the constructions in the bridge building contest (Engineering Encounters, 2016). CAE becomes ideal tools when a large number of constructions are evaluated to determine the best solution subjected to given constraints. As a summary, integrated computer aided techniques are essential parts of engineering designs for modern products and systems.

CAE refers to the techniques of using computers and information technologies (ITs) in the design analysis and synthesis. CAE tools are expected to empower designers to solve more challenging open-ended and/or integrated real-life design problems (Kim and Rezaei, 2008). CAE is a kind of artificial intelligence for the automation of intelligent behaviors. However, creative thinking is for human designers, that is, beyond the scope of design automation. Design automation involves the automation of routine works to free human designers (Johansson, 2011). Siemens (2016) have summarized the main advantages of CAE as follows.

■ Design candidates can be evaluated and refined using computer simulation rather than physical prototyping to save money and time.
■ CAE can provide performance insights in an earlier development process when design changes are less expensive to make.
■ CAE helps engineering teams manage risk and understand the performance implications of designs.

- Integrated CAE data and process management extend the ability to effectively leverage performance insights and improve designs to a broader community.
- Warranty exposure is reduced by identifying and eliminating potential problems. When properly integrated into product and manufacturing development, CAE can enable earlier problem resolutions, which can dramatically reduce the costs associated with the product life cycle.

1.4   FINITE ELEMENT ANALYSIS

FEA is one of the most important techniques among different CAE tools. This section gives a brief overview of FEA.

1.4.1   Classification of problem solving tools

Human beings are defined by their tool making, designing, and engineering skills, and the socialization and communication to make inventions, innovations, and technology transformation. Engineering relates to the development, acquisition, and application of technical, scientific, and mathematical knowledge for understanding, design, development, and use of materials, machines, structures, systems, and processes to meet human being's needs (Morjoram and Zhong, 2003). Therefore, engineering is all about solving problems using math, science, and technical knowledge.

After an engineering problem is formulated as a mathematical model, *graphic methods*, *experimental methods*, or *computational methods* can be used to solve the problem. Before computers were invented, graphic methods used to be very important. However, they can only tackle with some simple problems; graphic design and calculation can be made by basic mechanical instruments such as slide rules (Chapra and Canale, 2010). Experimental methods are still very popular, and they are the most reliable ways to solve engineering problems. However, some obvious disadvantages of experimental methods are (1) experiments have to be performed on physical systems, which are not always available at any design stage; (2) experiments involve high cost, because additional designs and implementations are required for measurement and instrumentation; it wastes time to prepare for and conduct experiments; (3) an engineering system usually involves a number of system inputs; there is a great deal of uncertainty about the conditions of system operations. It is unrealistic or impractical to predict the system performance at any circumstance. A computational method can be *analytical* or *numerical*. It solves an engineering problem by computers. As shown in Fig. 1.8, an analytical method refers to the case when an analytical solution exists and the designer is able to obtain the solution directly.

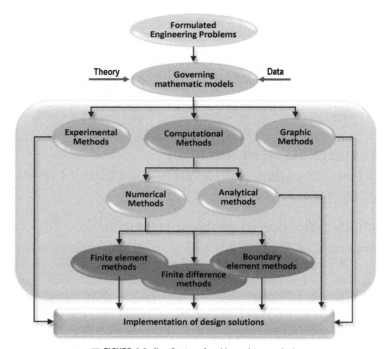

■ **FIGURE 1.8** Classification of problem-solving methods.

This applies only in simple problems occurring to objects with basic geometry. Numerical methods are by far the most widely used computational tools in engineering practices. In comparison with experimental or graphic methods, computational methods are becoming more and more important due to a number of reasons. (1) Most of the modern engineering products are so complicated that it becomes impossible or cost-forbidden to evaluate products or components thoroughly by experimental methods; (2) A computational method usually finds an answer to an engineering problem effectively. Before prototyping a physical system, the computational tool is capable of predicting system performance for a number of application scenarios. It can shorten the cycle time of product development greatly. (3) Modern CAD and CAE tools are very powerful and popular; engineers are able to use these tools by their engineering intuitions to solve a wide scope of engineering problems without sophisticated trainings.

As shown in Fig. 1.8, numerical methods include a number of varieties such as *finite element methods* (FEA), *finite difference methods*, and *boundary element methods*. The similarities of these methods are twofold: (1) they are generalized tools for various engineering problems; (2) they all use the *divide and conquer* strategy to deal with the variety and complexity of engineering problems. However, an FEA method differs from a finite difference method in the approximation of derivatives in a mathematical model. Derivatives are evaluated by the integration in FEA; although these derivatives are evaluated by finite differences in a finite difference method. An FEA method differs from a boundary element method in the discretization of a continuous domain. An FEA model has elements and nodes in the entire domain, although a boundary element method has elements and nodes only on the boundaries of domain.

1.4.2 **Top-down and bottom-up approaches**

The superior capability of solving diversified problems for complex geometries is one of the most significant attributes of FEA. FEA employs the piecewise approximation where a continuous domain of interest is divided into discrete subregions called *finite elements*. Each element is represented by a set of nodes; the behaviors of elements are represented by those of nodes, which are governed by one or a set of partial differential equations (PDEs) in respective physical disciplines. The behaviors of elements are represented by element models, and a system model can then be assembled from element models. By incorporating boundary and load conditions, the system model can be solved to obtain a numerical solution to original engineering problem.

The underlying idea of FEA is module-based design. In a module-based design, a complex problem can be treated as a set of simple subproblems, which can be solved easily; the solution to an original problem can be found by assembling subsolutions with the consideration of constraints of assemblies (Bi and Zhang, 2000). Module-based design has been widely applied in object-oriented design (Booch, 1994), axiomatic design theory (Suh, 2001), and the design of modularized systems (Bi et al., 2008a,b).

The implementation of a module-based design usually consists of two procedures, i.e., top-down approach for the decomposition and bottom-up approach for assembling and problem solving. Fig. 1.9 is modified from the well-known verification and validation (V&V) diagram (US Department of Transportation, 2009; Suh, 2001) to illustrate the top-down and bottom-up approaches in FEA modeling. The engineering design of a vehicle is used as an example of the decomposition in the illustration.

A top-down approach is applied to break down the complexity of system. *First*, the boundaries of the system are modeled to determine system

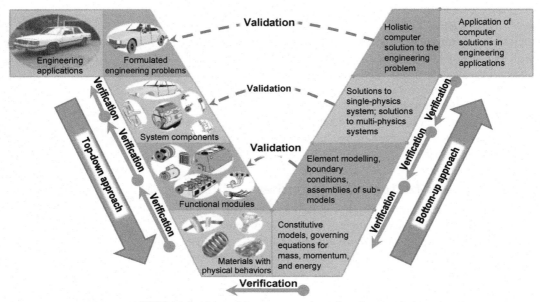

■ **FIGURE 1.9** Top-down and bottom-up approaches in finite element analysis.

parameters, design variables, constraints, and objectives. *Second*, the system is decomposed into components such as engine, chassis, exhauster, transmission, and body and trims of vehicle. *Third*, multiple layers of decompositions can be performed, so that the function of each module and its relations with others can be explicitly defined. *Finally*, the behaviors of the materials in part or components are analyzed to identify the physical phenomena, which have been studied and described by mathematical equations, such as the stress–strain relation in mechanical elements, mass, momentum, and energy conservations for heat transfer elements or fluid flow.

A bottom-up approach is to obtain the system solution based on the subsolutions to elements, parts, and components. *First*, element types and analysis types are selected to ensure that governing equations can be specified to describe physical phenomena of subdomains adequately. *Second*, the mathematic models with governing equations are solvable in order to define submodels of elements. *Third*, the submodels are assembled into a system model based on the mating relations of physical objects. *Finally*, the boundary conditions and loads are imposed, and the system model is solved to obtain the solution to the physical system. Finally, because an FEA tool is a type of software product, V&V must be performed throughout the modeling process to ensure the fidelity of FEA results.

1.4.3 **FEA—brief history**

FEA theories and methods were originated from the need to solve complex elasticity and structural analysis problems in engineering. Its origin can be traced back to the work (Hrennikoff, 1941; Courant, 1943). Hrennikoff (1941) discretized a continuous domain by using a lattice analogy. Courant (1943) calculated the torsional deflection of a hollow shaft by dividing a cross-section into triangles and used piecewise shape functions for the interpolation. It is worthy to note that earlier results by some mathematicians including Rayleigh, Ritz, and Galerkin for solving PDEs contributed significantly to the theory of finite element methods. Even though the published approaches at that period of time were not systematic and vary different from one to another, these works shared an essential attribute of FEA, i.e., a continuous domain was discretized into a mesh, and the system model was from an assembly of element submodels. Clough (1960) coined the concept of FEA; he was motivated to use FEA to analyze the distribution of plane stress over the wings on aircraft. Zienkiewicz and Cheung (1967) published the first book in the field of FEA.

The first patent related to FEA was filed for a number of sorting systems by the Applied Data Research, and the first prototype of software was developed in 1965. It was the origin of the commercial software NASA Structural Analysis (NASTRAN) (Multiphysics, 2014). Commercial FEA codes (e.g., Abaqus and Ansys) at their first generations were developed in the 1970s. However, FEA on a desktop computer was infeasible with the first (8086) Personal Computers (PCs), owing to excessive processing time and the constraints of limited memory and disk capacity. FEA was only made possible by the advance of the electronic, digital, computer, which at the time was making its entry into the field of large arithmetic processing (Zienkiewicz, 1995).

Until recent decades, the development of computer hardware and software brought desktop FEA within the range of reach even at the smallest companies (Woods, 2003). Nowadays, FEA has been widely used to find approximate solutions of a broad scope of applications such as stress distributions, fluid flow, heat transfer, and fluid structural interfaces.

1.4.4 **FEA applications**

With the maturity of commercial FEA codes and rapidly development of IT, FEA software codes tend to be essential tools to support product designs and developments. As long as computer hardware is powerful enough, FEA is capable of generating detailed engineering data relating to state variables and associate scalars about structure or system with any level of complexity. An FEA tool can run a numerical simulation on any design concept and predict behaviors and responses of systems in any imaginable operation conditions. It allows designers to refine design concepts prior to physical prototyping. FEA on computer models helps to identify design flaws and predict the performance of new designs cost-effectively.

For whatever types of engineering problems, an FEA package can be used as the main or assistive design tool to solve design problems. Fig. 1.10 shows some examples where FEA has been conducted to solve various problems in engineering designs.

1. **Structural analysis** is to determine the displacement and stress under static loading conditions. It can be either *linear* or *nonlinear* analysis. A linear model analysis assumes that the material works at its elastic region and there is no plastic deformation. Materials in a nonlinear model involve plastic deformation. Material properties change and stresses in the material will vary with the amount of deformation.

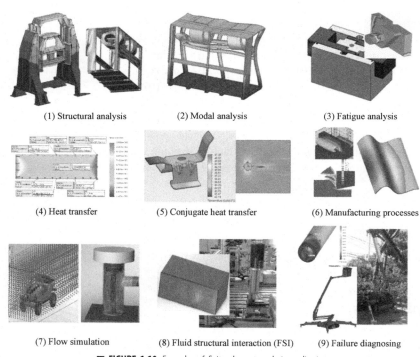

(1) Structural analysis   (2) Modal analysis   (3) Fatigue analysis

(4) Heat transfer   (5) Conjugate heat transfer   (6) Manufacturing processes

(7) Flow simulation   (8) Fluid structural interaction (FSI)   (9) Failure diagnosing

■ **FIGURE 1.10** Examples of finite element analysis applications.

2. **Modal analysis** is to evaluate the natural frequencies and mode shapes of a structure, different modal extraction methods can be used to solve the characteristic equations of structure. The structure tends to fail when an external excitation has a frequency close to one of its natural frequencies.

3. **Fatigue analysis** is to predict the life of a part or component in a model; a fatigue analysis predicts the effect of the cyclic loading on the part/assembly. Fatigue is responsible for approximately 80% of all structural failures. The simulation predicts the fatigue life and the safety factor against fatigue failure for an object of interest.

4. **Heat transfer analysis** is to consider the heat transfer behaviors including conduction, convection, and radiation in a system. A heat transfer analysis can be either *steady* or *transient*. Steady-state transfer refers to constant thermal properties that yield linear heat diffusion, which does not change with time. In contrast, transient heat transfer concerns the history of temperature changes. A heat transfer analysis can also run on a nonlinear model that involves time, radiation, and/or changing thermal properties in the material or through convection.

5. **Conjugate heat transfer** is a multiphysics problem because energy is transferred between a solid domain and fluid flow. Due to the coexisting of multiphysics behaviors, the corresponding governing equations have to be considered simultaneously to find the solution to meet the constraints in multiple disciplines. Many real-world applications such as quenching processes fall into this category.

6. **Analysis of manufacturing process** is to look into the response of materials in forming processes, for example, material removing processes, injecting, forging, and deforming. This analysis type involves the geometric change of an object, and the deformation on materials is dominated by plastic deformation. Nonlinear material properties must be defined, and the analysis process is mostly transient since the state of materials (including geometry) changes with respect to time. Such analysis requires intensive computation.

7. **Flow simulation** is to simulate liquid or gas flow in real-world conditions. Flow simulation predicts the distribution of pressure and velocity under external loads. Such type of information is very critical to the design of surrounding objects. Taking an example of a car body design, a flow simulation can be used to determine the dragging force by air flow; as a result, the body shape can be optimized to improve fuel efficiency.

8. **Fluid structural interaction** (FSI) is the interaction of a movable or deformable structure with an internal or surrounding fluid flows; for example, a piston in engine, a sealing part at a reciprocating interface.

FSI is another typical example of multiphysics systems. The solid and fluid domains must be modeled simultaneously because some state variables (such as pressure) are involved in the governing equations of both disciplines. FSI problems are in general too complex to be solved analytically. Therefore, either experimental or FEA methods have to be used to solve an FSI problem.

9. **Failure diagnose** is to determine the causes of failure when a failure occurs to a part, component, or structure. Not all of the design flaws can be successfully eliminated during the product development. This is in particular true for low-volume products from small- or medium-sized enterprises (SMEs). It is impractical and cost forbidden for SMEs to conduct a thorough V&V for new products; the remedies on failed parts or components are more economic options for noncrucial products. FEA can be used as a diagnosing tool to identify design flaws when failure occurs to products.

Note that the aforementioned examples are not an exhaustive list of where FEA can be applied to support engineering analysis. In contrast, the potential applications of FEA are unbounded. FEA can virtually be applied in any engineering design. For example, FEA has been widely used as a tool for the simulation-based optimization. Most of the commercial FEA codes are equipped with the functionalities of design studies. The design variables of an FEA model can be varied in the specified ranges, FEA processes can be executed repeatedly to find the combination of design variables for the best system performance.

1.4.5 **FEA—advantages and disadvantages**

FEA is one of the most important techniques for CAE tools. The advantages and disadvantages have been well known by researchers, and Table 1.5 gives a summary about the advantages and disadvantages of FEA (de Weck and Kim, 2004; Kim and Sankar, 2009).

1.5 **FEA—GENERAL MODELING PROCEDURE**

Before FEA modeling, a user has to construct a virtual model of the part or assembly to be analyzed and specify the material properties for every object in the model. The user must understand the physical phenomena occurring to the model, so that the analysis type and element types can be specified appropriately. In FEA modeling, the *divide and conquer* strategies is applied to obtain the system model from an assembly of submodels of simple units. As shown in Fig. 1.11, the continuous domain of the model is decomposed into small parts called "*elements*". Elements are assembled through the

Table 1.5 Advantages and Disadvantage of Finite Element Analysis

Advantages	Disadvantages
■ Insensitive to the complexity of geometries ■ Capable of analysing a wide variety of engineering problems such as solid mechanism, dynamics, heat problems, fluid flows, and electrostatic problems ■ Capable of dealing with complex boundary conditions such as over constraints in solid mechanics ■ Capable of dealing with complex loading conditions such as concentrated or distributed load, and time-dependent loads	■ Approximate solution instead of closed-form analytical solutions ■ Including inherent errors such as those caused by numerical computations and the idealization of models ■ Rely on designers to validate and verify the simulation results

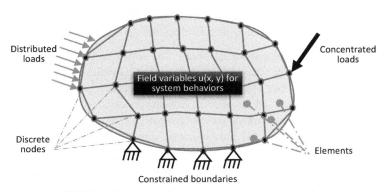

Distributed loads

Concentrated loads

Field variables u(x, y) for system behaviors

Discrete nodes

Elements

Constrained boundaries

■ **FIGURE 1.11** Discretization of a continuous domain into nodes and elements.

interconnection of points, which are called as "*nodes*". Accordingly, each element is associated with a set of nodes, and the element behavior is modeled by the behaviors of discrete nodes.

Generally, the procedure of FEA modeling consists of the following steps:

Step 1—Decomposition. The continuous domain is discretized into a collection of shapes or elements, the nodes for each element are specified. The assembling relations of elements as well as element—node relations are defined clearly. Because state variables on nodes are defined as design variables of elements, interpolation functions are specified to describe the response of any position in an element by nodal values.

Step 2—Develop element models. An analysis type is selected based on the physical behaviors of the model. Design variables are selected, and the mathematic model with governing differential equations is defined. The mathematic model is then converted into element models by the approximation methods such as *direct methods*, *minimum potential energy methods*, or *weighted residual methods*.

Step 3—Assembly. Based on the stored assembly relations in the decomposition, element models in local coordinate systems are transformed into element models in a global coordinate system, and they are assembled into a system model under the global coordinate system.

Step 4—Apply boundary conditions and loads. The interactions of the physical system with its application environment are defined, they are represented as boundary conditions or load conditions in the model.

Step 5—Solve for primary unknowns. Sufficient boundary conditions ensure the system model solvable. A system model usually consists of a large number of linear equations. A number of well-developed algorithms can be utilized to solve unknown variables from the system model.

Step 6—Calculate dependent variables. The design variables in an engineering system can be classified into independent variables and dependent variables. For example, stress and strain are dependent with each other, either stress or strain can be selected as independent one, and the other can be determined based on the constitutive model of materials. After independent variables are solved, postprocessing can be performed to evaluate dependent variables.

In the implementation of any FEA code, the above steps are essential to obtain the final solution to an FEA model; however, most of the activities in these steps are automatically accomplished by software. Users are only required to provide minimal information as the inputs of model to run a numerical simulation. A commercial FEA code provides a graphic user interface (GUI) only for the tasks where users provide inputs. As shown in Fig. 1.12, GUIs of a commercial FEA code are provided for the manual intervention at three solving stages, i.e., *preprocessing*, *processing*, and *postprocessing*.

Although most of the activities are automatically performed by an FEA code, users are responsible to formulate design problems, provide the correct and sufficient inputs for every steps, and interpret and verify the results from the software code adequately. If the inputs of an FEA model are wrongly given, the obtained results from the simulation could mislead users.

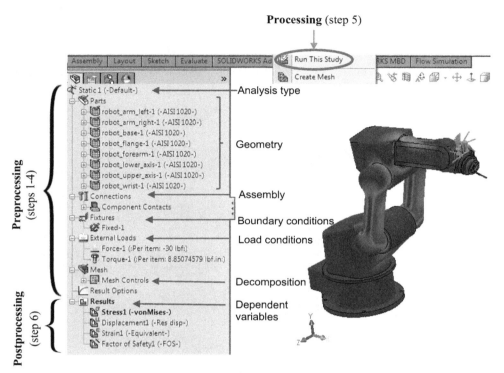

1.6 ORGANIZATION OF BOOK

This book covers both fundamental concepts and practical skills of FEA. It is designed as an introduction course of FEA suitable for undergraduate students, graduate students, and designers who are new to FEA but have the needs of using FEA as an engineering design tool. FEA is presented as a numerical tool to find approximate solutions to various engineering problems. Fig. 1.13 shows the organization of the book. It consists of 12 chapters, and the FEA fundamentals are introduced from element level, system level to application level in Chapters 3—6 and Chapter 11, the decomposition, computer implementation, and V&V are introduced in Chapters 2, 7, and 12, respectively. The abstracts of the rest of chapters are briefed as below.

2. Decomposition. This chapter introduces (1) a system approach to describe engineering problems and (2) the methodologies to deal with system complexity. To describe an engineering problem, system parameters and design variables are identified, physical principles are converted into governing differential equations as mathematical models; mathematic models are

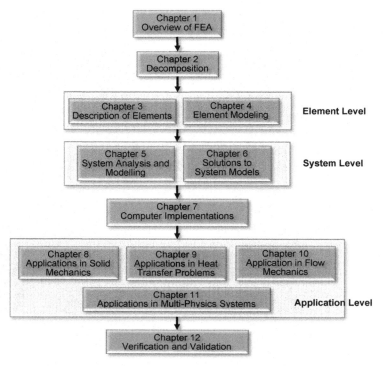

■ FIGURE 1.13 Organization of book.

used to represent the relational models of design variables and system outcomes. To deal with the complexity, it is discussed from the perspective of governing equations, geometries, time dependence, the nature of multidisciplinary, and uncertainties. The strategies to deal with complexity are explored, and they are adopted to accomplish the activities at the preprocessing phase of FEA modeling. The meshing process for the decomposition of geometry is especially discussed.

3. Description of Elements. FEA converts a continuous domain into a discretized domain, the behavior of an element is fully represented by the values of associated nodes. In this chapter, the characteristics of elements are discussed, and the elements are classified in terms of geometric shapes, dimensions, as well as linearity or nonlinearity nature. To describe elements, a number of coordinate systems (CS) are introduced, and the coordinate transformation from one CS to another is discussed and modeled. The focus is the interpolation within an element to describe the behavior at an arbitrary position based on nodal variables. In addition, because an FEA tool is basically a set of generic solvers to differential equations, integral operations are frequently performed in solving processes. Therefore, the methods of numerical integration are also covered.

4. Solutions of Element Models. An element model describes the relations of nodal variables and loads over the element, and the solution to an element model is to determine nodal variables based on given nodal loads. In Chapter 4, the solutions of element models are discussed. (1) Mathematic models are developed to represent physical behaviors of elements; (2) the techniques to model element behaviors are introduced to obtain state variables on nodes. Generally, an engineering problem can be modeled as PDEs. Therefore, generic solutions to PDEs are applicable to a broad scope of engineering problems. The governing equations of various engineering problems are reviewed, and a few of the commonly used methods are introduced to solve corresponding models at element levels.

5. System Analysis and Modeling. Spatial and geometric shapes are not the only source of complexity. A common misunderstanding for naïve designers is the underestimation of necessary manual interventions at the stage of system modeling. We argue that manual intervention is one of the most critical steps to build a good FEA model. In Chapter 5, we discuss the complexity other than geometric shapes. In particular, the complexity by time domain, part assembling, boundary conditions, and loads are investigated, the objective is to develop a conceptual model at the system level.

6. Solutions to System Models. The notations of vectors and matrices are widely adopted to describe an FEA modeling procedure and system models, and different design problems are eventually modeled as systems of linear equations. In Chapter 6, vectors and matrix algebra are reviewed, the mathematical models related to FEA formulations are discussed, and commonly used computer algorithms to solve systems of linear equations are reviewed. The focus is on the computation and efficiency of numerical simulation.

7. Integrated Engineering Analysis Environment. Computer programs are used to replace manual computations to find the solutions to various engineering problems. In Chapter 7, the basics about computer programs and programming techniques are introduced. Pros and cons of different programming techniques are compared, and an object-oriented (OO) approach is adopted to illustrate the computer implementation of FEA modeling. Types and topology of classes are defined accordingly based on the activities at each step of FEA modeling. Design and programming of an FEA for a generic spring system is provided to apply the object-oriented method for FEA computer implementation. In contrast to the introduction of computer programming in other textbooks, the context in this book focuses on the core idea of FEA, i.e., "*divide and conquer*". The program design is directly associated with the data and methods required in defining and solving an FEA model. The implementation of FEA has a modular structure, which allows students to create their own classes and FEA models. It is reusable and expendable as a workable CAD tool for a variety of engineering designs.

8. Application—Solid Mechanical Problems. The FEA theory was sprouted from the needs of solving structural design problems in solid mechanics. FEA has been proven as a vital solution to a variety of design problems in solid mechanics. In Chapter 8, the design analysis of mechanical structure is discussed. Mechanical design problems are classified as structural analysis, modal analysis, and fatigue analysis; they are formulated as FEA models. Commercial FEA tools are used to determine stress distribution, deformation, stability, and fatigue lives in the case studies.

9. Application—Heat Transfer Problems. Different types of heat transfer are introduced and the corresponding governing equations are discussed. The Galerkin method is applied to develop element models for triangle and rectangle elements. In addition, the solution of transient heat transfer problems is also presented.

10. Application—Fluid Mechanics. This chapter is to review the basic equations of fluid mechanics. The governing equations of fluid mechanics are introduced, and these equations are applied to define and solve basic fluid mechanics problems including pipe flows, steady incompressible fluid flow, and groundwater. In particular, the solution for fluid flow governed by Poisson's equations is modeled.

11. Applications—Multiphysics Simulation. Multiphysics is the study of systems with coupled physical behaviors in multiple disciplines. Numerical simulation of a multiphysics system deals with a set of governing differential equations from multiple single-physics subsystems in corresponding domains simultaneously. In Chapter 11, the generalized description of a multiphysics system is discussed, multiphysics systems in different applications are classified in terms of coupling natures. Thermal-structural systems, conjugate heat transfer, and fluid structural interaction are used as the case studies to illustrate FEA applications to model multiphysics systems.

12. Verification and Validation. The errors and sources of approximated solutions from an FEA are explored, the necessity and importance of V&V are discussed. The similarity and differences of V&V are analyzed. The method of quantifying the errors of FEA solutions is introduced; quantified errors are used as the criteria for V&V. The guidance and common techniques of performing V&V in the FEA-based simulation are introduced.

1.7 **SUMMARY**

A brief overview is given on FEA as well as its roles in modern engineering practices. The general process of an engineering design process is introduced and a few of the design problems have been used to illustrate how to convert customer's requirements into the functional requirements of engineering designs from the perspective of system complexity. The methods for performing main design activities, in particular, design analysis and design synthesis, are discussed. The strengths and weaknesses of computers and human designers are compared, and the numerical simulation is introduced to assist human designers in evaluating and optimizing design solutions. Different problem-solving tools are classified, and the focus is on FEA. FEA is viewed as an implementation of top-down and bottom-up approaches to deal with the complexity of engineering problems. The evolution of FEA technique is covered, the advantages and disadvantages are explored, and the general process of FEA modeling is presented.

REFERENCES

Bi, Z.M., Zhang, W.J., 2000. Concurrent optimal design of modular robotic configuration. Journal of Robotic Systems 18 (2), 77–87.

Bi, Z.M., Gruver, W.A., Zhang, W.J., Lang, S.Y.T., 2006. Automated modelling of modular robotic configurations. Robotics and Autonomous Systems 54, 1015–1025.

Bi, Z.M., Lang, S.Y.T., Verner, M., Orban, P., 2008a. Development of reconfigurable machines. Int. J. of Advanced Manufacturing Technology 39 (11-12), 1227–1251.

Bi, Z.M., Lang, S.Y.T., Shen, W.M., 2008b. Reconfigurable manufacturing systems: the state of the art. Int. J. of Production Research 46 (4), 967–992.

Bi, Z.M., Lin, Y., Zhang, W.J., 2010. The general architecture of adaptive robotic systems for manufacturing applications,. Robotics and Computer Integrated Manufacturing 26 (5), 461–470.

Booch, G., 1994. Object-Oriented Analysis and Deisgn. Addison Wesley Longman, Inc., ISBN 0-8053-5340-2

Chakrabarti, A., Shea, K., Stone, R., Cagan, J., Campbell, M., Hernandez, N.V., Wood, K.L., 2011. Computer-based design synthesis research: an overview. Journal of Computing and Information Science in Engineering 11, 021003–1 to 021003–10.

Chapra, S.C., Canale, R.P., 2010. Numerical Methods for Engineers, sixth ed. McGraw-Hill, 1221 Avenue of the Americas, New York, NY 10020, ISBN 978-0-07-340106-5.

Clough, R.W., 1960. The finite element method, in plane stress analysis. In: Proceedings, 2nd Conference on Electronic Computation. ASCE, Pittsburgh, PA.

Courant, R., 1943. Variational methods for the solutions of problems of equilibrium and vibrations. Bulletin of American Mathematical Society 49, 1–23.

de Weck, O., Kim, I.Y., 2004. Finite Element Method. http://web.mit.edu/16.810/www/16.810_L4_CAE.pdf.

DesignWorks, 2016. Types of Design. http://www-eng.lbl.gov/~dw/services/TypesOfDesign/TypesOfDesign.htm.

Dorst, K., 2016. The Problem of Design Problems — Creativity & Cognition. http://www.creativityandcognition.com/cc_conferences/cc03Design/papers/23DorstDTRS6.pdf.

Engineering Encounters, 2016. Bridge Building Contest. https://bridgecontest.org/.

Goel, K., June/July 1997. Design, analogy, and creativities. AI in Design, IEEE Expert 62—70.

Hrennikoff, A., 1941. Solution of problems of elasticity by the framework method. Journal of Applied Mechanics 8.4, 169—175.

Johansson, J., 2011. Automated Computer Systems for Manufacturability Analyses and Tooling Design: Applied to the Rotary Draw Bending Process (Doctorial thesis). Department of Product and Production Development, Chalmers University of Technology, Goteborg, Sweden.

Kim, K.-J., Rezaei, A., 2008. Efficient use of computational tools in machine design. In: Proceedings of the 2008 American Society for Engineering Education Pacific Southwest Section Conference March 27—28, 2008. Northern Arizona University, Flagstaff, AZ.

Kim, N.-H., Sankar, B.V., 2009. Introduction to Finite Element Analysis and Design. John Wiley & Sons, Inc., ISBN 978-0-470-12539-7

Knight, C.E., 1994. A Finite Element Method Primer for Mechanical Design. PWS Publishing Company, Boston, ISBN 0-534-93978-3.

Leger, C., 1999. Automated Synthesis and Optimization of Robot Configurations: An Evolutionary Approach, CMU-RI-TR-99-43 (Doctorial thesis). Carnegie Mellon University.

Morjoram, T., Zhong, Y., 2003. Engineering: Issues, Challenges, and Opportunities for Development. UNESCO Publishing. http://unesdoc.unesco.org/images/0018/001897/189753e.pdf.

Multiphysics, 2014. A Brief History of Finite Element Analysis. http://blog.multimechanics.com/brief-history-finite-element-analysis.

Seo, K., Fan, Z., Hu, J., Goodman, E.D., Rosenberg, R.C., 2003. Toward an Automated Design Method for Multi-Domain Dynamic Systems Using Bond Graphs and Genetic Programming. https://cse.sc.edu/~jianjunh/paper/mechatronics_gpbg.pdf.

Siemens, 2016. CAE/Computer Aided Engineering. https://www.plm.automation.siemens.com/en_us/plm/cae.shtml.

Suh, N.P., 2001. Axiomatic Design — Advances and Applications. Massachusetts Institute of Technology, Oxford University Press, New York. http://www.axiomaticdesign.com/technology/ADSChapter5.html.

Suh, N.P., 2005. Complexity: Theory and Applications, first ed. Oxford University Press, New York.

US Department of Transportation, 2009. System Engineering Guidebook for Intelligent Transportation Systems Version 3. Federal Highway Administration, California Division. http://www.fhwa.dot.gov/cadiv/segb/files/segbversion3.pdf.

Vanderplaats, G.N., 2006. Structural optimization for statics, dynamics and beyond. Journal of the Brazilian Society of Mechanical Sciences & Engineering XXVIII (3), 317—322.

Woods, R.I., 2003. The Application of Finite Element Analysis to the Design of Embedded Retaining Walls (Doctoral thesis). School of Engineering, University of Surrey.

Zienkiewicz, O.C., 1995. Origins, milestones and directions of the finite element method — a personal view. Archives of Computational Methods in Engineering 2 (1), 1—48.

Zienkiewicz, O.C., Cheung, Y.K., 1967. The Finite Element Method in Structural Mechanics. McGraw—Hill, p. 272.

FURTHER READING

Bickford, W.B., 1990. A First Course in the Finite Element Method. Rickard D. Irwin, Inc., ISBN 0-256-07973-0

Chandrupatla, T.R., Belegundu, A.D., 2012. Introduction to Finite Elements in Engineering, fourth ed. Pearson. ISBN-10: 0-13-216274-1.

Gupta, K.K., Meek, J.L., 1996. A brief history of the beginning of the finite element method,. International Journal for Numerical Methods in Engineering 39, 3761—3774.

Logan, D.L., 1985. A First Course in the Finite Element Method. PWS-KENT Publishing Company, ISBN 0-534-05394-7.

McCarthy, J.M., 2011. Kinematics, Polynomials, and Computers—A Brief History. http://synthetica.eng.uci.edu/mechanicaldesign101/JMR-Editorial-Feb2011-rev.pdf.K.

Moaveni, S., 2015. Finite Element Analysis — Theory and Application with ANSYS, fourth ed. Peason. ISBN-10: 0-13-384080-8.

Musa, S.M., Kulkarni, A.V., Hanvnur, V.K., 2014. Finite Element Analysis: A Primer. Mercury Learning and Information, ISBN 978-1-938549-34-2.

Schmit, L.A., 1960. Structural design by systematic synthesis. In: Proceedings, 2nd Conference on Electronic Computation. ASCE, New York, pp. 105—132.

Systems Management College, 2003. Department of Defense: Systems Engineering Fundamentals. ISBN-10: 1484129555, 2013.

Szabo, B., Babuska, I., 2011. Introduction to finite element analysis—formulation, verification and validation. In: Wiley Series in Computational Mechanics. Wiley, ISBN 978-0-470-97728-6.

Thompson, E.G., 2005. Introduction to the Finite Element Method — Theory, Programming, and Applications. John Wiley & Sons, Inc., ISBN 0471-45253-X

Turner, M.J., Clough, R.W., Martin, H.C., Topp, I.J., 1956. Stiffness and deflection analysis of complex structures. Journal of Aeronautical Sciences 23, 805—824.

Decomposition

2.1 INTRODUCTION

As generic software tools, commercial finite elemental analysis (FEA) codes are capable of simulating various physical phenomena such as structural deformation, heat transfer, fluid flow, acoustics, and electromagnetic waves. FEA has been applied widely in industry, science, and engineering for marvelously diverse purposes. In this chapter, the functional requirements (FRs) of a generic FEA software tool are discussed. A special attention is paid on the capabilities of an FEA code in solving various engineering problems with different levels of complexity. The decomposition of geometric complexity is used as an example; thus the meshing process in the FEA modeling is discussed.

2.2 GENERAL DESCRIPTION OF SYSTEM

In formulating an engineering problem, the first step is to isolate the system of interest from its application environment. In such a way, design variables are identified, and the boundaries of system can be clarified. As shown in Fig. 2.1, all of the connections of the system with its application environment will be represented by two types of interactions. The first type is *passive interactions* by which the system is influenced by the factors in the environment passively. For example, boundary conditions and loads on the system by other entities are passive interactions for the system of interest. The second type is *active interactions* by which the system makes its influence on the objects in the environment actively. For example, in a heat transfer problem, the heat flux from an object of system will affect the environment because heat energy is dissipated at the interaction. Every engineering system has its design purpose, and it is reflected by the desired active interactions from the system to its application environment. On the other hand, every engineering system has its boundaries, which are imposed by the application environment. Therefore, the boundaries of the system with the interactions specify the scope of an engineering problem.

Finite Element Analysis Applications. http://dx.doi.org/10.1016/B978-0-12-809952-0.00002-9

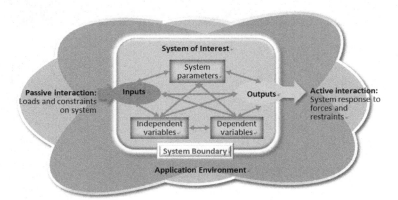

■ **FIGURE 2.1** Description of an engineering problem from system perspective.

A system model describes its constitutive components and their relations. As shown in Fig. 2.1, the passive and active interactions have to be converted and quantified as inputs and outputs of an FEA model, respectively. Design variables in the system can be generally classified into *system parameters*, *independent variables*, and *dependent variables*. System parameters represent the features and characteristics of the system to be modeled. For example, material properties and geometric shapes and dimensions are system parameters. Independent variables are some quantities to represent the states of the system. For example, the displacements are usually independent variables in a structural analysis, and temperature distributions over a continuous domain are independent variables in a heat transfer problem. In addition, independent variables are treated as primary unknowns, which will be solved in a solving process of FEA model. Dependent variables are also some quantities to describe the states of a system. For example, strains and stresses in a structural analysis are dependent variables; however, they are determined by system variables. Once independent variables have been solved from a system model, dependent variables are then calculated based on their dependencies on independent variables. In a system model, the governing differential equations describe the relations of system parameters, dependent variables, and independent variables. The distinguishing of independent and dependent variables is not clear and absolute. In solving a system model, it depends on the solver to choose a set of design variables as independent variables; while the rest of variables become dependent variables.

2.3 **SYSTEM COMPLEXITY**

One can see a system model describes the relations of inputs, outputs, system properties, and design variables. Therefore, the complexity of a system model relates not only to the application environment but also to system components and their connections. According to Suh (2005), the system complexity depends on two main factors: (1) the number and types of design variables for system components and their interactions and (2) their dynamic characteristics with respect to time. In addition, uncertainties contribute a great deal of complexity in modern products and systems (Bi and Kang, 2014).

A commercial FEA code is developed in such a way that the manual intervention is minimized in defining and solving an FEA model. The impact of the uncertainty and complexity on the simulation of an engineering problem can be easily underestimated by naïve designers. He or she tends to leave most of the important parameters as default settings in an FEA model, and this brings the risk of obtaining faulty results to original engineering problems. In this section, we discuss some important aspects that an FEA user needs to take into consideration in developing an FEA model.

To formulate an FEA model, the number and types of design variables are defined; the relations of design variables and system parameters are represented by governing equations. However, design variables and their relations are affected by many aspects such as *uncertainties*, *geometric shapes*, *assemblies*, *dynamics*, and *physical behaviors*, which are shown in Fig. 2.2. One has to take into consideration of these aspects carefully in the formulation.

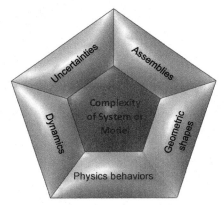

■ **FIGURE 2.2** Complexity of a finite element analysis model.

2.3.1 **Uncertainties**

There are many sources of uncertainties in the formulation of an engineering problem as an FEA model. (1) *Clarification of system boundaries*. When an object or system is isolated from its application environment, uncertainties exist in the definition of boundary conditions or loads. For example, if a boat is analyzed to determine its loading capacity, it is not an easy task to define the interaction between the boat and surrounding water. In addition, there are some uncertainties in applying appropriate supports and loads. (2) *Definition of system parameters*. A continuous domain can be made of multiple materials, each of them is represented by its own system parameters. However, the characterization of materials involves many uncertainties. For example, if a composite material is applied, the FEA modeler needs to decide how the heterogeneous properties are represented in the constitutive model. (3) *Interfaces at contacts*. Uncertainties exist at the interfaces of multiple objects. As shown in Fig. 2.3, if the drag force at a reciprocating interface of a sealing part has to be analyzed, uncertainties exist to characterize the friction at the interface, because three types of lubrications are possible in different operating conditions: under a slow reciprocating speed and light

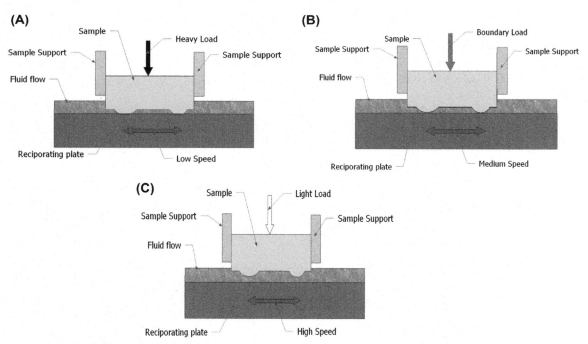

■ **FIGURE 2.3** Example of uncertainty of friction at contact surfaces. (A) Dry friction. (B) Boundary friction. (C) Shear by viscosity.

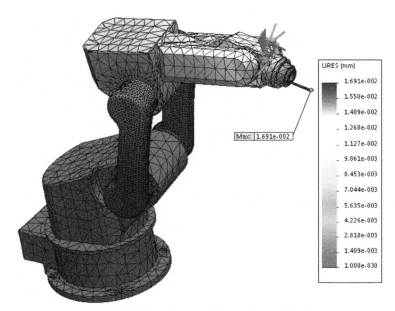

■ **FIGURE 2.4** Example of uncertainty caused by analysis goal.

external force, it is a dry contact dominated by the friction between two solids. If the reciprocating speed is very high, there will be a liquid film between two interacting surfaces and the drag force comes from the viscosity of fluid. It may also be a case in between where the drag force consists of the friction of two solids and the viscosity of fluid at a gap. Therefore, it is risky to formulate an FEA problem without a thorough understanding of the governing laws of mechanics in an engineering problem. (4) *Uncertainties of analysis goals*. Uncertainties exist if an analysis goal is not specific or directly relevant to state variables. Take an example of a robot in Fig. 2.4, one of the primary performances of a robot is the accuracy, or, the deflection of the structure under given loading conditions. FEA can be used to evaluate the displacement distribution of a robot under the given loads. However, FEA is only able to predict the deformation when both of the robot configuration (e.g., the structure and joint positions in this case) and external loads are given. FEA is not able to evaluate the accuracy of robot directly over its workspace.

2.3.2 **Assembling**

The majority of products are assembled from parts. If an engineering problem is defined for an assembled product or system, the complexity and

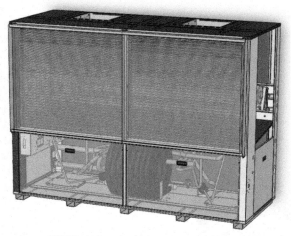

■ **FIGURE 2.5** Example of uncertainty in assembled product.

uncertainties exist in defining the contact relations of parts. The more the number of parts in an assembled product, the higher is the level of modeling complexity. Fig. 2.5 shows an assembled model for the support structure of ventilating device. A model analysis has to be conducted to ensure that the operating frequencies are far away from one of the natural frequencies of the structure. The model includes many insignificant parts, such as circuits, fasteners, and preventive panels. Therefore, the model has to be simplified to eliminate all of problematic parts or contacts in a valid FEA mode, in particular, for circuits, pipe components, or fasteners.

If an assembled product includes a large number of parts; it is suggested that the product be broken down into a number of components. FEA at component levels can be run on the detailed models, whereas FEA at the system level can be run with its components at the abstract level. By all means, the complexity of FEA model should be manageable, and it directly relates to the number and details of the parts in a model. As shown in Table 2.1, Hubka and Eder (1988) classified the system complexity into four levels based on the number and complexity of parts.

Few textbooks on FEA have discussed the modeling challenges caused by the complexity of assembling mates; note that any commercial FEA code can run a simulation on an assembled product. However, the product is usually treated as a bonded structure in its default. It often leads to some suspicious results if a contact interface turns in a critical area of interest.

Table 2.1 Level of Complexity Based on Assembly Relations

Level of Complexity	Engineering System	Characteristics	Examples
I	Parts, structure	Integral objects with no interface of assembly	Machine elements such as fasteners, shafts, and frames
II	Mechanisms, components	Simple product from assembling to fulfill specific functions	Gear boxes, hydraulic drives, spindles, braking systems
III	Machines, apparatuses, devices	Systems with a number of parts and subassembled components to perform complex tasks	Lathes, vehicles, motorcycles, robots
IV	Plants, integrated systems, complex systems	Integrated systems to fulfill multiple functions; a system consists of machines, work cells, and structures to constitutes a functional and spatial unity	Product lines, machining transfer lines, flexible manufacturing systems

2.3.3 **Dynamics**

FEA deals with the approximation of state variables in a continuous domain. Depending on the characteristics of a design problem, state variables can be *steady* or *transient* in a time domain. If the time has to be considered in defining load or boundary conditions, the FEA model will be needed to add another dimension upon the dimensions for the space domain. Therefore, the dynamic behaviors of state variables increase the complexity of modeling. Fig. 2.6 shows an example of a fatigue analysis on a mechanical component. The dynamics comes from the loading conditions of a mechanical structure; these vary with time. Such loads have to be represented as fluctuated loads. Fatigue analysis is usually performed as an additional analysis beyond a static analysis. The definition of the load conditions involves a high level of uncertainty. Dynamic characteristics must be taken into account when the performance of an engineering system relies greatly on the historical changes of state variables such as transient heat transfer or conjugate heat transfer problems.

2.3.4 **Multiphysics problems**

In many applications, the behaviors of a system involve a number of the laws of mechanics in different disciplines. The state variables related to an individual discipline are governed by corresponding mathematical models. However, some state variables happen to be in several mathematical models in multidisciplines. In other words, these variables are

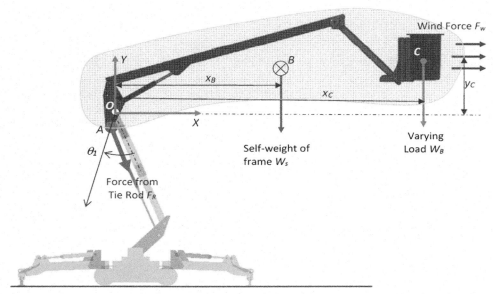

■ **FIGURE 2.6** Example of complexity by dynamic factors.

decoupled with each other due to the fact that the same set of state variables has to meet the constraints by governing laws of mechanics for multiphysics behaviors. The FEA model becomes more complex because multiple mathematic models in respective disciplines have to be solved simultaneously. Fig. 2.7 shows an example of a pressure vessel. The stress distribution in the shell of vessel depends on both the mechanical loads and thermal loads. If these loads are transient, the simulations on both of solid mechanics and heat transfer must be performed iteratively to evaluate stress distributions with respect to time.

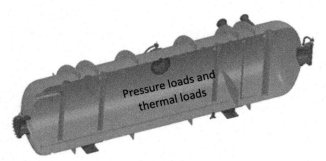

■ **FIGURE 2.7** Example of modeling complexity due to multiphysics.

2.3.5 **Geometric shapes**

Design problems in engineering textbooks usually have the objects with a simple geometric shape; this makes it possible to obtain analytical solutions to these problems by hand calculation. However, real-world design problems are usually very different in the sense that the geometric shapes of objects are very complex. The geometric complexity and varieties are the most critical reasons why general FEA code becomes so attractive in today's engineering practices. However, even though a generic FEA code can cope with an object with an arbitrary geometric shape technically, it is worthy to mention that the strategies to decompose geometric shapes affect the complexity of models and modeling processes greatly.

Some insignificant features over an object may affect the model complexity adversely in terms of model sizes, the efficiency of modeling processes, and the accuracy of simulation results. Take an example of a simple object in Fig. 2.8, the small fillets can increase the number of elements and nodes greatly; the example shows that small fillets in the object demand approximately three times of elements and nodes to represent the same object without these fillets. Note that these features have an ignorable impact on the stress distribution when the loads are normal forces on surfaces. Therefore, it is desirable to examine small features in a model to see if the model should be simplified before this model can be adopted as the computational domain in the simulation.

2.4   **FEA AS PROBLEM-SOLVING TOOL**

FEA is a problem-solving tool to obtain approximate solutions to various engineering problems. These problems should be formulated as one or a set of partial differential equations (PDEs) with given boundary conditions.

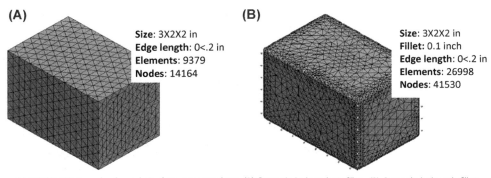

(A)

Size: 3X2X2 in
Edge length: 0<.2 in
Elements: 9379
Nodes: 14164

(B)

Size: 3X2X2 in
Fillet: 0.1 inch
Edge length: 0<.2 in
Elements: 26998
Nodes: 41530

■ **FIGURE 2.8** Example of complexity from geometric shape. (A) Rectangle body without fillets. (B) Rectangle body with fillets.

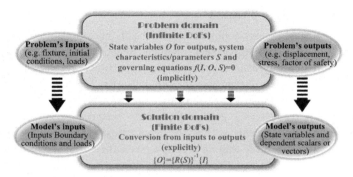

■ **FIGURE 2.9** Finite element analysis code as a problem-solving tool.

Fig. 2.9 describes the relation of an FEA model with its original engineering problem. A formulated FEA model is different from the original problem in the sense that (1) an FEA model has a discretized domain to represent a continuous domain in the original problem; (2) an FEA model has a finite number of degrees of freedom, or the number of unknowns to be solved; the behavior at any location in the continuous domain is approximated by the interpolation in the corresponding element; (3) an FEA model must be solvable to find state variables *explicitly* based on given boundary conditions and loads. From this perspective, the solution to an FEA model is a conversion from the given inputs to the explicit outputs for the given engineering problem. (4) Inputs or outputs in the original problem must be reformatted, so that the system inputs can be formulated as boundary conditions and loads can be applied to discretized nodes, and the outputs are approximated as the scalars or vectors derived from a finite number of state variables.

The complexity and variety of engineering problems have been discussed in Section 2.5. A further question to be explored here is "how can an FEA code be so generic that it can be applied to solve diversified engineering problems with different level of complexity?" An intuitive answer to this question is that an FEA code is a kind of design toolbox indeed. It includes all of the software functional modules with model templates, algorithms, and solvers for any anticipated engineering problems. If an FEA code is selected to solve a specific type of engineering problem, for example, a fluid structural interaction (FSI) model is selected for a multiple disciplinary problem, one must ensure that the selected code has such capabilities in modeling and solving FSI behaviors.

From an FEA user's perspective, an FEA modeling process is to customize a set of tools for solving a given problem optimally. To this end, an FEA user must understand well about what is to be actually modeled and what the software tool is good at. A user may ask some questions: (1) What is the complexity of an engineering problem? (2) How to decompose the complexity, so that functional modules in an FEA code can be applied to perform submodeling tasks effectively? (3) How can boundary conditions and loads be defined appropriately? (4) What outcomes can one expect from the simulation? At what level of accuracy, and with what acceptable time frame for the computation? To answer these questions, the axiomatic design theory (ADT) can be used to understand the complexity of an engineering problem and simplify a computable model when an FEA simulation is developed.

2.5 AXIOMATIC DESIGN THEORY

In dealing with the complexity, ADT can be used as a systematic approach to guide the decomposition (Suh, 1990). ADT lays the foundation for system analysis and design. ADT uses two basic axioms as well as several other corollaries and theorems to support the design theory. *The first axiom* is to maintain the independence of FRs and *the second axiom* is to minimize the information content of the resulting design solution. The act of design consists of the mapping of FRs to *design parameters* (DPs), and eventually, to arrive at a solution, which satisfies these two axioms.

Development of a complete solution to a given problem is proceeded by mapping FRs from *a functional or modal domain* to DPs in *a solution domain*. The mapping begins from a high, general level to a lower, increasingly detailed level. As shown in Fig. 2.10, the decomposition from a high level to a low level of FRs and DPs is the design process called a "*zig-zagging*" process, in which, the system design refers to the mappings

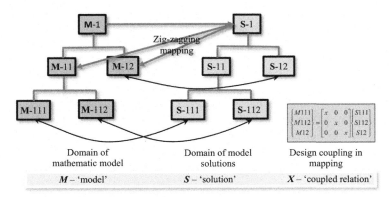

■ **FIGURE 2.10** Axiomatic design theory for decomposition of system complexity.

between the functional and solution domains. To develop a better FEA model, the design problem must be formulated in such a way that the high-level system representation is decomposed into the detailed level representation. In such a way, corresponding functional modules or settings in an FEA code can be directly selected or specified appropriately.

As shown in Fig. 2.10, in the hierarchy of a decomposition, the relations between $\{M\}$ and $\{S\}$ are represented in the form of a matrix $[DC]$ for coupling relations; $\{M\}$ and $\{S\}$ are the vectors of unknowns in the mathematic model and the solution model. By checking the elements in $[DC]$, the levels of decoupling can be justified based on the first design axiom. Depending on the characteristics of $[DC]$, a design may be classified as *uncoupled*, *decoupled*, or *coupled*. An uncoupled decomposition has nonzero elements at diagonal positions and zero elements at all off-diagonal positions. A decoupled decomposition has a $[DC]$ with all of nonzero elements in a triangular region of matrix, and a coupled decomposition has a $[DC]$ where nonzero elements exist off the principal diagonal line.

An uncoupled design satisfies the first axiom because each element in $[M]$ maps to one and only one element in $[S]$ to maintain the functional independence. If ADT is applied to an FEA modeling process, one FEA submodel is devoted to each mapping pair of elements in $[M]$ and $[S]$. *A decoupled design* may still satisfy the first axiom if the design is proceeded in certain order. With a triangular matrix, the independence of elements in $[M]$ may be maintained if the elements in $[S]$ are determined in a precedential sequence: one element in $[M]$ that is determined later does not affect predetermined elements. *A coupled design* exists whenever the adjustment of one element in $[S]$ affects more than one element $[M]$ regardless of how other elements in $[S]$ are adjusted. In such a case, all elements of $[M]$ have to be assembled as one FEA model, so that the obtained $[S]$ can satisfy the constraints by design couplings among the elements of $[M]$.

Fig. 2.11 gives a decoupled example about what level of the decomposition a structure should have to analyze a specific part of the structure (Oliver et al., 1997). It is an assembled model of a pocket knife and the designer is interested in the maximized torque one could apply on the cork screw subjected to the given safety of factor. The decomposition of the design problem can be made based on the assembly structure of the product. Based on ADT, the cork screw can be separated from the assembled model and used in the FEA modeling process to find the answer to the concerned problem independently.

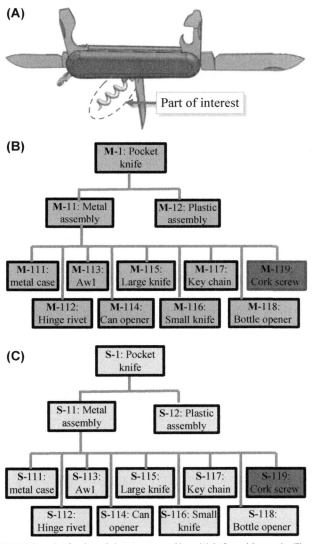

■ **FIGURE 2.11** Example of a decoupled engineering problem: (A) knife model example; (B) model domain; (C) solution domain.

The decomposition may not be always possible if system components are functioned together to achieve a given objective of the system. In such a case, the entire model has to be treated as an integral computable domain of an FEA model. Fig. 2.12 gives an example where the FEA simulation has to be run on the whole model (Bi and Wang, 2012; Zhang et al., 2015).

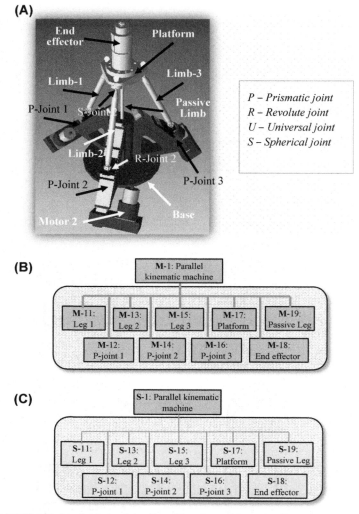

■ **FIGURE 2.12** Example of a coupled problem: (A) product model example; (B) model domain; (C) solution domain.

It is a parallel kinematic machine tool, and the design is to ensure the stresses over the structure are below the yield strengths of materials for the character-ized loads at its end effector. Applying ADT in Fig. 2.12 shows that the stress distribution in one component is affected by that of other components in the structure. Therefore, such a structure is not decomposable.

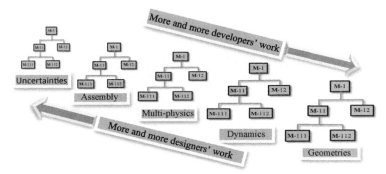

■ **FIGURE 2.13** Designer's role in formulating an finite element analysis model.

Fig. 2.2 has shown that five main factors contribute to the complexity of an engineering problem. The corresponding FEA model should be as specific as possible so that the level of uncertainty in the model can be minimized. FEA users are mainly responsible to prepare for data in formulating an engineering problem as an FEA simulation model. Note that an FEA code is functioning to convert system inputs to system outputs; if inputs are uncertain, it is unrealistic to expect a certain answer to the original problem. Therefore, an FEA user should have a good understanding of FEA theory and engineering knowledge to formulate FEA problems. More specifically, Fig. 2.13 illustrates the roles of FEA users and developers in tackling the complexity from different aspects. The more uncertainties an engineering problem has, the more critical the users' inputs are. Computer programs by developers can take over design activities to find the right solutions to engineering problems only if these problems are well formulated.

2.6 MODULAR ARCHITECTURE OF FEA CODE

In applying a commercial FEA code, an engineering problem is formulated to clarify correct inputs to corresponding simulation models, although the detailed implementation of commercial code is hidden from users. However, the organization of graphic user interfaces in the majority of commercial FEA tools is modularized. As shown in Fig. 2.14, every aspect of the complexity is associated with a set of inputs and settings for an FEA model. For an example, if an engineering problem is to find the strength of the structure under static loading conditions, the user should specify "*static analysis*" as an analysis type. The software will adopt the corresponding model template and solver to match the mathematic model of static analysis. The complexities related to uncertainties, assemblies, dynamics, and geometries are

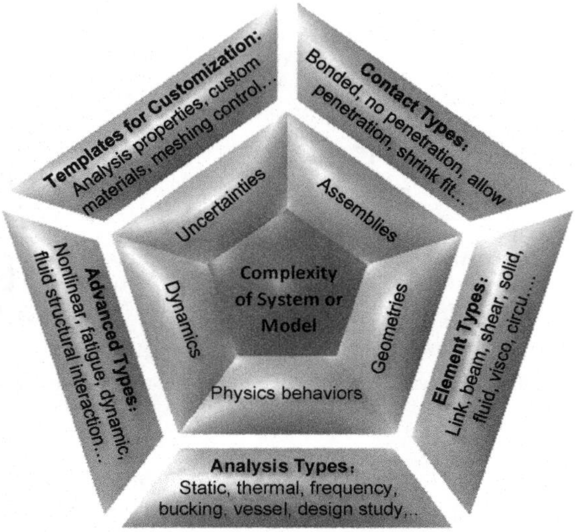

■ **FIGURE 2.14** Architecture of finite element analysis code for system complexity.

treated in a similar way; users are required to specify the modules or settings based on their understanding on engineering problems. Therefore, the importance of understanding the engineering problem itself can never be overemphasized.

2.7 DECOMPOSITION

The complexity of an FEA model is discussed in this section. An FEA user tends to focus more on the complexity by geometric shapes embedded in a CAD model. A typical CAD model usually represents the shape and structure in details, and it may also contain cosmetic features or manufacturing details that can prove to be extraneous in the FEA simulation. The user should make his or her engineering judgment into examining CAD models and deciding if some features should be removed or simplified before the decomposition. Starting with a simple model and adding complexity is almost always easier than starting with a complex model and simplifying it later.

Fig. 2.14 shows major options of FEA model setups at the abstract level. Once a high-level option is selected, additional inputs can be provided to tailor the FEA model to the given problem. For example, if an object is to be discretized as beam elements, and beam element types will be used where governing equations of elements are for beams. In addition, the number and sizes of elements and nodes are not determined automatically. An FEA code provides the interfaces for users to provide more inputs. Even though the software has its default settings (e.g., default mesh size) to ensure that the simulation generates the result with minimal inputs, the obtained result should be refined.

Here, we discuss how a continuous domain is decomposed into discretized elements and nodes; the result of such a decomposition is called *a mesh*. This procedure is commonly known as a *meshing process*, which is covered mostly by other textbooks.

A meshing process serves two primary purposes: (1) to discretize a continuous domain into a set of elements and nodes, so that the problem has a finite number of unknowns; (2) it approximates the system solution by assembling a set of simple submodels for elements. Even though the meshing process is performed by the computer automatically. Manual intervention is needed to generate a good mesh.

2.7.1 **Preparation of model**

A model with complex geometric shapes requires simplification before it can be meshed. For example, a product model may include some features used for manufacturing processes. These features should be defeatured if their impact on physical behaviors is ignorable. As some general rules, the features in an FEA model must (1) be meshable, (2) allow for the creation of a mesh that will correctly model the data of interest, (3) allow for the creation of a mesh solvable within a reasonable time. FEA software suppliers recommended the following tips for the preparation of FEA simulation (Andrei, 2017).

Defeature. A product model often contains the detailed information necessary to make and assembly parts. Not all of the details are useful in FEA modeling. Less important parts or features should be suppressed. An example of defeaturing is shown in Fig. 2.15, many products are assembled by mechanical fasteners. However, the detailed model on fasteners or threads on bolts or nuts is unnecessary for FEA simulation, and they should be removed before a meshing process.

Idealization. Idealization is to convert a detailed product model into an ideal model for a mesh with certain element types. For examples, (1) if truss or beam members reflect the behaviors of a structure better, all of the structural members should be remodeled accordingly so that the structure can be modeled by truss or beam elements. (2) If the product is made of sheet metal or it mainly consists of thin-wall panels, solid models should be converted into surface models so that shell elements can be used in the meshing process. Fig. 2.16 shows the case for the idealization from a solid model to

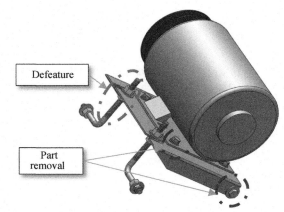

■ **FIGURE 2.15** Suppressed features or parts.

(A) **(B)**

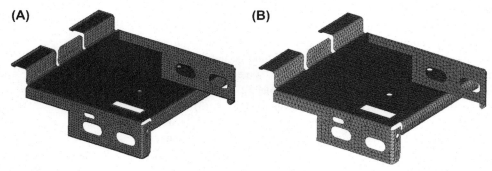

■ **FIGURE 2.16** Idealization of shell elements for solid model. (A) Mesh with solid elements; (B) mesh with shell elements.

the surface model. Such an idealization creates an abstract geometry for a plane stress or plane strain analysis. If the thickness is thin, plane stress elements can be used. On the other hand, if the dimension along one axis is extremely large in contrast to those in other two axes, plane strain elements can be used.

Clean-up is to address geometric quality issues that may affect the completion of meshing process and/or solving process. The most common issues to an assembly model occur to the interferences of solid objects. One has to ensure that any interference must be removed before the model is meshed. Features that are designed for manufacturing processes may not affect state variables of interest. However, a large number of elements are required to represent these small features. They should be cleaned before the meshing process performs. There are some other features, which are desirable to be

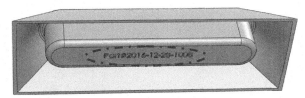

■ FIGURE 2.17 Example of unnecessary features to be cleaned-up.

cleaned, for an example, some drafts in casting. Fig. 2.17 shows an example of a part with the feature for a product number, which should be cleaned up if an FEA model is developed.

An automated meshing process can fail if a model has some quality issues such as duplicated entities, floating objects, or some tangent edges. These quality issues have to be addressed in the preparation of FEA simulation.

2.7.2 **Mesh and element types**

In regards to geometric complexity, a continuous domain is discretized as a mesh with a set of finite elements and nodes. The process for the discretization is referred as a *meshing process*. Even though all of the calculations for a mesh are performed by computers automatically, users need to specify the settings for meshing. On the one hand, the elements in a mesh should be small enough to generate results with an acceptable level of accuracy; on the other hand, element sizes should be large enough to make the amount of computation in meshing and solving processes manageable. To improve the fidelity of simulation, small elements, or element types with high orders, are usually required when state variables are very sensitive to spatial locations and their values are changed rapidly. To reduce the computation, large elements are encouraged where state variables are insensitive to spatial changes.

As shown in Fig. 2.18, based on how a continuous domain is discretized, a mesh can be categorized as one of three types, i.e., *structured*, *unstructured*, and *hybrid* mesh. In a structured mesh, the number of elements around an internal node is constant and the connectivity of the mesh can be calculated rather than stored explicitly. A structured mesh is simpler and consumes less computer memory; but it lacks the flexibility to deal with geometric complexity. In an unstructured mesh, the numbers of elements around an internal node are different from one node to another. An unstructured mesh is very flexible to deal with the geometric complexity; but it is relatively expensive in terms of time and memory needs. A hybrid mesh includes several regions where either structured or unstructured mesh is used for the spatial

(A) **(B)** **(C)**

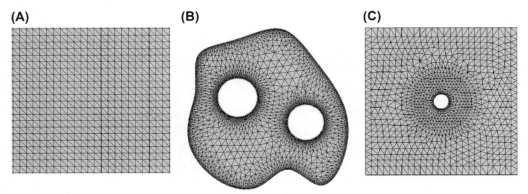

■ **FIGURE 2.18** Types of meshes in discretization. (A) Structured mesh; (B) unstructured mesh; (C) hybrid mesh.

decomposition. It is a trade-off discretization to balance the requirements of flexibility and computation. The regions of the continuous domain that are regular can adopt structured meshes and the other regions with complex boundaries can use unstructured ones.

Meshes can also be classified based on other criteria, for example, analysis type (static analysis, thermal, modal analysis, nonlinear), time dependence (steady and transient), and spatial dimensions. In terms of the dimensions, meshes can be classified into four categories: *one-dimensional, two-dimensional, three-dimensional,* and *other types* whose behaviors are irrelevant to spatial displacements. These mesh categories and types are listed in Table 2.2.

2.7.3 **Mesh quality**

A mesh must satisfy nearly contradictory requirements: (1) it must conform to the shape of the object or the computational domain in a simulation; (2) its elements may be neither too large nor too numerous; (3) it should not vary from small to large elements in very short distance; and (4) it must be composed of elements that are of right shapes and sizes. An element with *a right shape* typically refers to an element that is nearly equilateral and equiangular.

Elements can be distorted when they are mapping to a geometric shape. If an excessive distortion occurs, the element will be degenerated. In a meshing process, this problem can be alleviated by controlling the default element sizes or applying *mesh control* at local regions. The quality of mesh must be overseen based on the following criteria:

Table 2.2 Classification of Meshes

Category	Element Shapes and Types	Applications
1D: one of the dimensions (x) in very large in comparison to other two dimensions (y and z). The information on the cross-section must be defined	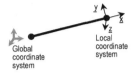 **Shape**: line **Types**: Link, beam, frame, and pipe.	Truss structures, long shafts, beams, construction frames, etc.
2D: one of the dimensions (z) in very small in comparison to other two dimensions (x and y). The information on the thickness direction z must be defined	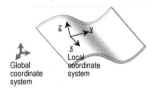 **Shape**: triangular and quad **Types**: plate, membrane, shell, plane stress, plane strain, axisymmetric elements.	Sheet metal parts, instrumental panels, roofs, autobodies, airplane bodies, axisymmetric problems, etc.
3D: the dimensions along three directions (x, y, and z) are comparable. The details of nodes provide complete information of a solid element in terms of its geometry.	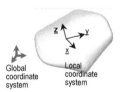 **Shape**: tetrahedral, hexahedral, pyramid, pentahedral, etc. **Types**: solids	Engines, machine beds, fixtures, furniture, appliances, structures, etc.
Others: the behaviors of elements are irrelevant to spatial shapes or displacements.	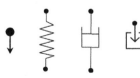 **Shape**: not applicable. **Types**: *mass* as a point element for centralized gravity load, *spring* as a point element for transitional or rational load, *damping* as a point element for damping coefficient, and *gap* as an element for distance, friction, and stiffness.	Idealization of corresponding mechanical or thermal elements or physical phenomena in a numerical simulation.

Aspect ratio: A mesh with uniform triangular or tetrahedral elements whose edges have similar lengths usually leads to the better accuracy of simulation results. If a model includes small edges, curved boundaries, thin features, and share corners, a computer-generated mesh may include some elements whose edges are much different in length with each other. As a consequence, the accuracy of simulation deteriorates. *The aspect ratio* is introduced to measure the distortions of elements. However, the definition of the aspect ratio varies with the types of elements. Table 2.3 shows the definitions for several element types.

Jacobian check: Nonlinear elements fit the curved geometry better than linear elements with the same size. Middle nodes on the edges of a nonlinear element are placed on actual geometry of model. However, if the boundaries are very sharp, middle nodes may result in the distorted elements; this causes the issue where edges are overlapped with each other. This problem can be identified the Jacobian check.

The Jacobian ratio at a point inside an element measures the degree of element distortion at that location. The Jacobian check is based on a number of Gaussian points (e.g., 4, 16, or 29) located within each element. In the Jacobian check, the Jacobian determinant of each point is calculated, and the Jacobian ratio is found by the ratio of the maximum and minimum determinant values. The Jacobian determinant of a two-dimensional (2D) element is calculated after it has been projected to a plane, and the determinant of three-dimensional (3D) elements is found by direct calculation. In either case, the Jacobian of an overdistorted element becomes negative. If an element with a negative Jacobian is found, the analysis program should be stopped before the mesh is fixed. An element with the perfect shape will be used to determine the Jacobian matrix for a point. Fig. 2.19 shows a case of a quadratic triangular element with a negative Jacobian ratio. This causes self-intersecting of geometry. After we learn the shape functions of elements, interested students can come back to look into how the Jacobian matrix for the mapping from a reference geometry to actual geometry of element can be determined.

Besides aspect ratio and Jacobian check, there are other criteria to check the mesh quality, such as *skewness*, *warpage*, and *taper*. Users can choose one or more criteria in the mesh control to ensure no abnormal element is included in an FEA model.

2.7.4 **Meshing methods**

The process to generate a mesh to be analyzed is referred as a meshing process. All of the calculation for a mesh is performed by the computer

Table 2.3 Definition of Aspect Ratios of Different Element Types

Element	Definition	Ideal Element	Bad element
Triangle element	The radius ratio of the circumscribe circle and inscribe circle: $a = \dfrac{R}{r}$		
Rectangle element	The ratio of the length L and height H ($L \geq H$): $a = \dfrac{L}{H}$		
Tetrahedral element	The ratio of the longest edge and the shortest normal projected from a vertex to the opposite face $a = \dfrac{L}{H}$		
Wedge element	The ratio of the longest edge and the shortest edge $a = \dfrac{\max\_L}{\min\_L}$		
Hexahedral element	The ratio of the longest edge and the shortest edge $a = \dfrac{\max\_L}{\min\_L}$		

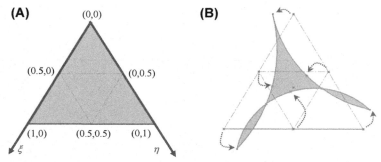

■ FIGURE 2.19 Example of Jacobian check in quadratic-triangular element. (A) Reference perfect quadratic-triangular element; (B) distorted element with negative Jacobian ratio.

automatically; users only need to specify the settings for meshing. However, it is a very common case that a meshing process runs into some problems to achieve the required quality of mesh. To solve a problem occurring in the meshing process, one often needs to refine the model and adjust meshing parameters. Some knowledge on the computation implementation of the mesh generation is helpful.

The mesh quality is ensured by the Delaunay property (Delaunay, 1934). In 2D, a *Delaunay triangulation* for a set P of points in the plane is a triangulation $DT(P)$, such that no point in P is inside the circumcircle of any triangle in $DT(P)$ except for three points of P that conform the triangle. Fig. 2.20A shows an example of the Delaunay triangulation as a mesh for a 2D region defined by a set of nodes. Fig. 2.20B and C show the case that two

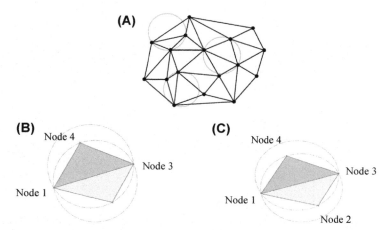

■ FIGURE 2.20 Illustration of the Delaunay property: (A) example of a Delaunay triangulation as a mesh in 2D space (van Kreveld, 2016); (B) no achieved; (C) achieved.

triangulations for a set of four nodes. The left triangulation does not follow the Delaunay property and the right one does. In 3D, the concept remains the same but replaces the circumcircle by the circumsphere. In particular, 2D Delaunay triangulation maximizes the minimum angle of all the triangles in the tessellation.

A Delaunay triangulation does not contain any illegal edge (such as the connection between nodes 1 and 3 in Fig. 2.20). Therefore, the algorithms of the Delaunay triangulation for a meshing process rely on the fast operations for detecting when a point is within a triangle's circumcircle and an efficient data structure to store triangles and edges. The most important algorithm to perform this operation is called as the flip algorithm, which is described as follows (van Kreveld, 2016):

Flip algorithm. *Input*. Given a triangulation T of a point set P. *Output*. A Delaunay triangulation.

```
while the set T contains an illegal edge pᵢpⱼ
    do (%flip edge pᵢpⱼ);

    let pᵢpⱼpₖ and pᵢpⱼpₗ be two triangles adjacent to pᵢpⱼ;

        remove pᵢpⱼ from T and add pₖpₗ instead;
update T and return while.
```

The flip algorithm is generalized to three or higher dimensions; however, it does not guarantee the convergence in these cases. In addition, the computation to achieve a Delaunay triangulation is high; it can take $O(n^2)$ flips (Hurtado et al., 1999). An alternative algorithm is *an incremental algorithm*: points in the set P are repeatedly added one by one to a triangulation, when a new point p is added, the affected parts of the graph are retriangulated by the flip algorithm; it takes $O(n)$ time for each added point. Because both of these algorithms have been well developed and implemented in commercial FEA codes, their implementations are not discussed here.

2.7.5 Mesh control

An FEA tool usually supports users to control meshing settings in local regions such as vertices, edges, faces, parts, and components. This is achieved by the *mesh control* tool. To define a mesh control, the applied entity is selected, and the meshing settings such as element sizes are specified (Fig. 2.21).

In meshing control, the mesh density should be determined based on the gradients of state variables; if the gradient is higher, a finer mesh should

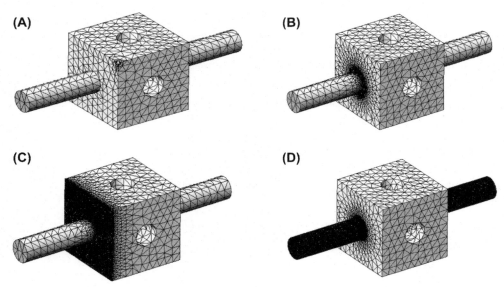

■ FIGURE 2.21 Example of mesh controls on different types of entities: (A) vertices; (B) edge; (C) surface; (D) part/components.

be used. The regions with a high gradient are often with the discontinuities of geometries, materials, and load conditions. Fig. 2.22 shows some examples of critical areas with a high gradient of state variables. If a region is far away from these critical areas, low-density mesh can be used to reduce the number of unknowns with little impact on the accuracy of system solution (Felippa, 2016).

2.8 SUMMARY

The complexity of various engineering problems is discussed; the sources of complexity are classified into five groups: uncertainties, dynamics, multidisciplinary behaviors, assemblies, and geometric shapes. It is argued that the modular architecture is desirable to facilitate the FEA implementation of the top-down and bottom-up approaches. In such a way, the complexity from five groups can be tackled by selecting right functional software modules and integrating them accordingly as holistic solutions to given engineering problems. The ADT is introduced to guide the decomposition of an engineering problem and formulate it as an FEA simulation model. In regards to the complexity by geometric shapes, the meshing process is introduced, some important measures of mesh quality are discussed.

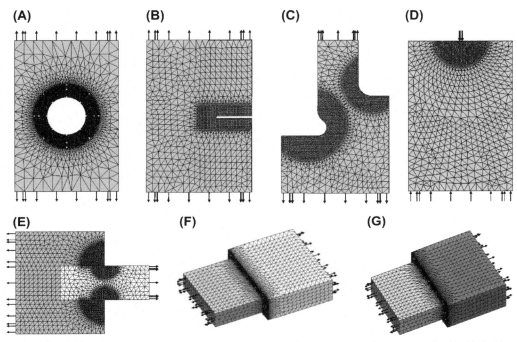

■ **FIGURE 2.22** Examples of common regions for a finer mesh: (A) cut-out; (B) seam; (C) entrant corner; (D) concentration; (E) welds; (F) shoulder; (G) bonded materials.

The Delaunay triangulation is used as an example to illustrate the meshing process. The main concepts covered in this chapter are

1. The complexity of an FEA model depends on five major aspects including uncertainties, dynamics, assemblies, multiphysics, and geometric shapes.
2. FEA is a problem-solving tool for explicit and approximated solutions to implicit mathematical models with PDEs.
3. Designers' roles are critical in developing good FEA models.
4. The systematic ADT can be applied to decompose system in FEA modeling.
5. A meshing process is to decompose the complexity from geometric shapes. Delaunay triangulation is a fundamental method to discretize a continuous domain into a mesh.
6. Meshing processes are highly automated; however, manual interventions are essential in preparing for good CAD models and set up mesh control in the FEA modeling process.

REFERENCES

Andrei, L., 2017. MECH4460: Mechanical Design. http://sydney.edu.au/engineering/aeromech/MECH4460/Course_Documents/FEA/FEA/Lecture%20and%20problem/tricks%20tips%20chpt.pdf.

Bi, Z.M., Wang, L., 2012. Energy modeling of machine tool for optimization of machine setup. IEEE Transaction on Automation Science and Engineering 9 (3), 607–613.

Bi, Z.M., Kang, B., 2014. Sensing and responding to the changes of geometric surfaces in flexible manufacturing and assembly. Enterprise Information Systems 8 (2), 225–245.

Delaunay, B., 1934. Sur la sphere vide. Izvestia Akademii Nauk SSSR, Otdelenie Matematicheskikh i Estestvennykh Nauk 7, 793–800.

Felippa, C., 2016. Introduction to Finite Element Methods (ASEN5007). University of Colorado, Boulder. http://www.colorado.edu/engineering/cas/courses.d/IFEM.d/.

Hubka, V., Eder, W.E., 1988. Theory of Technical Systems. Springer, New York. ISBN 978-3-642-52121-8.

Hurtado, F., Noy, M., Urrutia, J., 1999. Flipping edges in triangulations. Discrete & Computational Geometry 22 (3), 333–346.

Oliver, D.W., Kelliher, T.P., Keegan Jr., J.G., 1997. Engineering Complex Systems with Objects and Models. McGraw Hill, NY.

Suh, N.P., 1990. The Principles of Design. Oxford University Press. ISBN 0-19-504345-6.

Suh, N.P., 2005. Complexity: Theory and Applications. Oxford University Press, New York. ISBN-13: 978-0195178760.

van Kreveld, M., 2016. Geometric Algorithms. http://www.cs.uu.nl/docs/vakken/ga/slides9alt.pdf.

Zhang, J., Zhao, Y.Q., Jin, Y., 2015. Elastodynamic modeling and analysis for an eechon parallel kinematic machine. Journal of Manufacturing Science and Engineering 138 (3), 031011.

FURTHER READING

Accel, 2016. Productivity Improvement. http://www.accel-team.com/productivity/productivity_01_what.html.

Arinez, J.F., 1998. An Equipment Design Approach for Achieving Manufacturing System Design Requirements. Mechanical Engineering. Massachusetts Institute of Technology.

Becker, S.H., Koziolek, H., Reussner, R., 2009. The Palladio component model for model-driven performance prediction. Journal of Systems and Software 82 (1), 3–22.

Latzko, W.J., 2016. Manage Knowledge Workers for Lean Process Quality. In: http://latzko-associates.com/Publications/Managing%20Knowledge%20Workers%20for%20Lean%20Process%20Quality(QUIP)Revised.pdf.

Law, A., 2006. Simulation Modeling and Analysis, fourth ed. McGraw Hill Higher Education.

Magee, C.L., de Weck, O.L., 2004. Complex System Classification. https://pdfs.semanticscholar.org/8ebb/f076136aaf0b1578db2fbf6be1d389099980.pdf.

Szabo, B., Babuska, I., 2011. Introduction to finite element analysis-formulation, verification and validation. In: Wiley Series in Computational Mechanics. Wiley, ISBN 978-0-470-97728-6.

■ PROBLEMS

2.1. Create a model for a cantilever beam with a dimension of 300 × 10 × 5 (mm), use 1D, 2D, and 3D members to mesh the model. Set the mesh size as 5 mm, and find the details of meshes for the corresponding models with beam elements, sheet elements, and solid elements.

	Number of nodes	Number of elements	Mesh display
CAD model			
1-D			
2-D			
3-D			

2.2. Create a cup model with its main dimensions shown in the drawing, and minor features can be customized. Do the mesh in Solidworks and find the details about the meshes in two settings: (1) the general mesh size of 0.2-in, (2) applied mesh control on the handle and annotation Indiana University − Purdue University Fort Wayne with the mesh size of 0.1-in. Compare the numbers of elements and nodes in two meshes; comment on the necessity when finer mesh is needed at the handle region (Fig. 2.23).

2.3. Analyze the design of a structural support for a wind turbine with the following specification:

Type	Horizontal-Axis Wind Turbine
Configuration	Stall-regulated, passive-yawing
Weight	500 lbf
Number of blades	3
Rotor diameter	13 ft
Rated power	3-kW
Rated rotor speed	250 rpm
Rate wind speed	22.5 mph
Rotor tip to wind speed ratio λ	4
Cut-off wind turbine speed	300 rpm

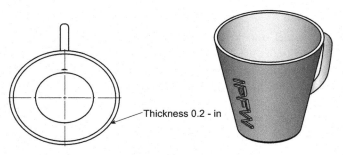

Thickness 0.2 - in

All lengths are inch and angle is in degree

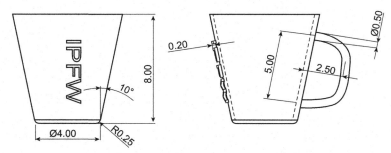

0.20

8.00

10°

Ø4.00

R0.25

5.00

2.50

Ø0.50

■ **FIGURE 2.23** Reference drawing of a cup model.

The structure must stand over 70 ft; it uses ASTM A572 structural steel. The foundation consists of piles which are made from reinforced concrete. The swept area of the turbine is 1.5×10^4 in^2. The structure must provide the safe support to the wind turbine, it must be operable for more than 10 years, and it must involve in a minimal cost. For this structural design, (1) formulate the design problem, (2) discuss the types of uncertainties involved in your problem definition, (3) identify FRs and design variables and build their correspondences, (4) decompose high-level FRs until you have a direct solution to an individual FR at the lowest level, (5) easy access for maintenance, (6) natural frequencies are far away from operational frequency, and (7) comment on how numerical simulations can assist you to achieve your design objectives (find the solution to FRs).

Chapter 3

Description of Elements

In a finite element analysis (FEA) approach, elements and nodes are basic units to build a system model. A continual domain is represented by a set of elements and nodes, and the behaviors of a continual domain are approximated by nodal values of state variables in a solution space. Elements in an FEA model are continuous subdomains, although the solution to the FEA model is for discretized state variables on nodes. Their correlations must be modeled. The behavior at an arbitrary location in an element has to be derived from the behaviors of associated nodes. In this chapter, nodes, elements, and coordinate systems (CS) are introduced, and the relations of these entities are modeled. As an application, the developed relations are applied to perform the numerical integration.

3.1 NODES AND ELEMENTS

After a continuous domain is discretized, it is represented by elements and nodes. *Elements* are continuous subdomains; *nodes* are a set of vertices used to represent elements. The behaviors of a physical object are described in terms of state or field variables. In an FEA model, a state variable at an arbitrary position depends on nodal variables in the element. State variables associated with the nodes include both of (1) quantities to represent spatial positions of nodes and (2) quantities to represent physical behaviors of elements. Note that the first type of variables depends on the type of used CS.

3.2 TYPES OF COORDINATE SYSTEMS (CS)

Nodal positions are specified as the coordinates in a selected CS. Different CS can be used to describe nodes and elements for different purposes. Three commonly used CS in FEA are *global CS* (GCS), *local CS* (LCS), and dimensionless *natural CS*. These CS will be discussed in this section.

Because different CS are applied alternatively, the representations of nodes in one CS should be transformed into the representation in another CS. In the next section, the coordinate transformation will be discussed.

Finite Element Analysis Applications. http://dx.doi.org/10.1016/B978-0-12-809952-0.00003-0

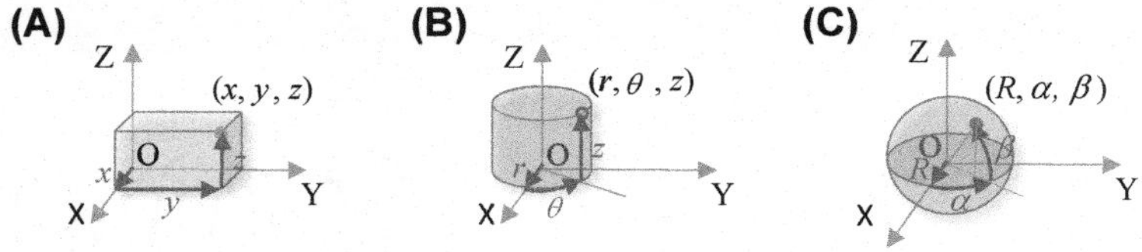

■ **FIGURE 3.1** Three commonly used coordinate systems (CS). (A) Cartesian CS; (B) cylindrical CS; (C) spherical CS.

A position within an element can be specified using different types of CS. Three common types of CS are illustrated in Fig. 3.1, namely, *Cartesian CS*, *cylindrical CS*, and *spherical* CS. Conventionally, the axes of coordinates in these CS are defined using the right-hand rule. A Cartesian CS consists of three axes, which are perpendicular to each other mutually. A position in the Cartesian CS is defined by its distance to the origin projected on three axes (X, Y, Z). A cylindrical CS consists of two linear axes (X and Z) and one rotational axis. Correspondingly, a position in the cylindrical CS is defined by two scalar variables and one angular variable, i.e. (r, θ, z). A spherical CS consists of two rotational axes and one translational axis; a position in the spherical CS is defined by two rotational variables and one translational variable, i.e. (R, α, β).

3.3 GLOBAL AND LOCAL CS

A GCS is an absolute reference frame. In a process of FEA modeling, a GCS is usually attached on the base component of object. GCS is commonly referred by all of the elements in an FEA model to describe system-level properties. The contribution to certain quantity from elements can be added subjected to the same reference of CS. For example, if the gravity force is taken into the consideration in an FEA, it is convenient to align the z-axis of GCS with the direction of the gravity.

In other cases, it is more convenient to evaluate state variables under a local CS. For example, if a designer is interested in strains and stresses in a truss member, the calculation can be simplified when nodal displacements in the element are given in an LCS because an LCS can align its main axis with the axial direction of truss. Both GCS and LCS can be Cartesian CS, although an LCS differs from GCS for its origin as well as the directions of axes. Another example for the difference of LCS and GCS is shown in Fig. 3.2. LCS and GCS are defined to describe the motion of an industrial robot. It is a common practice to attach LCS on individual motion axes or

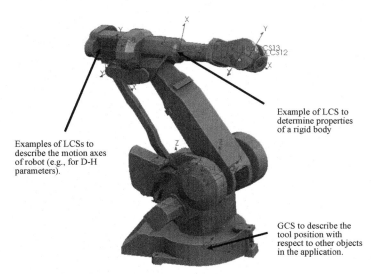

Examples of LCSs to describe the motion axes of robot (e.g., for D-H parameters).

Example of LCS to determine properties of a rigid body

GCS to describe the tool position with respect to other objects in the application.

■ **FIGURE 3.2** Difference of global and local coordinate systems.

rigid bodies. These LCS help in evaluating the dynamic properties of rigid bodies as well as the motions on actuated axes including Denavit—Hartenberg (D—H) parameters (Bi et al., 2006).

3.4 **COORDINATE TRANSFORMATION**

Different CS are established at different stages of an FEA modeling process. However, these CS are established on the same physical model, and the coordinates with respect to one CS are correlated to the coordinates with respect to another CS. The coordinate transformation from one CS to another CS is called as *coordinate transformation*.

3.4.1 **Coordinate transformation of GCS**

Section 3.2 gives three common types of GCS, and the coordinates in two GCS can be transformed with each other. It should be noted that the coordinate transformation from one CS to another CS is not always one-to-one mapping.

Cylindrical CS $(r, \theta, z) \leftrightarrow$ Cartesian CS (x, y, z):

$$\left. \begin{aligned} x &= r \cos \theta \\ y &= r \sin \theta \\ z &= z \end{aligned} \right\} \text{ and } \left. \begin{aligned} r &= \sqrt{x^2 + y^2} \\ \theta &= \tan^{-1} \frac{y}{x} \\ z &= z \end{aligned} \right\} \tag{3.1}$$

Spherical CS (R, α, β) ↔ Cartesian CS (x, y, z):

$$\left.\begin{aligned} x &= R \cdot \cos\beta \cdot \cos\alpha \\ y &= R \cdot \cos\beta \cdot \sin\alpha \\ z &= R \cdot \sin\beta \end{aligned}\right\} \quad \text{and} \quad \left.\begin{aligned} R &= \sqrt{x^2 + y^2 + z^2} \\ \alpha &= \tan^{-1}\frac{y}{x} \\ \beta &= \tan^{-1}\frac{z}{\sqrt{x^2 + y^2}} \end{aligned}\right\} \tag{3.2}$$

Spherical CS (R, α, β) ↔ Cylindrical CS (r, θ, z)

$$\left.\begin{aligned} r &= R \cdot \cos\beta \\ \theta &= \alpha \\ z &= R \cdot \sin\beta \end{aligned}\right\} \quad \text{and} \quad \left.\begin{aligned} R &= \sqrt{r^2 + z^2} \\ \alpha &= \theta \\ \beta &= \tan^{-1}\frac{z}{r} \end{aligned}\right\} \tag{3.3}$$

Note that multiple solutions occur in the second group of relations in Eqs. (3.1)–(3.3).

3.4.2 **From LCS to GCS**

To transform the coordinates from an LCS to GCS, the spatial relations of origins and coordinate axes of two CSs are found out, and then, a homogenous transformation is determined accordingly. Here, a transformation between two Cartesian CSs is used as the example. Such a transformation is often required in an FEA modeling.

3.4.2.1 CS transformation in one-diemsnional space

As shown in Fig. 3.3, a one-dimensional element such as a truss member only involves one coordinate to define its nodes or any point within the element. Therefore, the only difference of a GCS and an LCS is reflected by an offset of their origins. In Fig. 3.4, the element has two nodes, i.e., nodes i and j. The coordinates of nodes i and j under GCS are X_i and X_j, respectively. If the origin of an LCS is set at node i, the offset of its origin o from that of GCS is the coordinate of node i with respect to GCS. If the coordinates of nodes are transformed from GCS to LCS, the coordinates of two nodes with respect to the LCS are 0 and l, respectively; where l is the length of the element, i.e., $l = X_j - X_i$. For any position A within the element, the coordinate transformation can be performed by,

From GCS (X) to LCS (x):

$$x = X - X_i \tag{3.4}$$

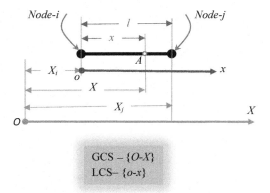

■ **FIGURE 3.3** Example of coordinate transformation for one-diemnsional element.

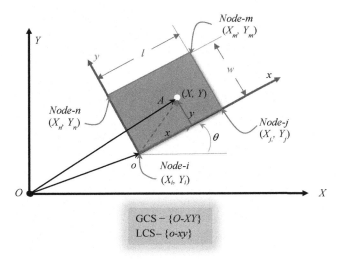

■ **FIGURE 3.4** Coordinate transformation for two-dimensional element.

From LCS (x) to GCS (X):

$$X = X_i + x \qquad (3.5)$$

3.4.2.2 *CS transformation in two-dimensional space*

As shown in Fig. 3.4, a two-dimensional rectangle element consists of four nodes (node i, j, m, and n). The coordinates are (X_i, Y_i) where $i = 1,2,3,4$, respectively, in the GCS $\{O\text{-}XY\}$. The length and width of the elements are l and w. An LCS {o-xy} is set at node i, whose x-axis is defined by an angle θ, which is measured from the length direction of the element with respect to the X-axis of GCS in the counterclock-wise direction. To transform an arbitrary point A (x, y) from the LCS to GCS (X, Y), Fig. 3.4 shows a vector loop for their relations:

$$\boldsymbol{OA} = \boldsymbol{Oo} + \boldsymbol{oA} = \boldsymbol{Oo} + x\cdot\boldsymbol{x} + y\cdot\boldsymbol{y} \tag{3.6}$$

where,

$\boldsymbol{OA}, \boldsymbol{oA}$ are the vectors of an arbitrary point A with respect to the origins of GCS and LCS, respectively.

$\boldsymbol{Oo}$ is the vector from the origin of GCS to that of LCS.

x, y are the coordinates of a point in the LCS.

$\boldsymbol{x} = \begin{Bmatrix} \cos\theta \\ \sin\theta \end{Bmatrix}; \boldsymbol{y} = \begin{Bmatrix} -\sin\theta \\ \cos\theta \end{Bmatrix}$ are the vectors of axes of the LCS represented in the GCS.

Eq. (3.6) can then be converted into a scalar formation as,

$$\begin{Bmatrix} X \\ Y \end{Bmatrix} = \begin{Bmatrix} X_i \\ Y_i \end{Bmatrix} + x\begin{Bmatrix} \cos\theta \\ \sin\theta \end{Bmatrix} + y\begin{Bmatrix} -\sin\theta \\ \cos\theta \end{Bmatrix} \tag{3.7}$$

For the computer implementation, it is an advantage to represent Fig. 3.4 as an expression with the homogeneous matrix T,

$$\begin{Bmatrix} X \\ Y \end{Bmatrix} = \begin{Bmatrix} X_i \\ Y_i \end{Bmatrix} + x\begin{Bmatrix} \cos\theta \\ \sin\theta \end{Bmatrix} + y\begin{Bmatrix} -\sin\theta \\ \cos\theta \end{Bmatrix}$$
$$\begin{Bmatrix} X \\ Y \\ 1 \end{Bmatrix} = [T]\cdot\begin{Bmatrix} x \\ y \\ 1 \end{Bmatrix} \tag{3.8}$$

where $[T]$ is the homogeneous matrix to transform a point from LCS to GCS, i.e.,

$$[T] = \begin{bmatrix} \cos\theta & -\sin\theta & X_i \\ \sin\theta & \cos\theta & Y_i \\ 0 & 0 & 1 \end{bmatrix} \tag{3.9}$$

If a reverse transformation is needed from GCS to LCS, Eq. (3.8) can be rewritten as,

$$\begin{Bmatrix} x \\ y \\ 1 \end{Bmatrix} = [T]^{-1}\cdot\begin{Bmatrix} X \\ Y \\ 1 \end{Bmatrix} \tag{3.10}$$

where $[T]^{-1}$ is the inverse matrix of $[T]$, i.e.,

$$[T]^{-1} = \begin{bmatrix} \cos\theta & \sin\theta & -X_i \\ -\sin\theta & \cos\theta & -Y_i \\ 0 & 0 & 1 \end{bmatrix} \tag{3.11}$$

Example 3.1

Fig. 3.5 shows a two-bar mechanism where the tool is attached on point B, the approaching direction (x) of the tool is aligned with the axial direction of the binary bar AB. The positions and lengths of two bars under the GCS {O-XY} are given as $L1 = 4$ (in) at $\theta_1 = 30$ degrees and $L2 = 3$ (in) at $\theta_2 = 60$ degrees. Find the homogenous matrix for the coordinate transformation from LCS {o-xy} to GCS {O-XY}.

Solution

In the GCS, the x-axis of LCS is $\theta = 60$ degrees along the counterclock-wise direction; the origin of local CS is at B whose coordinates in GCS can be found as,

$$X_i = L1 \cdot \cos 30° + L2 \cdot \cos 60° = 4.6 \text{ in}$$
$$Y_i = L1 \cdot \sin 30° + L2 \cdot \sin 60° = 4.96 \text{ in} \tag{3.12}$$

Therefore, the transformation matrix from LCS to GCS is,

$$[T] = \begin{bmatrix} \cos\theta & -\sin\theta & X_i \\ \sin\theta & \cos\theta & Y_i \\ 0 & 0 & 1 \end{bmatrix} = \begin{bmatrix} 1/2 & -\sqrt{3}/2 & 4.6 \\ \sqrt{3}/2 & 1/2 & 4.96 \\ 0 & 0 & 1 \end{bmatrix} \tag{3.13}$$

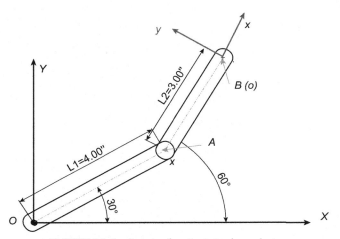

■ **FIGURE 3.5** Coordinate transformation in two-bar mechanism.

3.4.2.3 CS transformation in three-dimensional space

As shown in Fig. 3.6, a three-dimensional brick element consists of eight nodes (node $i, j, k, l, m, n, o,$ and p), whose coordinates are (X_i, Y_i, Z_i) where $i = 1, 2, 3,...8$ in GCS $\{O\text{-}XYZ\}$, respectively. An LCS $\{o\text{-}xyz\}$ is attached on node i, whose x-axis is aligned with the length direction (a unit vector n_x from node i to node j) of the element. Accordingly, y-axis is aligned with the width direction (a unit vector n_y from node i to node l), and z-axis is aligned with the height direction (a unit vector n_z from node i to node m). To transform an arbitrary point A (x, y, z) from LCS to GCS (X, Y, Z), Fig. 3.6 illustrates a vector loop for their relations:

$$\boldsymbol{OA} = \boldsymbol{Oo} + \boldsymbol{oA} = \boldsymbol{Oo} + x \cdot \boldsymbol{n_x} + y \cdot \boldsymbol{n_y} + z \cdot \boldsymbol{n_z} \tag{3.14}$$

where,

> $\boldsymbol{OA}, \boldsymbol{oA}$ are the vectors of the arbitrary position A with respect to the origins of GCS and LCS, respectively.
> $\boldsymbol{Oo}$ is the vector from the origin of GCS to that of LCS.
> x, y, z the coordinates of a point in the LCS.
> $\boldsymbol{n_x}, \boldsymbol{n_y}, \boldsymbol{n_z}$ are the vectors of axes of the LCS with respect to the GCS.

Eq. (3.14) can then be converted into the scalar formation as,

$$\begin{Bmatrix} X \\ Y \\ Z \end{Bmatrix} = \begin{Bmatrix} X_i \\ Y_i \\ Z_i \end{Bmatrix} + x \cdot \begin{Bmatrix} n_{x,x} \\ n_{x,y} \\ n_{x,z} \end{Bmatrix} + y \cdot \begin{Bmatrix} n_{y,x} \\ n_{y,y} \\ n_{y,z} \end{Bmatrix} + z \cdot \begin{Bmatrix} n_{z,x} \\ n_{z,y} \\ n_{z,z} \end{Bmatrix} \tag{3.15}$$

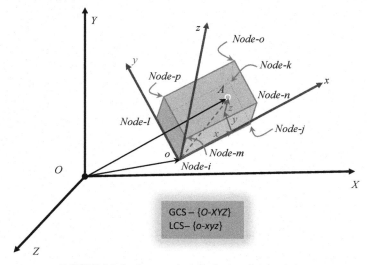

■ **FIGURE 3.6** Coordinate transformation for three-dimensional element.

where

$$n_x = \begin{Bmatrix} n_{x,x} \\ n_{x,y} \\ n_{x,z} \end{Bmatrix}; n_y = \begin{Bmatrix} n_{y,x} \\ n_{y,y} \\ n_{y,z} \end{Bmatrix}; n_z = \begin{Bmatrix} n_{z,x} \\ n_{z,y} \\ n_{z,z} \end{Bmatrix}$$ are the expansion of vectors

of axes of the LCS.

The homogeneous representation of the coordinate transformation from LCS to GCS from Eq. (3.15) is defined as,

$$\begin{Bmatrix} X \\ Y \\ Z \\ 1 \end{Bmatrix} = [T] \cdot \begin{Bmatrix} x \\ y \\ z \\ 1 \end{Bmatrix} \tag{3.16}$$

where $[T]$ is the homogeneous matrix to transform a point from LCS to GCS, i.e.,

$$[T] = \begin{bmatrix} n_{x,x} & n_{y,x} & n_{z,x} & X_i \\ n_{x,y} & n_{y,y} & n_{z,y} & Y_i \\ n_{x,z} & n_{y,z} & n_{z,z} & Z_i \\ 0 & 0 & 0 & 1 \end{bmatrix} \tag{3.17}$$

Eq. (3.17) can be used to transform the point from GCS (X, Y) back to LCS (x, y) as,

$$\begin{Bmatrix} x \\ y \\ z \\ 1 \end{Bmatrix} = [T]^{-1} \cdot \begin{Bmatrix} X \\ Y \\ Z \\ 1 \end{Bmatrix} \tag{3.18}$$

where $[T]^{-1}$ is the inverse matrix of $[T]$, i.e.,

$$[T]^{-1} = [T]' = \begin{bmatrix} n_{x,x} & n_{x,y} & n_{x,z} & -X_i \\ n_{y,x} & n_{y,y} & n_{y,z} & -Y_i \\ n_{z,x} & n_{z,y} & n_{z,z} & -Z_i \\ 0 & 0 & 0 & 1 \end{bmatrix} \tag{3.19}$$

3.5 NATURAL CS AND SHAPE FUNCTIONS

So far, we have discussed how an element is described by different CS, and how the expression in one CS can be transformed into the expression in another CS. These types of the transformations are critical in FEA modeling. For examples, coordinate transformations are required to transform element models from LCS to GCS in system modeling. The simulation results have to be transformed from GCS to LCS in the postprocessing to determine dependent variables in elements. We will revisit these coordination transformations in coming chapters.

In the rest of the chapter, we are especially interested in deriving a state variable at any point in a continuous domain by nodal values of the variable in an element. For an arbitrary point in an element, it is interesting to know what its relative position is by taking nodal positions as references. To achieve this goal, we define shape functions. A *shape function* is a weight function to represent the influence of state variables at nodes on the state variable of an arbitrary point in element. The number of shape functions equals to the number of nodes that the element has. We will illustrate the meanings and significance of shape functions through examples.

A natural CS is first established to define the shape functions in an element. *A natural CS* is a dimensionless coordinate system to represent relative positions of a spatial point with respect to the nodes of element. The natural CS for one-dimensional element is illustrated in Fig. 3.7. The shape of one-dimensional linear element is normalized in such a way that

- the natural axis ξ is aligned with x-axis of an LCS; but it is dimensionless;
- the reference origin of this element is the geometric center of the element o;
- the total length of element is normalized as "2" units.

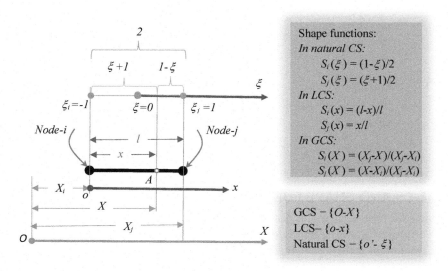

Shape functions:
In natural CS:
$$S_i(\xi) = (1-\xi)/2$$
$$S_j(\xi) = (\xi+1)/2$$
In LCS:
$$S_i(x) = (l-x)/l$$
$$S_j(x) = x/l$$
In GCS:
$$S_i(X) = (X_j-X)/(X_j-X_i)$$
$$S_j(X) = (X-X_i)/(X_j-X_i)$$

GCS – $\{O\text{-}X\}$
LCS– $\{o\text{-}x\}$
Natural CS – $\{o'\text{-} \xi\}$

■ **FIGURE 3.7** Coordinate transformation for one-dimensional element.

As the result, the relative distance from "0" to node i is "1" unit in the negative direction, and its coordinate under the natural CS is set as $\xi_i = -1$. The relative distance from "0" to node j is "1" unit in the positive direction, and its coordinate under the natural CS is $\xi_j = 1$.

The relative importance of the state variable of a certain node on the point of interest can be determined by the distance of the point to this node. The larger the distance is, the less importance of the state variable of this node on the state variable of the point is.

Denote the natural coordinate of an arbitrary point A as ξ, the state variable on A can be approximated by the values of the same state variable at node i and node j. The relative distances from A to nodes i and j are $(\xi + 1)$ and $(\xi - 1)$, respectively. The closer A is to node i, the more influence of the value of the same variable at node i is. Assume that the total weight from both nodes is set as "1", the weights from two individual nodes can be defined as,

$$\left.\begin{array}{l} S_i(\xi) = \dfrac{(1 - \xi)}{2} \\[3mm] S_j(\xi) = \dfrac{(1 + \xi)}{2} \end{array}\right\} \tag{3.20}$$

It is interesting to look into the properties of shape functions. For two shape functions $S_i(\xi)$ and $S_j(\xi)$ of one-dimensional element, one has,

1. For a shape function of a certain node, its value is 1 when the arbitrary point is specified at the same node, i.e.,

$$S_i(\xi) = 1 \text{ at node } i \text{ and } S_j(\xi) = 1 \text{ at node } j$$

2. For a shape function of a certain node, its value is 0 when the arbitrary point is specified at the other node, i.e.,

$$S_i(\xi) = 0 \text{ at node } j \text{ and } S_j(\xi) = 0 \text{ at node } i$$

3. The sum of all shape functions at any position is 1, i.e.,

$$S_i(\xi) + S_j(\xi) = 1 \text{ for any } \xi.$$

The above properties of the shape functions can be illustrated in Fig. 3.8. Later on, these properties of the shape functions will be utilized to derive the interpolation formula for high-order elements.

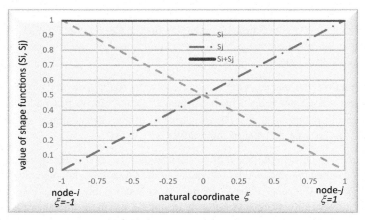

■ **FIGURE 3.8** Properties of shape functions.

Once the shape functions of an element are defined, they can be applied to interpolate state variables of any point when the values of state variables on nodes are given. For example, if the coordinates of a position within the element in different CSs are viewed as state variables, one can find the coordinate transformation from natural CS to LCS readily as,

$$x = S_i(\xi) \cdot (0) + S_j(\xi) \cdot (l) = \frac{(1 + \xi)l}{2} \tag{3.21}$$

From GCS to LCS, the coordinate transformation becomes,

$$\xi = \frac{2x}{l} - 1 \tag{3.22}$$

The coordinate transformation from the natural CS to GCS is,

$$X = S_i(\xi) \cdot X_i + S_j(\xi) \cdot X_j = \frac{(1 - \xi) \cdot X_i}{2} + \frac{(1 + \xi) \cdot X_j}{2} \tag{3.23}$$

From GCS to natural CS, the coordinate transformation is,

$$\xi = \frac{2X - (X_j + X_i)}{X_j - X_i} \tag{3.24}$$

where ξ, x, and X are the coordinates of an arbitrary point in element with respect to natural CS, LCS, and GCS, respectively.

3.6 INTERPOLATION

Because a continuous domain is represented by a set of discrete nodes, the relation of the behavior in any position of an element with those of nodes

Table 3.1 Example of State Variables in Different Applications

Application	State Variables
Solid mechanics	Displacement, stress, strain
Heat transfer	Temperature, temperature gradient
Fluid dynamics	Velocity, pressure, entropy
Electromagnetics	Voltage, amplitude, frequency

must be modeled. Such relational models are approximated by the interpolation in FEA modeling. In this section, the Taylor Exploration is first reviewed as the approximation approach; it is then applied to derive the interpolation equations for elements.

The coordinates give the spatial information over nodes; because physical behaviors of an element are represented discretely by nodes, state variables are associated with the coordinates of nodes. For example, if the temperature distribution is a concern in an FEA simulation, the temperature will be treated as a state variable at each node. State variables are directly associated with physical behaviors of the object under investigation, and they are changed with respect to positions. Table 3.1 gives an example of some common state variables in different applications.

As shown in Fig. 3.9, a basic question about the discretization is as follows: *if the behaviors of an element are represented by its behaviors on a set of discrete nodes, how to find the state variable at any point based on the given values at nodes of the same variable*? Here, we use two different methods to address this issue. The first one is based on the Taylor expansion formula and the second one is based on the properties of shape functions.

3.6.1 **Taylor expansion**

A *Taylor series* is a series of polynomial terms to approximate the function ($f(x)$) at a point ($x = a$). A one-dimensional Taylor series is given by

$$f(x) = f(a) + f'(a)(x - a) + \frac{f''(a)}{2!}(x - a)^2 + \cdots + \frac{f^{(n)}(a)}{n!}(x - a)^n + \cdots$$

$$(3.25)$$

If a is set as 0, the corresponding Taylor expansion is known as a Maclaurin series,

$$f(x) = f(0) + f'(0)x + \frac{f''(0)}{2!}x^2 + \cdots + \frac{f^{(n)}(0)}{n!}x^n + \cdots \qquad (3.26)$$

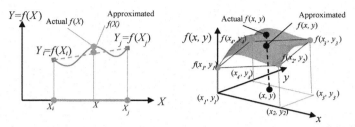

■ **FIGURE 3.9** Interpolations in one-dimensional and two-dimensional elements: (A) one-dimensional element; (B) two-dimensional element.

The *Taylor's theorem* states that any function $f(x)$ satisfying certain conditions can be expressed as a Taylor series: assume $f^{(n)}(0)$ $(n = 1, 2, 3...)$ is finite and $|x| < 1$, the term of $\dfrac{f^{(n)}(0)}{n!}x^n$ becomes less and less significant in contrast to the terms when n is small.

Therefore, a function over the continuous domain of x can be approximated as,

$$f(x) \cong C_0 + C_1 x + C_2 x^2 + \cdots + C_n x^n \tag{3.27}$$

where $C_i = f^{(i)}(0)$ $(i = 1, 2, ...n)$ are the constants determined by initial conditions, and n is an integer for the number of constraints applied on the function.

3.6.2 Interpolation in one-dimensional linear element

The approximation Eq. (3.27) is used as an interpolation function to determine the state variable at an arbitrary point based on the state variables on nodes. However, to define an interpolation function explicitly, one has to determine the values of n and C_i $(i = 1, 2, ...n)$ based on the constraints.

A one-dimensional linear element in Fig. 3.10 has two nodes. If the values of state variables are known at these two nodes, this gives two constraints on the interpolation. Therefore, n in Eq. (3.27) should be set as $n = 1$. The corresponding approximation equation becomes,

$$Y = f(X) \cong C_0 + C_1 X \tag{3.28}$$

Because the state variables at node i and j are assumed to be known as Y_i and Y_j, the constants in Eq. (3.28) must satisfy,

$$\left. \begin{array}{l} Y_i = C_0 + C_1 X_i \\ Y_j = C_0 + C_1 X_j \end{array} \right\} \tag{3.29}$$

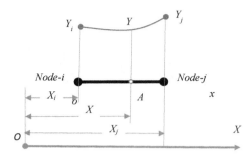

■ **FIGURE 3.10** Interpolation in one-dimensional element.

The solution to Eq. (3.29) gives,

$$\left.\begin{array}{c} C_0 = \dfrac{Y_i X_j - Y_j X_i}{X_j - X_i} \\[2mm] C_1 = \dfrac{Y_j - Y_i}{X_j - X_i} \end{array}\right\} \qquad (3.30)$$

As a result, the interpolation in a one-diemnsional linear element is,

$$Y = C_0 + C_1 X = \frac{Y_i X_j - Y_j X_i}{X_j - X_i} + \frac{Y_j - Y_i}{X_j - X_i} X \qquad (3.31)$$

Eq. (3.31) can be further formatted as an interpolation from the state variables on two nodes as,

$$Y = \frac{X_j - X}{X_j - X_i} Y_i + \frac{X - X_i}{X_j - X_i} Y_j = S_i(X) \cdot Y_i + S_j(X) \cdot Y_j \qquad (3.32)$$

where $S_i(X)$ and $S_j(X)$ are the shape functions of one-diemnsional linear element in GCS.

The similar procedure can be performed when the LCS is used, as a result,

$$Y = \frac{l - x}{l} Y_i + \frac{x}{l} Y_j = S_i(x) \cdot Y_i + S_j(x) \cdot Y_j \qquad (3.33)$$

where $S_i(x)$ and $S_j(x)$ are two shape functions of one-dimensional linear element in LCS.

As a matter of factor, the shape functions in Eqs. (3.20), (3.32), and (3.33) under different CS can be converted each other directly using the equations in the coordinate transformations expressed in Eqs. (3.4), (3.5), and (3.23). Fig. 3.11 illustrates these conversions.

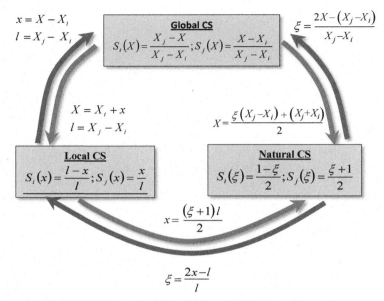

$x = X - X_i$
$l = X_j - X_i$

Global CS
$S_i(X) = \dfrac{X_j - X}{X_j - X_i}; S_j(X) = \dfrac{X - X_i}{X_j - X_i}$

$\xi = \dfrac{2X - (X_j - X_i)}{X_j - X_i}$

$X = X_i + x$
$l = X_j - X_i$

$X = \dfrac{\xi(X_j - X_i) + (X_j + X_i)}{2}$

Local CS
$S_i(x) = \dfrac{l - x}{l}; S_j(x) = \dfrac{x}{l}$

Natural CS
$S_i(\xi) = \dfrac{1 - \xi}{2}; S_j(\xi) = \dfrac{\xi + 1}{2}$

$x = \dfrac{(\xi + 1)l}{2}$

$\xi = \dfrac{2x - l}{l}$

■ **FIGURE 3.11** Conversion of shape functions under different coordinate systems.

It is interesting to see that the state variable at an arbitrary position is a weighted sum of the values on nodes and the weights on these nodes are determined by corresponding shape functions at this position.

Example 3.2

Fig. 3.12 shows a cantilever beam which is made of plain carbon steel ($E = 3.045 \times 10^7$ psi). It has a cross-section with $I_z = 2.75$ in⁴. The left end is completely fixed, and the right side carries 50-lb load. An FEA model consists of five elements and six nodes, and the Y-deflections on node-4 and node-5 are found as 0.0107-in and 0.0173-in, respectively. Find the Y-deflection of cross-section at 32.5-in from the fixed end.

Solution

The question is about how to use nodal values to obtain the state variable of an arbitrary point within an element. The element type is one-dimensional linear element, and Eq. (3.32) can be directly applied. Because the point of interest is within *element* 4 and it has node 4 and node 5 to represent its behaviors. Let $i = 4$ and $j = 5$.

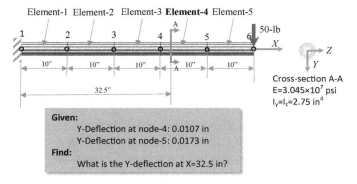

Given:
> Y-Deflection at node-4: 0.0107 in
> Y-Deflection at node-5: 0.0173 in

Find:
> What is the Y-deflection at X=32.5 in?

■ **FIGURE 3.12** Using shape functions to approximate state variable in element.

Under the GCS, the coordinates of two nodes are,

$$\left.\begin{array}{l} \text{for node } i, \\ \quad X_i = 30 \text{ in}, Y_i = 0.0107 \text{ in} \\ \text{for node } j, \\ \quad X_j = 40 \text{ in}, Y_j = 0.0173 \text{ in} \end{array}\right\}$$

The arbitrary position is at $X = 32.5$-in; therefore,

$$\left.\begin{array}{l} S_i(X) = \dfrac{X_j - X}{X_j - X_i} = \dfrac{(40) - (32.5)}{(40) - (30)} = 0.75 \\[2ex] S_j(X) = \dfrac{X - X_i}{X_j - X_i} = \dfrac{(32.5) - (30)}{(40) - (30)} = 0.25 \\[2ex] Y|_{X=32.5} = S_i(X) \cdot Y_i + S_j(X) \cdot Y_j = (0.75) * (0.0107) + (0.25) * (0.0173) = 0.0124 \text{ in} \end{array}\right\}$$

3.7 ONE-DIMENSIONAL QUADRATIC AND CUBIC ELEMENTS

The accuracy of the discretization in an FEA simulation depends on two critical factors. *The first factor* is sizes and shapes of elements. The finer and more regular elements are, the better accuracy of simulation results one can expect. *The other factor* is the polynomial order of elements. We have discussed that state variables in a continuous domain are approximated and generally represented by an algebra function with polynomial terms. However, the accuracy of the approximation within an element depends on the polynomial order. The higher polynomial order the function has, the better accuracy of the approximation one can obtain, and the better nonlinear behaviors can be represented. In FEA modeling, the approach to increase the simulation accuracy by reducing the sizes of elements in a mesh is called an *h-refinement*, and the approach to increase the simulation accuracy by increasing the polynomial order of elements for the interpolation is called a *p-refinement*. When a state variable demonstrates high nonlinearity and rapid changes over a spatial or temporal domain, the *p*-refinement is more effective and can achieve better accuracy. A *p*-refinement is implemented by introducing more nodes in each element. In this section, one-dimensional elements are used as examples to develop the shape functions for nonlinear elements.

3.7.1 One-dimensional quadratic element

When a Taylor series is used as a function for an approximation, the polynomial order is determined by the number of the constraints from discrete nodes. Therefore, one needs to increase the number of discrete nodes within an element to define more polynomial terms in the approximation. Fig. 3.13 illustrates a one-dimensional quadratic element where an additional node k is inserted in the middle of the linear element. As a result, a one-dimensional quadratic element has three nodes, i.e., node i, node j, and node k. Here, we are interested in determining the state variable at an arbitrary point in the element based on the given values of state variable at three nodes.

A quadratic element in Fig. 3.13 has three nodes. If the values of a state variable are known at three nodes, this gives three constraints on an interpolation equation. Therefore, n in Eq. (3.28) should be set as $n = 2$. The corresponding approximation equation becomes,

$$Y = f(X) \cong C_0 + C_1 X + C_2 X^2 \tag{3.34}$$

Because values of state variable at nodes i, j and k are assumed to be known as Y_i, Y_j, and Y_k, the constants in Eq. (3.34) must satisfy,

$$\left.\begin{array}{l} Y_i = C_0 + C_1 X_i + C_2 X_i^2 \\ Y_j = C_0 + C_1 X_j + C_2 X_j^2 \\ Y_k = C_0 + C_1 X_k + C_2 X_k^2 \end{array}\right\} \tag{3.35}$$

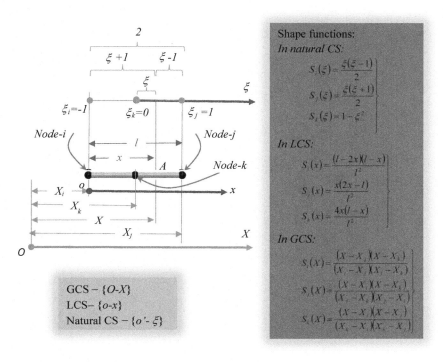

Shape functions:

In natural CS:

$$S_i(\xi) = \frac{\xi(\xi-1)}{2}$$

$$S_j(\xi) = \frac{\xi(\xi+1)}{2}$$

$$S_k(\xi) = 1 - \xi^2$$

In LCS:

$$S_i(x) = \frac{(l-2x)(l-x)}{l^2}$$

$$S_j(x) = \frac{x(2x-l)}{l^2}$$

$$S_k(x) = \frac{4x(l-x)}{l^2}$$

In GCS:

$$S_i(X) = \frac{(X-X_j)(X-X_k)}{(X_i-X_j)(X_i-X_k)}$$

$$S_j(X) = \frac{(X-X_i)(X-X_k)}{(X_j-X_i)(X_j-X_k)}$$

$$S_k(X) = \frac{(X-X_i)(X-X_j)}{(X_k-X_i)(X_k-X_j)}$$

GCS – {*O-X*}
LCS– {*o-x*}
Natural CS – {*o'- ξ*}

■ **FIGURE 3.13** Coordinate transformation for one-dimensional element.

As a result, the solution to Eq. (3.35) gives,

$$\left.\begin{aligned}
C_0 &= \frac{X_j X_k (X_k - X_j)}{|\Delta|} Y_i + \frac{X_i X_k (X_i - X_k)}{|\Delta|} Y_j + \frac{X_i X_j (X_j - X_i)}{|\Delta|} Y_k \\
C_1 &= \frac{\left(X_j^2 - X_k^2\right)}{|\Delta|} Y_i + \frac{\left(X_k^2 - X_i^2\right)}{|\Delta|} Y_j + \frac{\left(X_j^2 - X_i^2\right)}{|\Delta|} Y_k \\
C_2 &= \frac{(X_k - X_j)}{|\Delta|} Y_i + \frac{(X_i - X_k)}{|\Delta|} Y_j + \frac{(X_j - X_i)}{|\Delta|} Y_k
\end{aligned}\right\} \qquad (3.36)$$

where $|\Delta| = \begin{vmatrix} 1 & X_i & X_i^2 \\ 1 & X_j & X_j^2 \\ 1 & X_k & X_k^2 \end{vmatrix} = (X_i - X_j)(X_j - X_k)(X_k - X_i)$

Thus, the interpolation in a one-dimensional quadric element is written as,

$$Y = C_0 + C_1 X + C_2 X^2 = S_i(X) \cdot Y_i + S_j(X) \cdot Y_j + S_k(X) \cdot Y_k \qquad (3.37)$$

where $S_i(X)$, $S_j(X)$, and $S_k(X)$ are the shape functions of the quadratic element in GCS, which can be found as,

$$\left. \begin{aligned} S_i(X) &= \frac{(X - X_j)(X - X_k)}{(X_i - X_j)(X_i - X_k)} \\ S_j(X) &= \frac{(X - X_i)(X - X_k)}{(X_j - X_k)(X_j - X_i)} \\ S_k(X) &= \frac{(X - X_i)(X - X_j)}{(X_k - X_i)(X_k - X_j)} \end{aligned} \right\} \qquad (3.38)$$

Let $X = X_i + x$ and $X_j - X_i = l; X_j - X_k = X_k - X_i = l/2$ in Eq. (3.38), the shape functions of a quadratic element in the LCS can be found as,

$$\left. \begin{aligned} S_i(x) &= \frac{(l - 2x)(l - x)}{l^2} \\ S_j(x) &= \frac{x(2x - l)}{l^2} \\ S_k(x) &= \frac{4x(l - x)}{l^2} \end{aligned} \right\} \qquad (3.39)$$

Let $x = (\xi + 1)l/2$ in Eq. (3.39), the shape functions of a quadratic element in the natural CS can be found as,

$$\left. \begin{aligned} S_i(\xi) &= \frac{\xi(\xi - 1)}{2} \\ S_j(\xi) &= \frac{\xi(\xi + 1)}{2} \\ S_k(\xi) &= 1 - \xi^2 \end{aligned} \right\} \qquad (3.40)$$

Now, we look into the properties of shape functions in a quadratic element, Eq. (3.39) turns into a plot for three shape functions with respect to the natural coordinate ξ.

Fig. 3.14 shows that the following properties about the shape functions are still valid:

1. For a shape function of a certain node, its value is 1 at the same node, i.e.,

$$S_i(\xi) = 1 \text{ at node } i, \ S_j(\xi) = 1 \text{ at node } j, \ S_k(\xi) = 1 \text{ at node } k$$

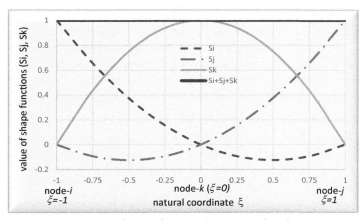

■ **FIGURE 3.14** Shape functions of quadratic element in natural coordinate system.

2. For a shape function of a certain node, its value is 0 at other node, i.e.,

$$S_i(\xi) = 0 \text{ at node } j \text{ and } k, \ S_j(\xi) = 0 \text{ at node } i \text{ and } k, \text{ and}$$
$$S_k(\xi) = 0 \text{ at node } i \text{ and } j.$$

3. The sum of all shape functions at any position is 1, i.e.,

$$S_i(\xi) + S_j(\xi) + S_k(\xi) = 1 \text{ for any } \xi \ desire$$

However, one difference of shape functions of a quadratic element from those of a linear element is that the shape functions of a quadratic element show nonlinearity. In particular, the value of a shape function is not always nonnegative anymore.

It is seen that the process of defining a shape function from Eq. (3.38) is trivial. It can be time consuming and error prone when more and more nodes are introduced in an element. Therefore, it is desirable to have an easy way to define shape functions. Here, a new approach based on the properties of the shape functions is illustrated through an example.

Example 3.3

Define shape functions of a quadratic element using the properties of shape functions under the natural coordinate system.

Solution

A quadratic element has three nodes with the coordinates of $\xi_i = -1$, $\xi_k = 0$, and $\xi_j = 1$, respectively. A Taylor series is used to represent each shape function. The polynomial order depends on the number and degrees of freedom of nodes. A quadratic element has three nodes. Therefore, a Taylor series for a shape function has the highest order of 2.

Let us define the shape function $S_i(\xi)$ at i at first.

1. According to property (2) of the shape function $S_i(\xi)$, i.e., $S_i(\xi) = 0$ at node j and k. We can assume that $S_i(\xi)$ can include the terms $(\xi - \xi_j)$ and $(\xi - \xi_k)$; so that the condition $S_i(\xi) = 0$ can be naturally satisfied at node j and k. In addition, because a quadratic element includes only three constraints for the Taylor series, the highest order of ξ must be 2. Therefore, $S_i(\xi)$ has no more term related to ξ other than $(\xi - \xi_j)$ and $(\xi - \xi_k)$, and $S_i(\xi)$ can be written as,

$$S_i(\xi) = k_i(\xi - \xi_j)(\xi - \xi_k) = k_i\xi(\xi - 1) \tag{3.41}$$

where k_i is a constant to be determined.

2. According to property (1) of the shape function $S_i(\xi)$, its function value is 1 at node i; therefore, using Eq. (3.41) gives,

$$S_i(\xi = -1) = k_i\xi(\xi - 1) = 1 \tag{3.42}$$

Eq. (3.42) finds $k_i = 1/2$; the complete expression of $S_i(\xi)$ is

$$S_i(\xi) = \frac{\xi(\xi - 1)}{2} \tag{3.43}$$

To define the shape functions $S_j(\xi)$ and $S_k(\xi)$ at j at k, the similar procedure can be applied.

Using the property (2), $S_j(\xi)$ and $S_k(\xi)$ can be defined as,

$$\left. \begin{aligned} S_j(\xi) &= k_j(\xi - \xi_k)(\xi - \xi_i) = k_j\xi(\xi + 1) \\ S_k(\xi) &= k_k(\xi - \xi_i)(\xi - \xi_j) = k_k(\xi^2 - 1) \end{aligned} \right\} \tag{3.44}$$

Using the property (1) in Eq. (3.44), we can find $k_j = 1/2$ and $k_k = -1$. The complete expression of $S_j(\xi)$ and $S_k(\xi)$ are,

$$\left. \begin{aligned} S_j(\xi) &= k_j(\xi - \xi_k)(\xi - \xi_i) = \frac{\xi(\xi + 1)}{2} \\ S_k(\xi) &= k_k(\xi - \xi_i)(\xi - \xi_j) = (1 - \xi^2) \end{aligned} \right\} \tag{3.45}$$

The results in Eqs. (3.42) and (3.45) are consistent with these in Eq. (3.40).

Eqs. (3.40), (3.42), and (3.45) are called *interpolation functions*. They satisfy three properties of shape functions, and they are referred as the *Lagrange interpolation functions.* The general expression of the Lagrange interpolation functions are,

$$S_k(x) = \prod_{M=1}^{M=N} \frac{(x - x_M) \text{omitting} (x - x_M)}{(x_k - x_M) \text{omitting} (x_k - x_M)}$$

$$= \frac{(x - x_1)...(x - x_{M-1}) \text{\rlap{}}(x - x_{M+1})...(x - x_N)}{(x_k - x_1)...(x_k - x_{M-1}) \text{\rlap{}}(x_k - x_{M+1})...(x_k - x_N)}$$

(3.46)

Note that the interpolation functions are applicable to any CS with any number of nodes in an element. In the next section, Eq. (3.46) will be used to find the shape functions of one-dimensional cubic element.

3.7.2 **One-dimensional cubic element**

A one-dimensional cubic element is shown in Fig. 3.15. Besides node i and j at two ends, it has another two nodes k and m in the middle section. The interpolation is for the determination of the value of state variable at an arbitrary position based on the values of state variable at these four nodes. The shape functions on these nodes represent the influence of nodal values on the point of interest.

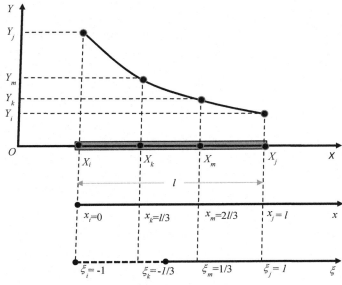

■ **FIGURE 3.15** A cubic element.

The Lagrange interpolation functions are used to define the shape functions of a cubic element directly. Here, the shape functions under a natural CS are used as an example.

The coordinates of four nodes in Fig. 3.15 are $\xi_i = -1$, $\xi_k = -1/3$, $\xi_m = 1/3$, and $\xi_j = 1$. Let $\xi \to x$, and $i, j, k, m \to 1, 2, 3,$ and 4, respectively, applying Eq. (3.46) gives,

$$
\left.
\begin{aligned}
S_i(\xi) &= \frac{(\xi - \xi_j)(\xi - \xi_k)(\xi - \xi_m)}{(\xi_i - \xi_j)(\xi_i - \xi_k)(\xi_i - \xi_m)} = \frac{(1 - \xi)(3\xi + 1)(3\xi - 1)}{16} \\
S_j(\xi) &= \frac{(\xi - \xi_i)(\xi - \xi_k)(\xi - \xi_m)}{(\xi_j - \xi_i)(\xi_j - \xi_k)(\xi_j - \xi_m)} = \frac{(\xi + 1)(3\xi + 1)(3\xi - 1)}{16} \\
S_k(\xi) &= \frac{(\xi - \xi_m)(\xi - \xi_i)(\xi - \xi_j)}{(\xi_k - \xi_m)(\xi_k - \xi_i)(\xi_k - \xi_j)} = \frac{9(\xi^2 - 1)(3\xi - 1)}{16} \\
S_m(\xi) &= \frac{(\xi - \xi_i)(\xi - \xi_j)(\xi - \xi_k)}{(\xi_m - \xi_i)(\xi_m - \xi_j)(\xi_m - \xi_k)} = \frac{9(1 - \xi^2)(3\xi + 1)}{16}
\end{aligned}
\right\}
\qquad (3.47)
$$

Once the shape functions are defined, they can be used for the interpolation within the element to find the state variable at any position ξ as,

$$
Y(\xi) = S_i(\xi) \cdot Y_i + S_j(\xi) \cdot Y_j + S_k(\xi) \cdot Y_k + S_m(\xi) \cdot Y_m \qquad (3.48)
$$

3.8 INTERPOLATION IN TWO-DIMENSIONAL ELEMENTS

The concept of shape functions can be expanded into two-dimensional elements. Fig. 3.16 shows two base types of two-dimensional elements, i.e., *rectangle element* and *triangle element* in Fig. 3.16A and B, respectively. The purpose of the interpolation is to derive the state variable in an arbitrary position from the nodal values of state variable within the element.

3.8.1 Two-dimensional rectangle element

As shown in Fig. 3.16, the interpolation in a two-dimensional rectangle element can be defined as suppose that nodal values of ψ_i, ψ_j, ψ_m, and ψ_n are given at the nodes of $(0, 0)$, $(l, 0)$, (l, w), and $(0, w)$, find the approximation of the variable ψ at the coordinate of (x, y).

The Taylor series is expanded in a two-dimensional space now, and its generic expression is given as,

$$
\psi(x, y) = C_0 + C_1 x + C_2 y + C_3 xy + C_4 x^2 + C_5 y^2 + C_6 x^2 y + \cdots \qquad (3.49)
$$

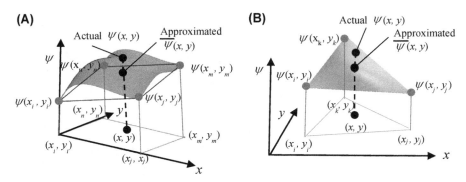

■ **FIGURE 3.16** Two basic two-dimensional elements: (A) rectangle element; (B) triangle element.

Similar to the approximation in one-dimensional space, the polynomial order is determined by the number of nodes or constraints given in the interpolation. The values of state variable at four nodes of a two-dimensional rectangle element bring four constraints. Therefore, the approximation in a rectangle element can be expressed as,

$$\psi(x, y) = C_0 + C_1 x + C_2 y + C_3 xy \tag{3.50}$$

The constants in Eq. (3.50) are solved from the following set of equations,

$$\left. \begin{aligned} \psi_i &= C_0 + C_1(0) + C_2(0) + C_3(0)(0) \\ \psi_j &= C_0 + C_1(l) + C_2(0) + C_3(l)(0) \\ \psi_m &= C_0 + C_1(l) + C_2(w) + C_3(l)(w) \\ \psi_n &= C_0 + C_1(0) + C_2(w) + C_3(0)(w) \end{aligned} \right\} \tag{3.51}$$

Solving Eq. (3.51) gives four constants in the approximation as,

$$\left. \begin{aligned} C_0 &= \psi_i \\ C_1 &= \frac{(\psi_j - \psi_i)}{l} \\ C_2 &= \frac{(\psi_n - \psi_i)}{w} \\ C_3 &= \frac{(\psi_i - \psi_j + \psi_m - \psi_n)}{lw} \end{aligned} \right\} \tag{3.52}$$

Substituting Eq. (3.52) into Eq. (3.50) and reformatting it with shape functions gives,

$$\psi(x, y) = S_i(x, y) \cdot \psi_i + S_j(x, y) \cdot \psi_j + S_m(x, y) \cdot \psi_m + S_n(x, y) \cdot \psi_n \tag{3.53}$$

where

$$
\left.\begin{aligned}
S_i(x, y) &= \left(\frac{l - x}{l}\right)\left(\frac{w - y}{w}\right) \\
S_j(x, y) &= \frac{x}{l}\left(\frac{w - y}{w}\right) \\
S_m(x, y) &= \frac{x}{l}\frac{y}{w} \\
S_n(x, y) &= \left(\frac{l - x}{l}\right)\frac{y}{w}
\end{aligned}\right\}
\tag{3.54}
$$

The shape functions are applicable to natural CS as well. Similar to one-dimensional elements, a natural CS can be introduced for a two-dimensional element by normalizing the dimensional lengths along x and y, respectively. Fig. 3.17 shows the normalization: the length along x- and y-axes is set as 2 units, and the origin of the natural CS is set at the geometric center of the rectangle element. As the result, the coordinates of four nodes in a natural CS are $(-1, -1)$, $(1, -1)$, $(1, 1)$, $(-1, 1)$ for i, j, m, and n, respectively.

To find the coordinate transformation from LCS to natural CS, we assume that,

$$
\left.\begin{aligned}
\xi &= C_{00} + C_{01}x \\
\eta &= C_{10} + C_{11}y
\end{aligned}\right\}
\tag{3.55}
$$

By transforming nodal coordinates from an LCS to a natural CS, we have

$$
\left.\begin{aligned}
Node\_i &\rightarrow & (-1) &= C_{00} + C_{01}(0) \\
Node\_j &\rightarrow & (1) &= C_{00} + C_{01}(l) \\
Node\_m &\rightarrow & (-1) &= C_{10} + C_{11}(0) \\
Node\_n &\rightarrow & (1) &= C_{10} + C_{11}(w)
\end{aligned}\right\}
\tag{3.56}
$$

Solving Eq. (3.56) gives four constants as,

$$
C_{00} = C_{10} = -1, \quad C_{01} = \frac{2}{l}, \quad C_{11} = \frac{2}{w}
\tag{3.57}
$$

The coordinate transformations from LCS to natural CS are,

$$
\left.\begin{aligned}
\xi &= -1 + \frac{2x}{l} \\
\eta &= -1 + \frac{2y}{w}
\end{aligned}\right\}
\quad \text{or} \quad
\left.\begin{aligned}
x &= \frac{(\xi + 1)l}{2} \\
y &= \frac{(\eta + 1)w}{2}
\end{aligned}\right\}
\tag{3.58}
$$

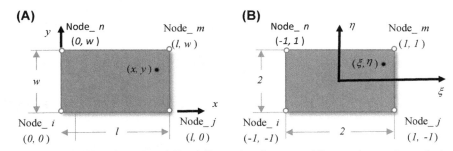

■ **FIGURE 3.17** (A) Local coordinate system (LCS) and (B) natural coordinate system (CS) in two-dimensional rectangle element.

Substituting Eq. (3.58) into (3.54) gives the interpolation equations in natural CS as,

$$\psi(\xi, \eta) = S_i(\xi, \eta) \cdot \psi_i + S_j(\xi, \eta) \cdot \psi_j + S_m(\xi, \eta) \cdot \psi_m + S_n(\xi, \eta) \cdot \psi_n \quad (3.59)$$

where

$$
\left.
\begin{aligned}
S_i(\xi, \eta) &= \frac{(1 - \xi)(1 - \eta)}{4} \\[6pt]
S_j(\xi, \eta) &= \frac{(1 + \xi)(1 - \eta)}{4} \\[6pt]
S_m(\xi, \eta) &= \frac{(1 + \xi)(1 + \eta)}{4} \\[6pt]
S_n(\xi, \eta) &= \frac{(1 - \xi)(1 + \eta)}{4}
\end{aligned}
\right\} \quad (3.60)
$$

3.8.2 Two-dimensional quadratic rectangle element

Nodes in a rectangle element can be increased to achieve the better accuracy in the approximation. One common element type is two-dimensional quadratic rectangle element with eight nodes. Under a natural CS in Fig. 3.18, the interpolation in such an element can be defined as giving nodal values of ψ_i, ψ_j, ψ_m, ψ_n, ψ_o, ψ_p, ψ_q, and ψ_r at the nodes of (ξ_i, η_i), (ξ_j, η_j), (ξ_m, η_m), (ξ_n, η_n), (ξ_o, η_o), (ξ_p, η_p), (ξ_q, η_q), and (ξ_r, η_r), approximating the state variable ψ at (ξ, η).

The interpolation in a quadratic rectangle element becomes,

$$\psi(\xi, \eta) = S_i \cdot \psi_i + S_j \cdot \psi_j + S_m \cdot \psi_m + S_n \cdot \psi_n + S_o \cdot \psi_o + S_p \cdot \psi_p + S_q \cdot \psi_q + S_r \cdot \psi_r$$
$$(3.61)$$

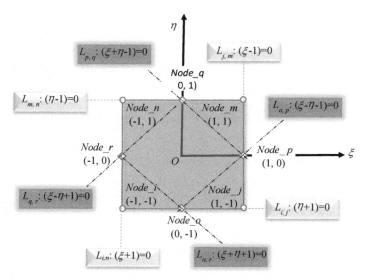

■ FIGURE 3.18 Description of a quadratic rectangle element in natural coordinate system.

Here, the properties of shape functions are utilized again to define these shape functions.

First, the shape function of a corner node is considered, and node i is used as an example. According to property (1), $S_i(\xi, \eta) = 0$ at other nodes j, m, n, o, p, q, and r. As shown in Fig. 3.18, these nodes are either on edge line j-m ($L_{j, m}$) and l-m ($L_{m, n}$) or on diagonal line o-r ($L_{o, r}$), and the equations for these lines can be easily defined; they are illustrated in the natural CS. Therefore, property (1) can be satisfied by assuming the expression of the shape function as,

$$S_i(\xi, \eta) = k_i \cdot (\xi - 1) \cdot (\eta - 1) \cdot (\xi + \eta + 1)$$

$$\underbrace{\qquad}_{L_{j,m}} \quad \underbrace{\qquad}_{L_{m,n}} \quad \underbrace{\qquad}_{L_{o,r}}$$

(3.62)

where k_i is a constant to be determined.

According to property (2), $S_i(\xi, \eta) = 0$ at node i ($\xi = -1$, $\eta = -1$). Substituting this condition about node i in Eq. (3.62) gives $k_i = -1/4$. $S_i(\xi, \eta)$ is then determined as,

$$S_i(\xi, \eta) = -\frac{(\xi - 1)(\eta - 1)(\xi + \eta + 1)}{4}$$

(3.63)

Following the same process in applying the properties of shape functions for other three corner nodes, one generates the shape functions on all four corner nodes as,

$$
\left.\begin{aligned}
S_i(\xi, \eta) &= -\frac{(\xi - 1)(\eta - 1)(\xi + \eta + 1)}{4} \\
S_j(\xi, \eta) &= -\frac{(\xi + 1)(\eta - 1)(\xi - \eta - 1)}{4} \\
S_m(\xi, \eta) &= \frac{(\xi + 1)(\eta + 1)(\xi + \eta - 1)}{4} \\
S_n(\xi, \eta) &= -\frac{(\xi - 1)(\eta + 1)(\xi - \eta + 1)}{4}
\end{aligned}\right\}
\tag{3.64}
$$

Second, the shape function of a middle node is considered, and node o is used as an example. According to property (1), $S_o(\xi, \eta) = 0$ at other nodes $i, j, m, n, p, q,$ and r. As shown in Fig. 3.18, these nodes are all on edge line j-m ($L_{j,\,m}$), l-m ($L_{m,\,n}$), or i-n ($L_{i,\,n}$) and the equations for these lines are illustrated in the natural CS. Therefore, property (1) can be satisfied by assuming

$$
S_o(\xi, \eta) = k_o \cdot \underbrace{(\xi - 1)}_{L_{j,\,m}} \cdot \underbrace{(\eta - 1)}_{L_{m,\,n}} \cdot \underbrace{(\xi + 1)}_{L_{i,\,n}}
\tag{3.65}
$$

where k_o is a constant to be determined.

According to property (2), $S_o(\xi, \eta) = 1$ at node o ($\xi = o, \eta = -1$). Substituting it in Eq. (3.65) gives $k_o = 1/2$. Therefore, $S_o(\xi, \eta)$ is found as,

$$
S_o(\xi, \eta) = \frac{(\xi^2 - 1)(\eta - 1)}{2}
\tag{3.66}
$$

Following the same process to applying the shape functions of middle nodes, one generates the shape functions on all four middle nodes as,

$$
\left.\begin{aligned}
S_o(\xi, \eta) &= \frac{(\xi^2 - 1)(\eta - 1)}{2} \\
S_p(\xi, \eta) &= -\frac{(\xi + 1)(\eta^2 - 1)}{2} \\
S_q(\xi, \eta) &= -\frac{(\xi^2 - 1)(\eta + 1)}{2} \\
S_r(\xi, \eta) &= \frac{(\xi - 1)(\eta^2 - 1)}{2}
\end{aligned}\right\}
\tag{3.67}
$$

3.8.3 **Two-dimensional linear triangle element**

The interpolation in a triangle element can be defined as: gives nodal values of ψ_i, ψ_j, and ψ_k at the nodes of (x_i, y_i), (x_j, y_j), and (x_k, y_k), find the approximation of the state variable ψ at any position (x, y) within the triangle element.

The polynomial order is determined by the number of nodes or constraints given in the interpolation. The given values of state variable at three nodes of a two-dimensional triangle element bring three constraints. Therefore, the approximation can be expressed as,

$$\psi(x, y) = C_0 + C_1 x + C_2 y \tag{3.68}$$

The constants in Eq. (3.68) are solved from the following equation set,

$$\left.\begin{aligned}
\psi_i &= C_0 + C_1(x_i) + C_2(y_i) \\
\psi_j &= C_0 + C_1(x_j) + C_2(y_j) \\
\psi_k &= C_0 + C_1(x_k) + C_2(y_k)
\end{aligned}\right\} \tag{3.69}$$

Solving Eq. (3.69) finds the three constants in the approximation as,

$$\left.\begin{aligned}
C_0 &= \frac{\alpha_i}{|\Delta|}\psi_i + \frac{\alpha_j}{|\Delta|}\psi_j + \frac{\alpha_k}{|\Delta|}\psi_k \\
C_1 &= \frac{\beta_i}{|\Delta|}\psi_i + \frac{\beta_j}{|\Delta|}\psi_j + \frac{\beta_k}{|\Delta|}\psi_k \\
C_2 &= \frac{\delta_i}{|\Delta|}\psi_i + \frac{\delta_j}{|\Delta|}\psi_j + \frac{\delta_k}{|\Delta|}\psi_k
\end{aligned}\right\} \tag{3.70}$$

where the constants in Eq. (3.70) are determined by the triangle shape as,

$$\left.\begin{aligned}
|\Delta| &= \begin{vmatrix} 1 & x_i & y_i \\ 1 & x_j & y_j \\ 1 & x_k & y_k \end{vmatrix} \\
\alpha_i &= x_j y_k - x_k y_j, \quad \alpha_j = x_k y_i - x_i y_k, \quad \alpha_k = x_i y_j - x_j y_i \\
\beta_i &= y_j - y_k, \quad \beta_j = y_k - y_i, \quad \beta_k = y_i - y_j \\
\delta_i &= x_k - x_j, \quad \delta_j = x_i - x_k, \quad \delta_k = x_j - x_i
\end{aligned}\right\} \tag{3.71}$$

where $|\bullet|$ is the determinant of a matrix.

Substituting Eq. (3.70) into Eq. (3.68) and reformatting it with shape functions gives,

$$\psi(x, y) = S_i(x, y) \cdot \psi_i + S_j(x, y) \cdot \psi_j + S_k(x, y) \cdot \psi_k \tag{3.72}$$

where

$$
\left.
\begin{aligned}
S_i(x, y) &= \frac{|\Delta_i|}{|\Delta|} = \frac{\alpha_i + \beta_i \cdot x + \delta_i \cdot y}{|\Delta|} \\
S_j(x, y) &= \frac{|\Delta_j|}{|\Delta|} = \frac{\alpha_j + \beta_j \cdot x + \delta_j \cdot y}{|\Delta|} \\
S_k(x, y) &= \frac{|\Delta_k|}{|\Delta|} = \frac{\alpha_k + \beta_k \cdot x + \delta_k \cdot y}{|\Delta|}
\end{aligned}
\right\}
\tag{3.73}
$$

Example 3.4

A triangle element is defined in a GCS as shown in Fig. 3.19. The element is under an external load, which causes the stresses 5-MPa, 6-MPa, and 7-MPa on nodes *i*, *j*, and *k*, respectively. (1) Determine the stress at the position *P* (5, 3); (2) determine a contour in the triangle element where the stress of any position of the contour equals to 5.5 MPa.

Solution to (1)

First, the nodal coordinates are listed in GCS as shown in Table 3.2.

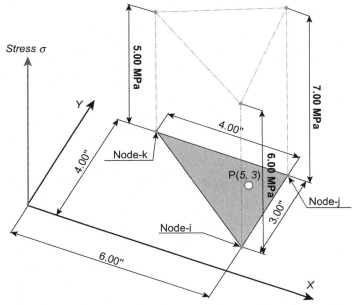

■ **FIGURE 3.19** Stress at a given position and a contour with given stress.

Table 3.2 The Nodal Coordinates in Global Coordinate System

Node	Node-i (x_i, y_i)	Node-j (x_j, y_j)	Node-k (x_k, y_k)
GCS	(6, 1)	(6, 4)	(2, 4)

Eq. (3.71) is applied to find all of the constants in the shape functions as,

$$|\Delta| = \begin{vmatrix} 1 & x_i & y_i \\ 1 & x_j & y_j \\ 1 & x_k & y_k \end{vmatrix} = \begin{vmatrix} 1 & 6 & 1 \\ 1 & 6 & 4 \\ 1 & 2 & 4 \end{vmatrix} = 12$$

$$\alpha_i = x_j y_k - x_k y_j = (6)(4) - (2)(4) = 16$$
$$\alpha_j = x_k y_i - x_i y_k = (2)(1) - (6)(4) = -22$$
$$\alpha_k = x_i y_j - x_j y_i = (6)(4) - (6)(1) = 18$$

$$\beta_i = y_j - y_k = (4) - (4) = 0$$
$$\beta_j = y_k - y_i = (4) - (1) = 3$$
$$\beta_k = y_i - y_j = (1) - (4) = -3$$

$$\delta_i = x_k - x_j = (2) - (6) = -4$$
$$\delta_j = x_i - x_k = (6) - (2) = 4$$
$$\delta_k = x_j - x_i = (6) - (6) = 0$$

Eq. (3.73) is applied for the evaluation of shape functions at P (5, 3) as,

$$S_i(5,3) = \frac{\alpha_i + \beta_i \cdot x + \delta_i \cdot y}{|\Delta|} = \frac{(16) + (0) \cdot (5) + (-4) \cdot (3)}{(12)} = \frac{1}{3}$$

$$S_j(5,3) = \frac{\alpha_j + \beta_j \cdot x + \delta_j \cdot y}{|\Delta|} = \frac{(-22) + (3) \cdot (5) + (4) \cdot (3)}{(12)} = \frac{5}{12}$$

$$S_k(5,3) = \frac{\alpha_k + \beta_k \cdot x + \delta_k \cdot y}{|\Delta|} = \frac{(18) + (-3) \cdot (5) + (0) \cdot (3)}{(12)} = \frac{1}{4}$$

To evaluate the state variable at P, Eq. (3.70) is applied for the interpolation,

$$\sigma(5,3) = S_i(5,3) \cdot \sigma_i + S_j(5,3) \cdot \sigma_j + S_k(5,3) \cdot \sigma_k$$

$$= \left(\frac{1}{3}\right)(6) + \left(\frac{5}{12}\right)(7) + \left(\frac{1}{4}\right)(5) = 6.1667 \text{ MPa}$$

Solution to (2)

A triangle element is a linear element whose state variable changes linearly in the domain with respect to its coordinates. Therefore, a contour with the same stress level is a straight line. If two points on a straight line can be specified, this line is completely determined. Because only node-k has a stress below 5.5-MPa; we can define two points on edge i-k and edge j-k, respectively, to define the contour line.

For a point A on edge i-k, if the stress on A is 5.5 MPa, it satisfies

$$\frac{x_A - x_i}{x_k - x_i} = \frac{y_A - y_i}{y_k - y_i} = \frac{5.5 - \sigma_i}{\sigma_k - \sigma_i} \qquad (3.74)$$

Eq. (3.74) yields the coordinates of A $(x_A, y_A) = (4, 2.5)$. Similarly, for a point B on edge j-k, if the stress on B is 5.5 MPa, it satisfies,

$$\frac{x_B - x_j}{x_k - x_j} = \frac{y_B - y_i}{y_k - y_i} = \frac{5.5 - \sigma_j}{\sigma_k - \sigma_j} \qquad (3.75)$$

Eq. (3.74) yields the coordinates of B $(x_B, y_B) = (3, 4)$. Therefore, the contour line with the stress of 5.5-MPa can be completely defined by A and B as shown in Fig. 3.20.

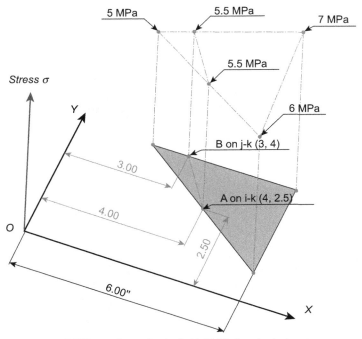

■ **FIGURE 3.20** Contour line A—B with 5.5 MPa in a triangle element.

3.8.4 **Two-dimensional linear triangle element under natural CS**

A close look in Eq. (3.73) finds that a shape function in a triangle element depends only on the triangle shape, i.e., an area ratio of two triangles. In addition, the value of a state variable on any point is interpolated based on how the given point divides the triangle into three subtriangles. One can utilize these properties to define a dimensionless natural CS for a triangle element.

As shown in Fig. 3.21, when a position $P(x, y)$ in the element is given, the whole area of the triangle element is divided into three subtriangles by connecting P with three nodes. Coincidentally, a shape function of the triangle element corresponds to an area ratio of a subtriangle and the whole triangle. Taking an example of $S_i(x, y)$, it corresponds to the ratio of A_i and A. Note that in Eq. (3.73), one has,

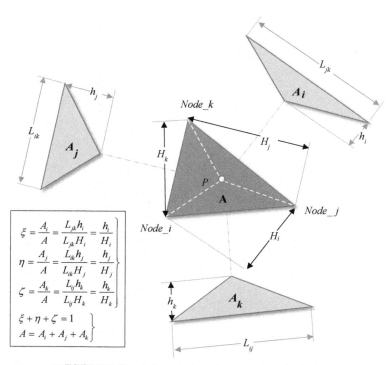

$$\xi = \frac{A_i}{A} = \frac{L_{jk}h_i}{L_{jk}H_i} = \frac{h_i}{H_i}$$

$$\eta = \frac{A_j}{A} = \frac{L_{ik}h_j}{L_{ik}H_j} = \frac{h_j}{H_j}$$

$$\zeta = \frac{A_k}{A} = \frac{L_{ij}h_k}{L_{ij}H_k} = \frac{h_k}{H_k}$$

$$\xi + \eta + \zeta = 1$$

$$A = A_i + A_j + A_k$$

■ **FIGURE 3.21** Dimensionless natural coordinate system in a triangle element.

$$|\Delta| = 2A = \begin{vmatrix} 1 & x_i & y_i \\ 1 & x_j & y_j \\ 1 & x_k & y_k \end{vmatrix} \quad |\Delta_i| = 2A_i = \begin{vmatrix} 1 & x & y \\ 1 & x_j & y_j \\ 1 & x_k & y_k \end{vmatrix},$$

$$|\Delta_i| = 2A_j = \begin{vmatrix} 1 & x_i & y_i \\ 1 & x & y \\ 1 & x_k & y_k \end{vmatrix}, \quad |\Delta_i| = 2A_k = \begin{vmatrix} 1 & x_i & y_i \\ 1 & x_j & y_j \\ 1 & x & y \end{vmatrix}$$

(3.76)

where $|\bullet|$ is the determinant of a matrix. Let $\xi = A_i/A$, $\eta = A_j/A$, and $\zeta = A_k/A$ are the natural coordinates of a triangle element, the shape functions in the GCS in Eq. (3.73) can be transformed into these in natural CS as,

$$\left. \begin{aligned} S_i(\xi, \eta) &= \xi \\ S_j(\xi, \eta) &= \eta \\ S_k(\xi, \eta) &= \zeta = 1 - \xi - \eta \end{aligned} \right\}$$

(3.77)

Obviously, the shape functions in Eq. (3.77) also possess three basic properties of a shape function as,

1. For a shape function of a given node, its value is 1 when the point is specified at the same node, i.e.,

$$S_i(\xi, \eta) = 1 \text{ at node } i, \ S_j(\xi, \eta) = 1 \text{ at node } j, \ S_k(\xi, \eta) = 1 \text{ at node } k$$

2. For a shape function of a given node, its value is 0 when the point of interest is at other nodes, i.e.,

$$S_i(\xi, \eta) = 0 \text{ at node } j \text{ and } k, \ S_j(\xi, \eta) = 0 \text{ at node } i \text{ and } k, \text{ and } S_k(\xi, \eta)$$
$$= 0 \text{ at node } i \text{ and } j.$$

3. The sum of all of the shape functions at any position within the triangle element is 1, i.e.,

$$S_i(\xi, \eta) + S_j(\xi, \eta) + S_k(\xi, \eta) = 1 \text{ for any position } (\xi, \eta).$$

3.8.5 Two-dimensional quadratic triangle element

In a quadratic triangle element, a middle node is added in each edge besides the nodes on three vertices. Under a natural CS in Fig. 3.22, the interpolation in such an element is defined as giving nodal values of $\psi_i, \psi_j, \psi_k, \psi_o, \psi_p,$ and ψ_q at the nodes $(\xi_i, \eta_i), (\xi_j, \eta_j), (\xi_k, \eta_k), (\xi_o, \eta_o), (\xi_p, \eta_p),$ and (ξ_q, η_q); find the approximation of the state variable ψ at (ξ, η).

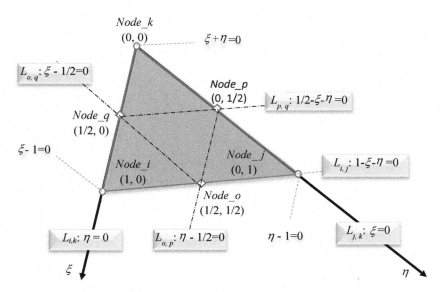

■ **FIGURE 3.22** Description of a quadratic rectangle element under natural coordinate system.

Taking into consideration the number of constraints by given values on six nodes, the interpolation of the state variable at (ξ, η) in the element should be,

$$\psi(\xi, \eta) = S_i \cdot \psi_i + S_j \cdot \psi_j + S_k \cdot \psi_k + S_o \cdot \psi_o + S_p \cdot \psi_p + S_q \cdot \psi_q \qquad (3.77)$$

The properties of the shape functions are utilized to define the shape functions themselves.

First, the shape function of a corner node is considered, and node i is used as an example. According to property (1), $S_i (\xi, \eta) = 0$ at node j, k, o, p, and q. As shown in Fig. 3.22, these nodes are either on edge line j-k ($L_{j, k}$) or on a middle line o-q ($L_{o, q}$), and the equations for these lines can be defined, which are illustrated with respect to natural CS. Therefore, property (1) can be satisfied by expressing $S_i (\xi, \eta)$ as,

$$S_i(\xi, \eta) = k_i \cdot (\xi - 1/2) \cdot \xi \qquad (3.78)$$

where k_i is a constant to be determined.

According to property (2), $S_i(\xi, \eta) = 1$ at node i ($\xi = 1$, $\eta = 0$). Substituting this condition in Eq. (3.78) gives $k_i = 2$. Therefore, $S_i(\xi, \eta)$ is found as,

$$S_i(\xi, \eta) = \xi \cdot (2\xi - 1) \tag{3.79}$$

Following the same process, one can obtain the shape functions of other three corner nodes as,

$$\left.\begin{aligned}
S_i(\xi, \eta) &= \xi \cdot (2\xi - 1) \\
S_j(\xi, \eta) &= \eta \cdot (2\eta - 1) \\
S_k(\xi, \eta) &= (1 - \xi - \eta) \cdot (1 - 2\xi - \eta)
\end{aligned}\right\} \tag{3.80}$$

Second, the shape function of a node in the middle of an edge is considered. Node o on edge i-j is used as an example. According to property (1), $S_o(\xi, \eta) = 0$ at node $i, j, k, p,$ and q. As shown in Fig. 3.22, all of these nodes are on edge line i-k ($L_{i,\,k}$), or line j-k ($L_{j,\,k}$). The algebra equations for these two lines are given in the natural CS. Therefore, property (1) can be easily satisfied by expressing $S_o(\xi, \eta)$ as,

$$S_o(\xi, \eta) = k_o \cdot \underbrace{\xi}_{L_{i,\,k}} \cdot \underbrace{\eta}_{L_{j,\,k}} \tag{3.81}$$

where k_o is a constant to be determined.

According to property (2), $S_o(\xi, \eta) = 1$ at node o ($\xi = 1/2$, $\eta = 1/2$). Substituting this condition in Eq. (3.81) gives $k_o = 4$. Therefore, $S_o(\xi, \eta)$ is found as,

$$S_o(\xi, \eta) = 4 \cdot \xi \cdot \eta \tag{3.82}$$

Following the same process for other two middle nodes, one can generate the shape functions of three middle nodes as,

$$\left.\begin{aligned}
S_o(\xi, \eta) &= 4 \cdot \xi \cdot \eta \\
S_p(\xi, \eta) &= 4 \cdot \xi \cdot (1 - \xi - \eta) \\
S_q(\xi, \eta) &= 4 \cdot (1 - \xi - \eta) \cdot \eta
\end{aligned}\right\} \tag{3.83}$$

3.9 NUMERICAL INTEGRATION

FEA is generally a solver for differential equations over a continuous domain; although integral operations are often performed over a domain to solve differential equations. Very few of differential equations have analytical solutions under given boundary conditions. As a generic solver of differential equations, integral operations over a continuous domain have to be performed numerically. A *numerical integration* approximates the integral about a state variable over a continuous domain based on the given values of state variable at nodes. To make the algorithms and programs generic, the nodes under a dimensionless natural CS are of special interest.

3.9.1 Integral over one-dimensional domain

The integral over a one-dimensional domain is described as follows: assume that there is a function $f(X)$ over a continuous domain $X \in [a, b]$, an integral of this function is to be found numerically by taking a weighted sum of the function values over N discrete points within the domain, i.e.,

$$I_{1-D} = \int_a^b f(X) \cdot dX \cong \sum_{i=1}^N w_i \cdot f(X_i) \tag{3.84}$$

where

I_{1-D} is the approximated integral of the function $f(X)$ over—one-dimensional domain,
N is the number of the discrete points in one-dimensional domain,
w_i is the weight for the function value at point i, and
X_i and $f(X_i)$ are the selected point i and the corresponding function value at this point.

$$I_{1-D} = \int_a^b f(X) \cdot dx \cong (b-a) \cdot f\left(\frac{a+b}{2}\right) \tag{3.85}$$

$$I_{1-D} = \int_a^b f(X) \cdot dx \cong \frac{(b-a)}{2} \cdot f(a) + \frac{(b-a)}{2} \cdot f(b) \tag{3.86}$$

Depending on the selections of N, w_i, and X_i, different approximations can be defined (Fig. 3.23). Before we derive the equations for the numerical integration, we transform the integral over GCS into a natural CS, so that the implemented program can be more flexible due to the dimensionless feature of the natural CS. The equations for the coordination transformation in Fig. 3.11 can be utilized here, i.e.,

$$\left.\begin{aligned}
\xi &= \frac{2X - (X_j + X_i)}{X_j - X_i} = \frac{2X - (b+a)}{b-a} \\
&\text{or} \\
X &= \frac{(b-a)\xi + (b+a)}{2}
\end{aligned}\right\} \tag{3.87}$$

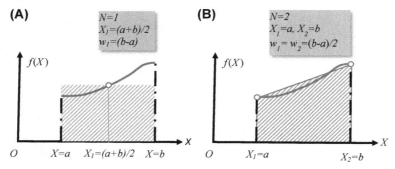

■ **FIGURE 3.23** (A) 1-Point and (B) 2-point approximation.

The bounds of one-dimensional domain in natural CS are determined by Eq. (3.87) as,

$$\text{when } X = a,$$

$$\xi = \frac{2X - (b + a)}{b - a} = -1$$

$$\text{when } X = b,$$

$$\xi = \frac{2X - (b + a)}{b - a} = 1$$

(3.88)

The one-dimensional domain in $[a, b]$ in the GCS corresponds to $[-1, 1]$ in the natural CS.

Substituting Eq. (3.87) into Eq. (3.84) gives,

$$
\begin{aligned}
I_{1-D} &= \int_a^b f(X) \cdot dX \\
&= \int_{-1}^1 f\left(\frac{(b - a)\xi + (b + a)}{2}\right) \cdot d\left(\frac{(b - a)\xi + (b + a)}{2}\right) \\
&= \frac{b - a}{2} \int_{-1}^1 f\left(\frac{(b - a)\xi + (b + a)}{2}\right) \cdot d\xi \\
&= \frac{b - a}{2} \int_{-1}^1 \overline{f}(\xi) \cdot d\xi = \frac{b - a}{2} \cdot \overline{I}_{1-D}
\end{aligned}
$$

(3.89)

where $\overline{f}(\xi) = f\left(\dfrac{(b - a)\xi + (b + a)}{2}\right)$ and $\overline{I}_{1-D} = \int_{-1}^1 \overline{f}(\xi) \cdot d\xi$

Now the original problem becomes an approximation of integration $\overline{I}_{1-D}$ over a normalized domain $[-1, 1]$,

$$\overline{I}_{1-D} = \int_{-1}^1 \overline{f}(\xi) \cdot d\xi \cong \sum_{i=1}^N w_i \cdot \overline{f}(\xi_i)$$

(3.90)

To use Eq. (3.90), one has to determine the number of points, the values, and the corresponding weights in the approximation.

The Taylor series has been adopted to approximate an unknown function. The higher polynomial order is used in a Taylor series, the better approximation one can obtain. However, using more polynomial terms needs more nodes whose function values are supposed known. Therefore, the number of polynomial terms should be tailored to match the amount of available information in the approximation. In addition, with the given number of points, the coordinates and weights of points should minimize the errors of the approximation of integration in the given domain.

3.9.1.1 Integral using 1-point approximation

If 1-point is used in Eq. (3.90), ξ_1 and w_1 are two constants that can be freely selected, but they can be selected to obtain a result with better accuracy for the integration of $\bar{f}(\xi)$ with the highest polynomial order of 1.

Let the polynomial term be $\bar{f}(\xi) = 1$,

$$\bar{I}_{1-D} = \int_{-1}^{1} \bar{f}(\xi) \cdot d\xi = \int_{-1}^{1} (1) \cdot d\xi = 2$$

$$= \sum_{i=1}^{N} w_i \bar{f}(\xi_i) = w_1 \bar{f}(\xi_1) = w_1(1) = w_1 \qquad (3.91)$$

Let the polynomial term be $\bar{f}(\xi) = \xi$,

$$\bar{I}_{1-D} = \int_{-1}^{1} \bar{f}(\xi) \cdot d\xi = \int_{-1}^{1} (\xi) \cdot d\xi = 0$$

$$= \sum_{i=1}^{N} w_i \bar{f}(\xi_i) = w_1 \bar{f}(\xi_1) = w_1(\xi_1) = w_1 \cdot \xi_1 \qquad (3.92)$$

Combining Eqs. (3.91) and (3.92) leads to $\xi_1 = 0, w_1 = 2$. Therefore, the approximated 1-point integration is,

$$\bar{I}_{1-D} = \int_{-1}^{1} \bar{f}(\xi) \cdot d\xi = 2\bar{f}(0) \qquad (3.93)$$

In GCS, it becomes,

$$I_{1-D} = \frac{(b-a)}{2} \cdot \bar{I}_{1-D} = (b-a) \cdot f\left(\frac{a+b}{2}\right) \qquad (3.94)$$

3.9.1.2 Integral using 2-point approximation

If the 2-point approximation is used in Eq. (3.90), ξ_1, ξ_2, w_1, and w_2 are four constants, which can be freely selected, and the highest polynomial order is 3.

The following conditions have to be satisfied to obtain the result for $\bar{f}(\xi)$ with best accuracy.

Let the polynomial term be $\bar{f}(\xi) = 1$,

$$\bar{I}_{1-D} = 2 = w_1 + w_2 \tag{3.95}$$

Let the polynomial term be $\bar{f}(\xi) = \xi$,

$$\bar{I}_{1-D} = 0 = w_1 \cdot \xi_1 + w_2 \cdot \xi_2 \tag{3.96}$$

Let the polynomial term be $\bar{f}(\xi) = \xi^2$,

$$\bar{I}_{1-D} = \frac{2}{3} = w_1 \cdot \xi_1^2 + w_2 \cdot \xi_2^2 \tag{3.97}$$

Let the polynomial term be $\bar{f}(\xi) = \xi^3$,

$$\bar{I}_{1-D} = 0 = w_1 \cdot \xi_1^3 + w_2 \cdot \xi_2^3 \tag{3.98}$$

Combining Eqs. (3.95)–(3.98) leads to the solution of four constants as $w_1 = w_2 = 1, \xi_1 = \frac{-1}{\sqrt{3}}, \xi_1 = \frac{1}{\sqrt{3}}$.

Therefore, the approximated 2-point integration is,

$$\bar{I}_{1-D} = \int_{-1}^{1} \bar{f}(\xi) \cdot d\xi = \bar{f}\left(\frac{-1}{\sqrt{3}}\right) + \bar{f}\left(\frac{1}{\sqrt{3}}\right) \tag{3.99}$$

In GCS, it becomes,

$$I_{1-D} = \frac{(b-a)}{2} \cdot \bar{I}_{1-D}$$

$$= (b-a) \cdot \left[f\left(\frac{(b-a)\left(\frac{-1}{\sqrt{3}}\right) + (b+a)}{2} \right) \right. \tag{3.100}$$

$$\left. + f\left(\frac{(b-a)\left(\frac{-1}{\sqrt{3}}\right) + (b+a)}{2} \right) \right]$$

The same procedure can be applied to derive the coordinates and weights if more points are needed in the numerical integration. Because the natural CS is dimensionless, the solutions to the coordinates and weights are fixed, and they can be used as a look up tables in the program implementation directly. The aforementioned approach for numerical integration is called as the **Gauss–Legendre integration**. The lookup table for the weights and coordinates up to six points are given in Table 3.3.

Table 3.3 Weights and Coordinates of Points for Gauss–Legendre Integration (Davis and Rabinowitz, 1956)

Approximation	Coordinates	Weights
1-point	$\xi_1 = 0$	$w_1 = 2$
2-point	$\xi_1 = -1/\sqrt{3} = -0.577350269$	$w_1 = 1$
	$\xi_2 = 1/\sqrt{3} = 0.577350269$	$w_2 = 1$
3-point	$\xi_1 = -\sqrt{3/5} = -0.774596669$	$w_1 = 5/9$
	$\xi_2 = 0$	$w_2 = 8/9$
	$\xi_3 = \sqrt{3/5} = 0.774596669$	$w_3 = 5/9$
4-point	$\xi_1 = -0.861136312$	$w_1 = 0.3478548$
	$\xi_2 = -0.339981044$	$w_2 = 0.6521452$
	$\xi_3 = 0.339981044$	$w_3 = 0.6521452$
	$\xi_4 = 0.861136312$	$w_4 = 0.3478548$
5-point	$\xi_1 = -0.906179846$	$w_1 = 0.2369269$
	$\xi_2 = -0.538469310$	$w_2 = 0.4786287$
	$\xi_3 = 0$	$w_3 = 0.5688889$
	$\xi_4 = 0.538469310$	$w_4 = 0.4786287$
	$\xi_5 = 0.906179846$	$w_5 = 0.2369269$
6-point	$\xi_1 = -0.932469514$	$w_1 = 0.1713245$
	$\xi_2 = -0.661209386$	$w_2 = 0.3607616$
	$\xi_3 = -0.238619186$	$w_3 = 0.4679139$
	$\xi_4 = 0.238619186$	$w_4 = 0.4679139$
	$\xi_5 = 0.661209386$	$w_5 = 0.3607616$
	$\xi_6 = 0.932469514$	$w_6 = 0.1713245$

As a summary, the procedure of using the Gauss–Legendre formula for the numerical integration over the one-dimensional domain is shown as follows:

1. Determine the range of integral $[a, b]$;
2. Determine the equation of coordinate transformation, i.e.,

$$x_i = \frac{(b-a)\xi_i + (b+a)}{2} \quad (i = 1, \dots N) \tag{3.101}$$

3. Select the number of points N to be used, and find the coordinate and weights ξ_i and w_i $(i = 1, \dots N)$ from Table 3.2;
4. Evaluate the integral using

$$I_{1-D} = \frac{(b-a)}{2} \cdot \bar{I}_{1-D} = \frac{(b-a)}{2} \sum_{i=1}^{i=N} w_i f(x_i) \tag{3.102}$$

where x_i is determined by ξ_i using Eq. (3.101).

Example 3.5

Using the Gauss–Legendre formula (2-point, 3-point, and 4-point) to evaluate the integral of

$$I_{1-D} = \int_0^4 xe^{-x}dx$$

Solution

Step 1: the range of the one-dimensional domain is $[a, b] = [0, 4]$;

Step 2: the coordinate transformation from the natural CS to GCS is found as,

$$x_i = \frac{((4) - (0))\xi_i + ((4) + (0))}{2} = 2\xi_i + 2 \quad (i = 1, ...N)$$

Steps 3 and 4: apply Eq. (3.102)

$$I_{1-D} = \frac{(b-a)}{2}\cdot \bar{I}_{1-D} = (2)\sum_{i=1}^{i=N} w_i f(x_i) \qquad (3.103)$$

Using the coordinates and weights in Table 3.2 gives the results as,

2-point:

$$\left.\begin{array}{l} x_1 = 2(-0.577350269) + 2 = 0.845529462 \\ x_2 = 2(0.577350269) + 2 = 3.154700538 \end{array}\right\}$$

$$I_{1-D} = (2)\cdot\{(1)(x_1 e^{-x_1}) + (1)(x_2 e^{-x_2})\} = 96$$

3-point:

$$\left.\begin{array}{l} x_1 = 2(-0.774596669) + 2 = 0.450806662 \\ x_2 = 2(0) + 2 = 2 \\ x_3 = 2(0.774596669) + 2 = 3.559193338 \end{array}\right\}$$

$$I_{1-D} = (2)\cdot\{(5/9)(x_1 e^{-x_1}) + (8/9)(x_2 e^{-x_2}) + (5/9)(x_3 e^{-x_3})\}$$
$$= 60.57777778$$

4-point:

$$\left.\begin{array}{l} x_1 = 2(-0.861136312) + 2 = 0.277727376 \\ x_2 = 2(-0.339981044) + 2 = 1.320037912 \\ x_3 = 2(0.339981044) + 2 = 2.679962088 \\ x_3 = 2(0.861136312) + 2 = 3.722272624 \end{array}\right\}$$

$$I_{1-D} = (2)\cdot\left\{\begin{array}{l} (0.3478548)(x_1 e^{-x_1}) + (0.6521452)(x_2 e^{-x_2}) \\ +(0.6521452)(x_3 e^{-x_3}) + (0.3478548)(x_4 e^{-x_4}) \end{array}\right\}$$
$$= 103.6119622$$

This example has shown that when the function is highly nonlinear and the domain range is large, the numerical integration could lead to a result with a considerable error. To evaluate the integral correctly, the domain has to be decomposed into small segments for multiple operations of integral.

3.9.2 Integral operation over two-dimensional domain

For a numerical integration over a two-dimensional domain, the weights and points along two axes can be selected separately, as a result,

$$
\begin{aligned}
\bar{I}_{2-D} &= \int_{-1}^{1}\int_{-1}^{1}\bar{f}(\xi,\eta)\cdot d\xi d\eta \\
&\cong \int_{-1}^{1}\left[\sum_{i=1}^{N} w_i\bar{f}(\xi_i,\eta)\right]d\eta \\
&\cong \sum_{i=1}^{N}\sum_{j=1}^{N} w_i w_j \bar{f}(\xi_i,\eta_j)
\end{aligned}
\tag{3.104}
$$

In a GCS, an integration over a two-dimensional domain is expressed as,

$$
I_{2-D} = \int_{c}^{d}\int_{a}^{b} f(x,y)dxdy
\tag{3.105}
$$

Applying Eq. (3.104), one can obtain its numerical integration as,

$$
\begin{aligned}
\bar{I}_{2-D} &\cong \left(\frac{b-a}{2}\right)\left(\frac{d-c}{2}\right)\sum_{i=1}^{N} \\
&\times \sum_{j=1}^{N} w_i w_j f\left(\frac{(b-a)\xi_i+(b+a)}{2},\frac{(d-c)\eta_j+(d+c)}{2}\right)
\end{aligned}
\tag{3.106}
$$

where w_i, w_j, ξ_i, and η_j are the constants given in Table 3.2.

The procedure of using the Gauss—Legendre formula for numerical integration over the two-dimensional domain can be summarized as follows.

1. Determine the range of integral $[a, b]$ and $[c, d]$;
2. Determine the equation of coordinate transformation, i.e.,

$$
\left.
\begin{aligned}
x_i &= \frac{(b-a)\xi_i+(b+a)}{2} \quad (i=1,...N) \\
y_i &= \frac{(d-c)\eta_i+(d+c)}{2} \quad (i=1,...N)
\end{aligned}
\right\}
\tag{3.107}
$$

3. Select the number of points N, and find the coordinate and weights of ξ_i and η_i with w_i ($i = 1, \dots N$) using Table 3.2;
4. Evaluate the integral using

$$\bar{I}_{2-D} \cong \left(\frac{b-a}{2}\right)\left(\frac{d-c}{2}\right) \sum_{i=1}^{N} \sum_{j=1}^{N} w_i w_j f(x_i, y_i) \qquad (3.108)$$

where x_i and y_i are determined by ξ_i and η_i using Eq. (3.107).

Example 3.6

Use the 2-point and 3-point approximation to evaluate the integral of

$$I_{2-D} = \int_0^6 \int_0^2 (2 + x + y + x^2 y)\, dx\, dy$$

Solution

When 2-point approximation is used, Table 3.2 shows that $w_1 = w_2 = 1$, $\xi_1 = -1/\sqrt{3}$, $\xi_2 = 1/\sqrt{3}$, $\eta_1 = -1/\sqrt{3}$, and $\eta_2 = 1/\sqrt{3}$ and the problem specifies the ranges of the two-dimensional domain as $a = 0$, $b = 2$, $c = 0$, $d = 6$. Therefore, applying Eq. (3.108) gives,

$$\bar{I}_{2-D} \cong \left(\frac{b-a}{2}\right)\left(\frac{d-c}{2}\right) \sum_{i=1}^{N} \sum_{j=1}^{N} w_i w_j f\left(\frac{(b-a)\xi_i + (b+a)}{2}, \frac{(d-c)\eta_j + (d+c)}{2}\right)$$

$$\cong (1)(3) \sum_{i=1}^{2} \sum_{j=1}^{2} w_i w_j f\left((\xi_i + 1), (3\eta_j + 3)\right)$$

$$= (1)(3)(1)(1)\left[2 + (\xi_1 + 1) + \left((\xi_1 + 1)^2 + 1\right)(3\eta_1 + 3)\right]$$

$$+ (1)(3)(1)(1)\left[2 + (\xi_1 + 1) + \left((\xi_1 + 1)^2 + 1\right)(3\eta_2 + 3)\right]$$

$$+ (1)(3)(1)(1)\left[2 + (\xi_2 + 1) + \left((\xi_2 + 1)^2 + 1\right)(3\eta_1 + 3)\right]$$

$$+ (1)(3)(1)(1)\left[2 + (\xi_2 + 1) + \left((\xi_2 + 1)^2 + 1\right)(3\eta_2 + 3)\right]$$

$$= 120$$

When 3-point approximation is used, Table 3.3 tells that $w_1 = w_3 = 5/9$, $w_2 = 8/9$, $\xi_1 = \eta_1 = -\sqrt{3/5}$, $\xi_3 = \eta_2 = 0$, $\xi_3 = \eta_3 = \sqrt{3/5}$. Applying Eq. (3.108) gives,

$$\bar{I}_{2-D} \cong \left(\frac{b-a}{2}\right)\left(\frac{d-c}{2}\right) \sum_{i=1}^{N} \sum_{j=1}^{N} w_i w_j f\left(\frac{(b-a)\xi_i + (b+a)}{2}, \frac{(d-c)\eta_j + (d+c)}{2}\right)$$

$$\cong (1)(3) \sum_{i=1}^{3} \sum_{j=1}^{3} w_i w_j f\left((\xi_i + 1), (3\eta_j + 3)\right)$$

$$= (1)(3)(5/9)(5/9)\left[2 + (\xi_1 + 1) + \left((\xi_1 + 1)^2 + 1\right)(3\eta_1 + 3)\right]$$

$$+ (1)(3)(5/9)(8/9)\left[2 + (\xi_1 + 1) + \left((\xi_1 + 1)^2 + 1\right)(3\eta_2 + 3)\right]$$

$$+ (1)(3)(5/9)(5/9)\left[2 + (\xi_1 + 1) + \left((\xi_1 + 1)^2 + 1\right)(3\eta_3 + 3)\right]$$

$$+ (1)(3)(5/9)(5/9)\left[2 + (\xi_2 + 1) + \left((\xi_2 + 1)^2 + 1\right)(3\eta_1 + 3)\right]$$

$$+ (1)(3)(5/9)(8/9)\left[2 + (\xi_2 + 1) + \left((\xi_2 + 1)^2 + 1\right)(3\eta_2 + 3)\right]$$

$$+ (1)(3)(5/9)(5/9)\left[2 + (\xi_2 + 1) + \left((\xi_2 + 1)^2 + 1\right)(3\eta_3 + 3)\right]$$

$$+ (1)(3)(5/9)(5/9)\left[2 + (\xi_3 + 1) + \left((\xi_3 + 1)^2 + 1\right)(3\eta_1 + 3)\right]$$

$$+ (1)(3)(5/9)(8/9)\left[2 + (\xi_3 + 1) + \left((\xi_3 + 1)^2 + 1\right)(3\eta_2 + 3)\right]$$

$$+ (1)(3)(5/9)(5/9)\left[2 + (\xi_3 + 1) + \left((\xi_3 + 1)^2 + 1\right)(3\eta_3 + 3)\right]$$

$$= 120$$

Both of the calculations lead to the accurate result because the integrated function is a polynomial equation itself with the highest polynomial order of 3.

3.10 **SUMMARY**

This chapter discusses mathematical representations of elements and their relations in simulation model. (1) Different types of CS are introduced to represent nodes, and establish and transfer element models from one stage to another stage of a modeling process. (2) Shape functions are developed to calculate state variables in any position from given nodal values in one element; the properties of shape functions are deployed to derive their expressions for elements with a high order. To facilitate the modeling and computation in elements; the natural CS is introduced, and it is adopted for an iso-parametric representation of any physical quantities in elements. (3) Shape functions have been utilized for programming of the numerical integration in a one-dimensional, two-dimensional, and three-dimensional domain.

REFERENCES

Bi, Z.M., Gruver, W.A., Zhang, W.J., Lang, S.Y.T., 2006. Automated modelling of modular robotic configurations. Robotics and Autonomous Systems 54, 1015−1025.

Davis, P., Rabinowitz, P., 1956. Abscissas and weights for Gaussian quadratures of high order. Journal of Research of the National Bureau of Standards 56 (1), 35−37. Research Paper 2645.

■ PROBLEMS

3.1. A rectangle element is shown in Fig. 3.24: (1) define the equation of coordinate transformation from LCS to natural CS; (2) define the equations of coordinate transformation from LCS to GCS; (3) fill in the coordinates for specified positions.

Position	Coordinates		
	Local CS (x, y)	GCS (X, Y)	Natural CS (ξ, η)
A			
B			
C			
D			
o	(0, 0)	(3.464, 2)	(−1, −1)

3.2. A triangle element is shown in Fig. 3.25: (1) draw out the axes and origin of natural CS in the figure; (2) define the equation of coordinate transformation from natural CS to GCS; (3) fill in the coordinates for the specified positions.

Position	Coordinates	
	GCS (X, Y)	Natural CS (ξ, η)
Node-i	(4, 2)	(1, 0)
Node-j		
Node-k		
A		
B		
C		
D		

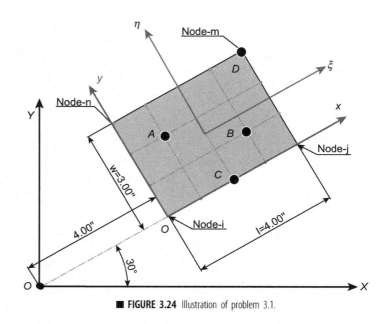

■ **FIGURE 3.24** Illustration of problem 3.1.

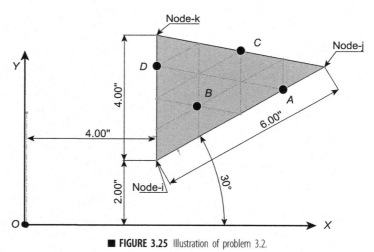

■ **FIGURE 3.25** Illustration of problem 3.2.

3.3. Fig. 3.26 shows a 3-bar mechanism where the tool is attached on point C, the approaching direction (x) of the tool is aligned with the axial direction of binary bar BC. The positions and lengths of 3 bars under the GCS $\{O\text{-}XY\}$ are given. Find the homogenous matrix for the coordinate transformation from LCS $\{o\text{-}xy\}$ to GCS $\{O\text{-}XY\}$.

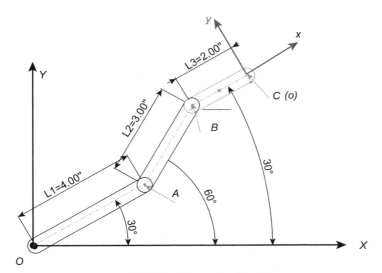

■ **FIGURE 3.26** Illustration of problem 3.3.

3.4. Use Lagrange interpolation equations to determine the shape functions of a cubic element under a GCS.

3.5. Use Lagrange interpolation equations to determine the shape functions of a cubic element under a local CS. Draw the shape functions with respect to coordinate x to verify if all properties of shape functions are valid.

3.6. As shown in Fig. 3.27, a rectangle element consists of nodes A, B, C, and D. The element is under an external load and the stresses on these nodes are illustrated in the figure, estimate the stress at the given position P.

3.7. As shown in Fig. 3.28, a triangle element consists of nodes A, B, and C. The element is under the thermal load and the temperatures on these nodes are illustrated, estimate the temperature at the given position P.

3.8. For the triangle element shown in Fig. 3.27, find the coordinates of position P under the natural CS. Use the shape functions in natural CS to determine the temperature at P based on the nodal temperatures.

3.9. Do the integral for the following expression analytically and use the Gauss-Legendre formula to determine how many points are needed to find an accurate integral.

$$I_{1-D} = \int_2^{10} (x^3 + x + 2)\,dx$$

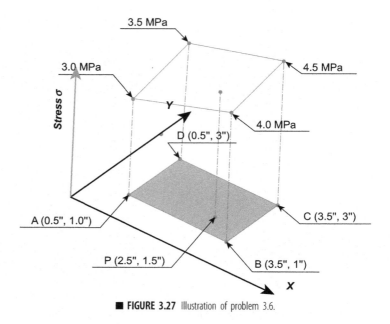

■ FIGURE 3.27 Illustration of problem 3.6.

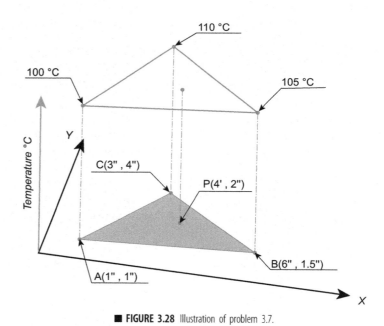

■ FIGURE 3.28 Illustration of problem 3.7.

3.10. Applying the Gauss-Legendre formula to evaluate the integral of the following expression with 2-, 3-, 4-, and 5- point approximation.

$$I_{1-D} = \int_0^2 \left(x^3 \sin(x) + 1\right)dx$$

3.11. Use the 2-point and 3-point approximation to evaluate the integral of

$$I_{2-D} = \int_0^3 \int_0^3 (2 + x + xe^{-y})dxdy$$

Chapter 4

Solutions of Element Models

4.1 INTRODUCTION

In the previous chapter, we discussed (1) how a continuous domain is decomposed into a set of elements with discrete nodes, and (2) how a state variable at any point is approximated based on the given values of state variable at nodes. In this chapter, we tackle with the solutions of state variables on nodes, and we correlate these variables to nodal loads. This process is called *element modeling*; it leads to a mathematic model as the relations of nodal state variables and loads.

Element models are derived from the governing equations of engineering problems. We first discuss some common laws of mechanics. Most of the physical phenomena in engineering systems are governed by the principles of conservation, such as mass conservation, momentum conservation, and energy conservation. These principles can be then interpreted as partial differential equations (PDEs). Therefore, the mathematical representations of these principles, their connections to engineering problems, and different solving methods to PDEs in elements are discussed.

4.2 GOVERNING EQUATIONS OF ENGINEERING PROBLEMS

For an engineering problem, an exact solution can only be obtained by analytical methods. However, the solution to an analytical model does not always exist. Even if it exists, it is not always practical to obtain the solution. Generally, an analytical solution is only applicable to simple geometry and boundary conditions (BCs). Therefore, the numerical simulation to solve engineering problems is desirable for a number of the reasons: (1) a continuous domain of a real-world design problem is often irregular and very complex, which becomes difficult to represent the boundaries of a domain accurately; (2) the domain may consist of a set of subdomains with different properties, or one material with anisotropic nature; (3) the engineering problem may involve multiphysics; (4) the engineering problem involves complex boundary or loading conditions, which are often cost forbidden for mock-up tests in real-life. Generally, a numerical simulation can be performed if the engineering problem can be formulated as a mathematic model where the system response is described by differential equations. Table 4.1 shows a few examples of engineering problems with the corresponding laws of mechanics, as well as differential equations.

Finite element analysis (FEA) is a generic approach to solve differential equations subjected to BCs. If two engineering systems can be formulated

Table 4.1 Examples of Engineering Problems and Governing Equations

Engineering Applications	Laws of Mechanics and Differential Equations
Solid mechanics 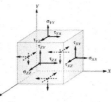 where $[\sigma]^T$ and $[\varepsilon]^T$ are stress and strain vectors as, $[\sigma]^T = [\sigma_{xx} \quad \sigma_{yy} \quad \sigma_{zz} \quad \tau_{xy} \quad \tau_{yz} \quad \tau_{xz}]$ $[\varepsilon]^T = [\varepsilon_{xx} \quad \varepsilon_{yy} \quad \varepsilon_{zz} \quad \gamma_{xy} \quad \gamma_{yz} \quad \gamma_{xz}]$	Hooke's Law $\varepsilon_{xx} = \dfrac{1}{E}[\sigma_{xx} - v(\sigma_{yy} + \sigma_{zz})]$ $\varepsilon_{yy} = \dfrac{1}{E}[\sigma_{yy} - v(\sigma_{xx} + \sigma_{zz})]$ $\varepsilon_{zz} = \dfrac{1}{E}[\sigma_{zz} - v(\sigma_{xx} + \sigma_{yy})]$ $\gamma_{xy} = \dfrac{\tau_{xy}}{G} \quad \gamma_{yz} = \dfrac{\tau_{yz}}{G} \quad \gamma_{xz} = \dfrac{\tau_{xz}}{G}$ where $\varepsilon_{xx} = \dfrac{\partial u}{\partial x} \quad \varepsilon_{yy} = \dfrac{\partial v}{\partial y} \quad \varepsilon_{zz} = \dfrac{\partial w}{\partial z}$ $\gamma_{xy} = \dfrac{\partial u}{\partial y} + \dfrac{\partial v}{\partial x} \quad \gamma_{yz} = \dfrac{\partial v}{\partial z} + \dfrac{\partial w}{\partial y} \quad \gamma_{xz} = \dfrac{\partial u}{\partial z} + \dfrac{\partial w}{\partial x}$ E and G are young's modulus and shear modulus, respectively.
Dynamics 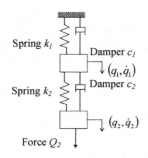	Lagrange's Equation where $\dfrac{d}{dt}\left(\dfrac{\partial T}{\partial \dot{q}_i}\right) - \dfrac{\partial T}{\partial q^i} + \dfrac{\partial \Lambda}{\partial q^i} = Q_i \quad (i = 1,2,3\cdots n)$ $t = $ time $T = $ kinetic energy of the system $q_i = $ coordinate system $\dot{q} = $ time derivate of coordinate system $\square = $ potential energy of the system $Q_i = $ nonconservative forces or moments

Table 4.1 Examples of Engineering Problems and Governing Equations *continued*

Engineering Applications	Laws of Mechanics and Differential Equations
Heat Transfer Problem 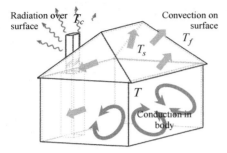	Fourier law of conduction $q = -k \cdot \nabla T$ Newton's law of convection (boundary condition or BC) $q = hA(T_s - T_f)$ Stefan–Boltzmann law of radiation (BC) $P = e\sigma A(T^4 - T_C^4)$ where q = heat flux k = conductivity coefficient T = temperature h = convection coefficient T_s = temperature on surface T_f = temperature in surrounding fluid P = net radiated power A = radiating area σ = Stefan's constant E = emissivity T_c = temperature of surrounding
Electromagnetics 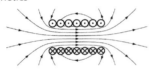 E = electric field B = magnetic field D = electric displacement H = magnetic field strength ρ = charge density ε_0 = permittivity μ_0 = permeability J = current density C = speed of light P = polarization I = electric current M = magnetization.	Maxwell's equations Gauss' law for electricity $\nabla \cdot (\varepsilon_0 E + P) = \rho$ Gauss's law for magnetism $\nabla \cdot B = 0$ Faraday's law of induction $\nabla \times E = \dfrac{\partial B}{\partial t}$ Ampere's law $\nabla \times H = J + \dfrac{\partial D}{\partial t}$ where $D = (\varepsilon_0 E + P)B = \mu_0(H + M)\nabla \cdot (\bullet)$ and $\nabla \times \cdot (\bullet)$ are the *divergence* and *curl* operations of vectors, respectively.
Fluid mechanics 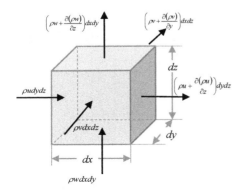	Law of conservation of mass $\dfrac{\partial \rho}{\partial t} + \dfrac{\partial \rho u}{\partial x} + \dfrac{\partial \rho v}{\partial y} + \dfrac{\partial \rho w}{\partial z} = 0$ where t = time ρ = density u, v, and w are the velocities along x, y, and z axis, respectively. For a steady flow: $\dfrac{\partial \rho}{\partial t} = 0$ From impressive flow where the density is constant: $\dfrac{\partial u}{\partial x} + \dfrac{\partial v}{\partial y} + \dfrac{\partial w}{\partial z} = 0$

as a mathematic model with the same set of differential equations; the same solver in the FEA implementation can be used to solve these problems. Many engineering problems have similar governing equations and even they are in different disciplines. Fig. 4.1 has shown a variety of engineering problems that are governed by the well-known Poisson equations (Chandrupatla and Belegundu, 2012). If an FEA tool includes a solver to find solution for the Poisson equations with given boundary values, it can be applied to solve all of these problems represented by the Poisson equations.

4.3 DIRECT FORMULATION

Governing equations for some basic engineering problems can be solved by *direct formulation*. In a direct formulation, the approximated solution in an element is derived from the governing equations directly. The procedure for the direction formulation is illustrated by modeling the elements of a spring system as shown in Fig. 4.2.

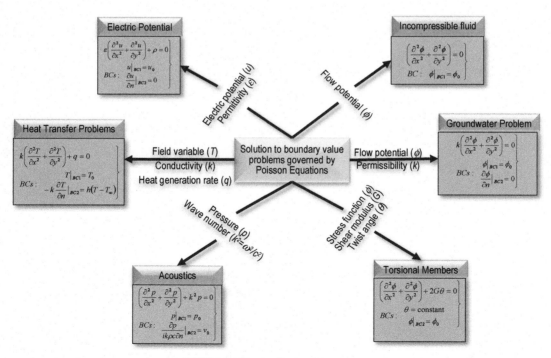

■ **FIGURE 4.1** Different applications of Poisson equations.

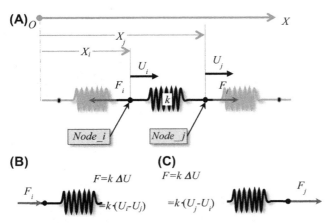

■ **FIGURE 4.2** Description of elements in a spring system. (A) An element in a spring; (B) force on node *i*; (C) force on node *j*.

As shown in Fig. 4.2, a spring element consists of two nodes, i.e., node *i* and node *j*. The physical behavior of the element can then be represented by the displacements at two nodes, which are denoted as U_i, and U_j at node *i* and *j*, respectively. The state variable of the spring element is displacement; although the external loads are F_i and F_j applied on two nodes. Fig. 4.2B and C shows the free body diagrams when the spring element is virtually cut into two parts. The physical behavior of a spring element is governed by the Hook's law as,

$$F = k \cdot \Delta U \qquad (4.1)$$

where

  F, the applied force at two ends,
  k, the spring coefficient,
  ΔU, the difference of the displacements of two ends.

After the discretization of a continuous domain, the behavior of an element in an FEA model is completely represented by the state variables of nodes. For the spring element in Fig. 4.2, its behaviors are represented by U_i and U_j. To model the relations of external loads F_i and F_j to state variables U_i and U_j, the force balances on two free bodies in Fig. 4.2B and C are considered. This leads to the force-balancing equations at two nodes as,

$$\left. \begin{array}{l} k(U_i - U_j) = F_i \\ k(U_j - U_i) = F_j \end{array} \right\} \qquad (4.2)$$

Eq. (4.2) consists of two linear equations for two-state variables U_i and U_j. Note that Eq. (4.2) is not solvable if there is no imposed BC. It is understandable that if no restraint is applied to the element, any rigid motion of spring is valid because it has no impact on the deformation of spring. However, such a motion changes the displacements of nodes. In other words, Eq. (4.2) only gives the solution about the difference of displacements caused by external load F. It is insufficient to determine absolute displacements on two nodes if there is a rigid motion. Therefore, an additional BC is required to obtain a unique solution. For example, if the left end of the spring system is fixed, the displacement of node i is given as $U_i = 0$, and U_j can be found as $U_j = F_j/k$ accordingly.

The above spring element model is generic and can be extended to solve a variety of simple engineering problems; as long as the system response to external loads can be treated as equivalent responses of springs. If it is the case, the solution to such an engineering problem can be defined by mapping their physical parameters, external loads, and state variables into spring rate (k), forces (F_i and F_j), and the displacement in a spring element (U_i and U_j). Table 4.2 gives the mappings of some basic engineering problems to equivalent spring systems.

Taking an example of a resistor element in Table 4.2, its element model can be defined by mapping all of physical properties and quantities to those of a spring element, i.e.,

$$
\begin{aligned}
\frac{1}{R} &\rightarrow k \\
V &\rightarrow U \\
I &\rightarrow F
\end{aligned}
$$

Replacing the corresponding parameters and variables in Eq. (4.2) gets the model of a heat transfer element as,

$$
\left.
\begin{aligned}
\frac{1}{R}(V_j - V_i) &= I_i \\
-\frac{1}{R}(V_j - V_i) &= I_j
\end{aligned}
\right\}
\tag{4.3}
$$

4.4 MINIMUM POTENTIAL ENERGY PRINCIPLE

The behavior of an engineering system may relate to one or a few of scalar quantities such as energy, momentum, and mass. For example, the energy change occurs to a computational domain must be conservative. Energy can be transferred from one form to another, but it is never vanished. Therefore, the minimum potential energy principle can be utilized to develop element solutions for corresponding differential equations. In this

Table 4.2 Similarity of a Spring System With Other Engineering Systems

Engineering Problems	Equivalent Terms in a Spring Element		
	Stiffness (k)	Load (F)	State Variable (U)
Axial Member E: Young's modulus A: Cross-section area L: length P: Axial load Y: Displacement	$\frac{EA}{L}$	P	Y
Torsional Member G: Shear modulus of rigidity J: Second moment of inertia T: Torque L: Length θ: Angular deflection	$\frac{GJ}{L}$	T	θ
Pipe Member A: Cross-section diameter L: Length μ: Viscosity P: Pressure Q: Flow rate	$\frac{\pi D^4}{128\mu}$	Q	P
Resistor R: Resistance I: Current V: Voltage	$\frac{1}{R}$	I	V
Heat Transfer A: Cross-section area K: Conductivity q: heat flow rate T: temperature L: Length	$\frac{kA}{L}$	q	T

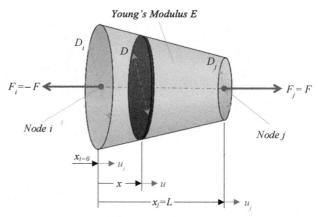

■ **FIGURE 4.3** Taped member under external axial loads.

section, the minimum potential energy principle is applied to model a tapered cylinder under axial loads to illustrate its application in FEA element modeling.

Fig. 4.3 gives the description of the tapered member under axial loads. Material properties is described by Young's modulus E, which is used to describe the stress–strain relation in the constitutive model. The tapered member consists of two nodes, i.e., node i and node j. The state variables are u_i and u_j on two nodes, respectively. The geometry of element is dimensioned by diameters D_i and D_j at two ends and the length L of the element. A local coordinate system (LCS) is established by select the position of node i as the origin of LCS. In this LCS, an arbitrary point within the element is represented by coordinate x, D for the diameter of the cross-section area, and $u(x)$ for the displacement.

The diameter of the cross-section varies linearly with respect to x. Therefore, $D(x)$ can be calculated from D_i and D_j as,

$$D(x) = D_i + x\frac{D_j - D_i}{l} \tag{4.4}$$

The total potential energy as well as the work by external loads on two nodes can be evaluated as,

$$\Pi = \int_V \frac{\sigma\varepsilon}{2}dV - (F_i \cdot U_i + F_j \cdot U_j)$$
$$= \int_0^L \frac{E}{2}\varepsilon^2 \left(\frac{\pi}{4}(D(x))^2\right)dx - (F_i \cdot U_i + F_j \cdot U_j) \tag{4.5}$$

where

$\prod$ is the sum of the strain energy and the work done by external force on the element,

$\int_V \frac{\sigma\varepsilon}{2} dV$ is the total strain energy caused by deformation,

$(F_i \cdot U_i + F_j \cdot U_j)$ is the work done by the external forces on node i and j, and the strain $\varepsilon(x)$ can be obtained by

$$\varepsilon = \frac{dU}{dx} = \frac{d(S_i \cdot U_i + S_j \cdot U_j)}{dx} = \frac{U_j - U_i}{l} \tag{4.6}$$

Eq. (4.5) for the total energy in the element involves in two state variables, i.e., U_i and U_j. According to the minimal potential energy principle, these two variables must meet the following conditions to achieve the energy conservation,

$$\frac{d\prod}{dU_i} = 0; \frac{d\prod}{dU_j} = 0 \tag{4.7}$$

Substituting Eqs. (4.4)–(4.6) into Eq. (4.7) gives two equations about U_i and U_j as,

$$\left.\begin{array}{l} \dfrac{\partial \prod}{\partial U_i} = \dfrac{\partial \left(\int_0^L \dfrac{E}{2} \left(\dfrac{U_j - U_i}{L} \right)^2 \left(\dfrac{\pi}{4} \left(D_i + x \dfrac{D_j - D_i}{L} \right) \right)^2 dx - (F_i \cdot U_i + F_j \cdot U_j) \right)}{\partial U_i} = 0 \\[4mm] \dfrac{\partial \prod}{\partial U_j} = \dfrac{\partial \left(\int_0^L \dfrac{E}{2} \left(\dfrac{U_j - U_i}{L} \right)^2 \left(\dfrac{\pi}{4} \left(D_i + x \dfrac{D_j - D_i}{L} \right) \right)^2 dx - (F_i \cdot U_i + F_j \cdot U_j) \right)}{\partial U_j} = 0 \end{array}\right\} \tag{4.8}$$

Simplifying Eq. (4.8) gets the relations of (U_i and U_j) with (F_i and F_j) as,

$$\left.\begin{array}{l} k_e \cdot U_i - k_e \cdot U_j = F_i \\ -k_e \cdot U_i + k_e \cdot U_j = F_j \end{array}\right\} \tag{4.9}$$

where k_e is equivalent spring rate calculated by

$$k_e = \frac{E \cdot \pi}{12} \left(D_i^2 + D_i D_j + D_j^2 \right) \tag{4.10}$$

To obtain U_i and U_j uniquely from Eq. (4.10), the BC has to be applied. For the tapered member in Fig. 4.3, the left side is restrained; therefore, we have $U_i = 0$. Using this BC in Eq. (4.8) gives the maximized displacement of the tapered member occurring to node j as,

$$U_j = \frac{F_j}{k_e} = \frac{12 F_j}{E \cdot \pi \left(D_i^2 + D_i D_j + D_j^2 \right)} \tag{4.11}$$

Physical behaviors of many mechanical systems are governed by differential equations based on the energy conservation. Therefore, the minimum potential energy principle can be applied to derive element models for a _variety of mechanical systems. In Chapter 8, it will be used to model a number of design problems in solid mechanics.

4.5 WEIGHTED RESIDUAL METHODS

Direct methods and minimum potential energy methods are able to find solutions to a limited number of engineering problems; but they lack the flexibility to solve a broad scope of PDEs. More generic methods are needed to develop element models for any PDEs. In this section, the weighted residual method is introduced as a generic way to approximate the solutions to any PDEs.

The solution to a PDE is approximated; to ensure the solution is acceptable, the measure of the approximation must be defined. Such a measure is defined in comparison with the exact solution of PDE.

4.5.1 Exact and weak solutions to PDEs

In this section, an engineering problem is formulated as a set of differential equations (N_D) subjected to a set of BCs (N_{BC}), which is also referred as a *boundary value problem*. Without losing the generality, the concepts of exact and weak solutions are introduced through the example of a boundary value problem as below.

$$\frac{d^2u}{dx^2} = p(x), a \leq x \leq b \tag{4.12}$$

where only one differential equation is involved in the 1D domain, i.e., $N_D = 1$.

The BCs of this example problem are,

$$\left. \begin{array}{c} u(a) = 0 \\ \dfrac{du}{dx}\bigg|_{x=b} = 1 \end{array} \right\} \tag{4.13}$$

where two BCs are defined, i.e., $N_{BC} = 2$.

Eq. (4.13) includes two types of BCs, i.e., essential BCs and natural BCs. The first BC is an *essential BC* where the value of a state variable is prescribed on boundary. The second BC is a *natural BC* where the derivate of the state variable is specified on boundary.

A residual function R(x) refers to a function for the error of an approximated solution $\bar{u}(x)$ to the original PDE by Eq. (4.12) for the element. $R(x)$ is defined as the difference of left and right sides of Eq. (4.12) when the exact solution $u(x)$ is approximated by $\overline{u}(x)$,

$$R(x) = \left(\frac{d^2\overline{u}(x)}{dx^2} - p(x)\right) \qquad (4.14)$$

Note that an *exact* solution to the differential equation Eq. (4.12) satisfies the condition that the residual function is $R(x) \equiv 0$ for $a \leq x \leq b$. In other words, an exact solution satisfies Eq. (4.12) strictly in any position of the element without an error.

A *residual* $\overline{R}$ refers to an integral of the error function of PDE over the element by using approximated solution $\overline{u}(x)$ in Eq. (4.12). Therefore, the *residual* $\overline{R}$ is defined as,

$$\overline{R} = \int_a^b R(x)dx \qquad (4.15)$$

A *weak* solution to the differential Eq. (4.12) satisfies the condition that the residual $\overline{R}$ is vanished $\left(\overline{R} = 0\right)$ in the element. Therefore, a *weak* solution satisfies Eq. (4.12) approximately in the sense that the summed error within the element is minimized even though $\overline{u}(x)$ is not the exact $u(x)$ to satisfy Eq. (4.12) strictly at any position.

A *weighted residual method* is to use a test function $w(x)$ in a weak solution in the evaluation of residual $\overline{R}$, i.e.,

$$\overline{R} = \int_a^b w(x) \cdot R(x)dx = 0 \qquad (4.16)$$

where $w(x)$ is a test function to be selected.

The selection of a test function is arbitrary. If Eq. (4.16) is satisfied for any test function, the solution $\overline{u}(x)$ is an exact solution. Moreover, different weighted residual methods can be developed based on different choices of $w(x)$.

4.5.2 **Procedure to find an approximated solution**

Because an exact solution to a PDE is hard or even impossible to find, an approximated solution should be defined to develop element models in FEA. The most wide approximation for an arbitrary function is based on the Taylor expansion. Here, the Taylor expansion is used to approximate an exact solution for Eq. (4.12). Because it is one-dimensional problem, the formula of the Taylor expansion is,

$$\overline{u}(x) \cong c_1 + c_2 x + c_3 x^2 + \cdots c_{N_T} x^{N_T - 1} \qquad (4.17)$$

Note that the expression of the approximation in Eq. (4.17) is partially known but incomplete because N_T and all of the coefficients are still unknown. To define a complete approximation, one has to determine (1) the number of polynomial terms (N_T) in the approximation and (2) all of the constants used in the expression $(c_1, c_2, \ldots c_{N_T})$ in Eq. (4.17).

Therefore, a weighted residual method is reduced to find the number of polynomial terms N_T and the corresponding coefficients $(c_1, c_2, \ldots c_{N_T})$ and it will be based on the conditions for a weak solution to the original PDE.

The procedure of using a weighted residual method to find an approximated solution consists of the following steps:

1. Determine the number of polynomial terms (N_T). This number must be larger than the sum of the numbers of the differential equation(s) and BCs, i.e.,

$$N_T \geq N_D + N_{\text{BC}} \tag{4.18}$$

Taking an example of a boundary value problem in Eqs. (4.12) and (4.13), it has one differential equation ($N_D = 1$) and two BCs ($N_{\text{BC}} = 2$), then the terms (N_T) of the Taylor series for the approximation must be equal to or larger than 3. The more terms the approximation has, the better the approximation that can be obtained, but much computation is needed.

2. BCs in Eq. (4.13) are applied as the algebra equations for the part of constraints for the approximated solutions. The number of the algebra equations equals the number of BCs (N_{BC}).

3. Select ($N_T - N_{\text{BC}}$) number of test functions and apply the condition Eq. (4.16) for a weak solution. This generates ($N_T - N_{\text{BC}}$) number of algebra equations. These ($N_T - N_{\text{BC}}$) algebra equations combined with N_{BC} equations from step 2 yields a system of N_T number of algebra equations about the constants $(c_1, c_2, \ldots c_{N_T})$ in Eq. (4.17).

4. Solve the system of N_T algebra equations to get $c_1, c_2, \ldots c_{N_T}$ and obtain the complete solution by substituting all of the coefficients back to Eq. (4.17).

In the aforementioned procedure, users are free to choose (1) the number of polynomial terms (N_T) and (2) the types of test functions. Different test functions lead to the different approximated solutions to original differential equations. Although test functions can be arbitrary functions, some commonly used test functions are selected by using *collocation methods*, *subdomain methods*, *Galerkin methods*, and *least square methods*.

4.5.3 A variety of weighted residual methods

The following value boundary problem is used as an example to illustrate the procedure of using the weighted residual method to find an approximated solution to PDEs.

The goal is to find an approximated solution $\bar{u}(x)$ for $u(x)$ governed by a PDE as follows,

$$\frac{d^2 u}{dx^2} = 6x - \sin(x), \ 0 \leq x \leq 1$$

subjected to :

$$\left. u(x) \right|_{x=1} = \sin(1) \\ \left. \frac{du}{dx} \right|_{x=0} = 0 \Bigg\}$$

$$(4.19)$$

Step 1. In the given boundary value problem, $N_D = 1$ and $N_{BC} = 2$. Therefore, Eq. (4.19) shows that the number of polynomial terms (N_T) must be larger than 3. We set $N_T = 4$ in the approximation of the Taylor series as,

$$\bar{u}(x) \cong c_1 + c_2 x + c_3 x^2 + c_4 x^3 \tag{4.20}$$

To define the approximation function completely, four unknown constants (c_1, c_2, c_3, c_4) have to be determined. Therefore, four algebra equations about these constants must be defined.

Step 2. The given problem has two BCs, substituting Eq. (4.20) into these two BCs gives,

$$\left. \bar{u}(x) \right|_{x=1} = c_1 + c_2(1) + c_3(1)^2 + c_4(1)^3 = \sin(1) \\ \left. \frac{d\bar{u}}{dx} \right|_{x=0} = c_2 + 2c_3(0) + 3c_4(0)^2 = 0 \Bigg)$$

$$(4.21)$$

Substituting Eq. (4.21) back into Eq. (4.20) for the simplification yields,

$$\bar{u}(x) \cong (\sin(1) - c_3 - c_4) + c_3 x^2 + c_4 x^3 \tag{4.22}$$

Taking the first and second derivative of Eq. (4.22) gives,

$$\frac{\bar{u}(x)}{dx} \cong 2c_3 x + 3c_4 x^2 \\ \frac{d\bar{u}^2(x)}{dx^2} = 2c_3 + 6c_4 x \Bigg\}$$

$$(4.23)$$

Step 3. Eq. (4.23) includes two unknown constants (c_3, c_4). Therefore, two different test functions have to be selected, so that another two equations can be defined based on the condition of weak solution (Eq. 4.16) to solve c_3 and c_4. Substituting Eq. (4.23) into Eq. (4.16) yields the expression of the residual as,

$$\bar{R} = \int_a^b w(x) \cdot R(x) dx = \int_0^1 w(x) \cdot ((2c_3 + 6c_4 x) - 6x + \sin(x)) dx = 0$$

$$(4.24)$$

Next, different methods are used in selecting test functions to evaluate residuals in Eq. (4.24).

4.5.3.1 Collocation method

In a collocation method, the test function is selected as,

$$w(x) = \begin{cases} 1 & x = x_i \\ 0 & x \neq x_i \end{cases} \tag{4.25}$$

where x_i is an arbitrary position in the domain of interest.

Because two additional equations are required to solve c_3 and c_4; the condition of the weak-solution has to be applied twice using the test function of Eq. (4.25) at two different positions. Let $x_1 = 1/3$ and $x_2 = 2/3$, using Eq. (4.25) in Eq. (4.24) twice gets,

$$\left.\begin{aligned}
\overline{R}_1 &= \left(\left(2c_3 + 6c_4\left(\frac{1}{3}\right)\right) - 6\left(\frac{1}{3}\right) + \sin\left(\left(\frac{1}{3}\right)\right)\right) = 0 \\
\overline{R}_2 &= \left(\left(2c_3 + 6c_4\left(\frac{2}{3}\right)\right) - 6\left(\frac{2}{3}\right) + \sin\left(\left(\frac{2}{3}\right)\right)\right) = 0
\end{aligned}\right\} \tag{4.26}$$

Solving Eq. (4.26) yields $c_3 = -0.01801$ and $c_4 = 0.8544$. The approximated solution becomes

$$\overline{u}(x) = 0.00508 - 0.01801 \cdot x^2 + 0.8544 \cdot x^3 \tag{4.27}$$

The plot of $R(x)$ by the collocation method is shown in Fig. 4.4. Substituting the exact solution $u(x)$ by $\overline{u}(x)$ has led to the error (or the residual) of the original PDE. However, the collocation method ensures that there is a zero residual at a position where a test function is defined.

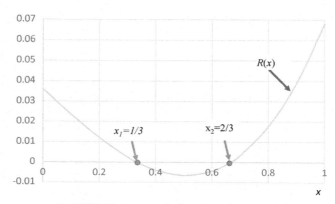

■ **FIGURE 4.4** Approximated solution from collocation method.

4.5.3.2 Subdomain method

In the subdomain method, the test function is selected as,

$$w(x) = \begin{cases} 1 & x \in (a_i, b_i) \\ 0 & x \notin (a_i, b_i) \end{cases} \tag{4.28}$$

where (a_i, b_i) is an arbitrary subdomain of interest.

Because two additional equations are required to solve c_3 and c_4; the condition of the weak solution has to be applied twice using the test function of Eq. (4.28) in two subdomains. Set two subdomains as (0, 0.5) and (0.5, 1.0), use Eq. (4.28) in Eq. (4.24), and set the corresponding residuals as zero,

$$\left. \begin{aligned} \overline{R}_1 &= \int_0^{0.5} (2c_3 + 6c_4 x - 6x + \sin(x))dx = c_3 + \frac{3}{4}c_4 + \frac{1}{4} - \cos\left(\frac{1}{2}\right) = 0 \\ \overline{R}_2 &= \int_{0.5}^1 (2c_3 + 6c_4 x - 6x + \sin(x))dx = c_3 + \frac{9}{4}c_4 - \frac{9}{4} + \cos\left(\frac{1}{2}\right) - \cos(1) = 0 \end{aligned} \right\} \tag{4.29}$$

Solving Eq. (4.29) yields $c_3 = -0.01499$ and $c_4 = 0.85676$ and the approximated solution becomes

$$\overline{u}(x) = 0.000299 - 0.01499 \cdot x^2 + 0.85676 \cdot x^3 \tag{4.30}$$

The plot of $R(x)$ by the subdomain method is shown in Fig. 4.5. Substituting the exact solution $u(x)$ by $\overline{u}(x)$ has led to the error (or residual) of the original PDE. However, the subdomain method ensures that there is an integrated zero residual over a subdomain where a test function is defined.

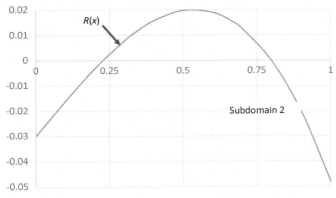

■ **FIGURE 4.5** Approximated solution from the subdomain method.

4.5.3.3 Least-squares method

In a least-square method, the test function is selected as,

$$w(x) = \frac{\partial \overline{R}(x)}{\partial (c_i)} \tag{4.31}$$

where c_i is an elected constant to be determined. The motivation of a least-square method is to minimize the absolute value of the residual integral of the whole domain. To achieve this, c_3 and c_4 must be chosen to minimize the residual integral,

$$\widetilde{R} = \int_0^1 \overline{R}^2(x)dx \tag{4.32}$$

To minimize $\widetilde{R}$, c_3 and c_4 have to satisfy the condition of,

$$\left. \begin{array}{l} \dfrac{\partial \widetilde{R}}{\partial c_3} = 2 \displaystyle\int_0^1 \dfrac{\partial \overline{R}}{\partial c_3} \cdot \overline{R}(x)dx = 0 \\[3mm] \dfrac{\partial \widetilde{R}}{\partial c_4} = 2 \displaystyle\int_0^1 \dfrac{\partial \overline{R}}{\partial c_4} \cdot \overline{R}(x)dx = 0 \end{array} \right\} \tag{4.33}$$

Eq. (4.33) shows if a test function in Eq. (4.30) is used, the residual based on the least-square method can be minimized. For the PDE in Eq. (4.24), the test functions are selected as

$$\left. \begin{array}{l} w_1(x) = \dfrac{\partial R(x)}{\partial c_3} = 2 \\[3mm] w_2(x) = \dfrac{\partial R(x)}{\partial c_4} = 6x \end{array} \right\} \tag{4.34}$$

Use Eq. (4.34) in Eq. (4.33) yields,

$$\left. \begin{array}{l} \overline{R}_1 = \displaystyle\int_0^1 (2)(2c_3 + 6c_4 x - 6x + \sin(x))dx = (2)(2c_3 + 3c_4 - 3 - \cos(1)) = 0 \\[3mm] \overline{R}_2 = \displaystyle\int_0^1 (6x)(2c_3 + 6c_4 x - 6x + \sin(x))dx = (6)\left(c_3 + 2c_4 - 2 + \displaystyle\int_0^1 x\sin(x)dx \right) = 0 \end{array} \right\} \tag{4.35}$$

where $\int_0^1 x\sin(x)dx = (\sin(x) - x\cos(x))|_0^1 = 0.301169$

Solving Eq. (4.35) yields $c_3 = -0.012462$ and $c_4 = 0.855076$ and the approximated solution becomes

$$\overline{u}(x) = -0.00114 - 0.012462 \cdot x^2 + 0.855076 \cdot x^3 \tag{4.36}$$

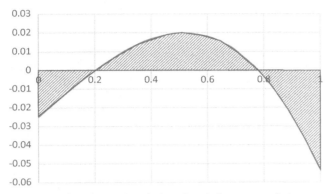

■ FIGURE 4.6 Approximated solution from the least square method.

The plot of $R(x)$ by the least square method is shown in Fig. 4.6. Substituting the exact solution $u(x)$ by $\bar{u}(x)$ has led to the error (or residual) of the original PDE. However, the least squares method ensures that the integral of absolute residual over the whole domain is minimized.

4.5.3.4 Galerkin method

In a Galerkin method, test functions are selected as constitutive functions in the expression of the approximated solution $\bar{u}(x)$. In Eq. (4.22), $\bar{u}(x)$ includes three constitutive functions, i.e., $\phi_1(x) = 1; \phi_2(x) = x^2; \phi_2(x) = x^3$.

Because only two equations are needed to determine c_3 and c_4, we choose $\phi_1(x)$ and $\phi_2(x)$ as test functions,

$$\left.\begin{array}{l} \bar{R}_1 = \int_0^1 \phi_1(x)(2c_3 + 6c_4 x - 6x + \sin(x))dx = (2c_3 + 3c_4 - 2 - \cos(1)) = 0 \\[2mm] \bar{R}_2 = \int_0^1 \phi_2(x)(2c_3 + 6c_4 x - 6x + \sin(x))dx = \left(\frac{2}{3}c_3 + \frac{3}{2}c_4 - \frac{3}{2} + \int_0^1 x^2 \sin(x)dx\right) = 0 \end{array}\right\}$$

$$(4.37)$$

where $\int_0^1 x^2 \sin(x)dx = (2 \cos(x) + 2x \sin(x) - x^2 \cos(x))\big|_0^1 = 0.223244$

Solving Eq. (4.37) yields $c_3 = -0.019815$ and $c_4 = 0.8599773$ and the approximated solution becomes

$$\bar{u}(x) = 0.00131 - 0.019815 \cdot x^2 + 0.85599773 \cdot x^3 \qquad (4.38)$$

The plot of $R(x)$ by the Galerkin method is shown in Fig. 4.7. The Galerkin method ensures that the integral of a weighted residual over the whole domain is minimized.

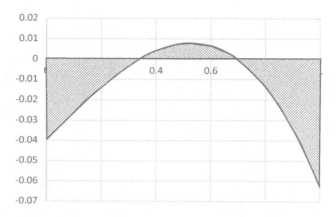

■ FIGURE 4.7 Approximated solution from the Galerkin method.

4.5.4 **Comments on weighted residual methods**

To implement a weighted residual method, test functions can be selected by the collocation method, the subdomain method, the least-square method, and the Galerkin method. The accuracy of the approximated solution $\bar{u}(x)$ from these methods varies from one boundary value problem to another. Therefore, it is no sense to judge the performance of weighted residual methods from the perspective of accuracy of approximation.

By all means, the accuracy of an approximation method can be validated by comparing the approximated solution with the exact one if it is available. The analytical solution to Eq. (4.19) can be found by taking two integrals and integrating with the consideration of the BCs. It is found as,

$$u = x^3 + \sin(x) - x \qquad (4.39)$$

Table 4.3 gives the absolute discrepancy of the approximated solutions from the exact solution over the whole domain of $x \in (0,1)$. Fig. 4.8 shows the closeness of exact functions and four approximated functions. It has shown that all of four weight-residual methods have obtained a good approximation of the function $u(x)$ with a reasonable error over the whole domain.

Among these four weighted residual methods, the Galerkin method is the most widely used in computer implementation of FEA code. Because the shape functions in the interpolation can be selected directly as test functions. In Sections 4.6—4.8, the Galerkin method will be used to develop element models for rectangle and triangular elements.

Table 4.3 Maximized Solution Errors by Different Weighted Residual Methods

x	Exact Solution $u(x)$	Maximized Errors by Different Methods $\|(\bar{u}(x) - u(x))\|$			
		Collocation Method	Subdomain Method	Least-Squares Method	Galerkin Method
0.00	0	0.005080985	0.000299015	0.001143015	0.005288
0.05	0.000104169	0.005038591	0.000333564	0.001171455	0.005242
0.10	0.000833417	0.004921868	0.000425572	0.001245976	0.005113
0.15	0.002813132	0.004746227	0.000557858	0.001350661	0.004918
0.20	0.006669331	0.004526454	0.000713866	0.001470218	0.004674
0.25	0.013028959	0.004276401	0.000877974	0.001590287	0.004396
0.30	0.022520207	0.004008678	0.001035802	0.00169775	0.004097
0.35	0.035772807	0.003734352	0.001174513	0.001781034	0.003789
0.40	0.053418342	0.003462642	0.001283118	0.001830414	0.003483
0.45	0.076090534	0.003200626	0.001352769	0.001838304	0.003188
0.50	0.104425539	0.002952946	0.001377054	0.001799554	0.002909
0.55	0.139062229	0.002721531	0.001352274	0.00171173	0.002649
0.60	0.180642473	0.002505311	0.001277729	0.001575393	0.002408
0.65	0.229811406	0.002299954	0.001155981	0.001394369	0.002183
0.70	0.287217687	0.002097598	0.000993122	0.001176014	0.001968
0.75	0.35351376	0.0018866	0.000799025	0.000931463	0.001753
0.08	0.429356091	0.001651294	0.000587586	0.000675874	0.001521
0.85	0.515405405	0.001371755	0.00037696	0.000428667	0.001256
0.90	0.61232691	0.001023575	0.000189785	0.000213741	0.000934
0.95	0.720790505	0.000577655	5.339E-05	5.96895E-05	0.000526
1.00	0.841470985	0	0	0	0
Maximized errors over the domain		**0.005080985**	**0.001377054**	**0.001838304**	**0.005288**

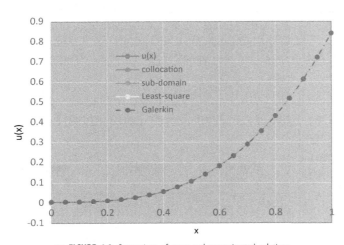

■ **FIGURE 4.8** Comparison of exact and approximated solutions.

4.6 TYPES OF DIFFERENTIAL EQUATIONS

Tables 4.1 and 4.2 have shown that PDEs are introduced as governing equations for different engineering problems. PDEs are about the constraints of state variables over a continuous domain. For a given problem, state variables are changing quantities over space (x, y, z) or both space and time domain (x, y, z, t). The highest derivative of state variables in a PDE is called the order of PDE. All of engineering problems in Tables 4.1 and 4.2 are governing by the first-order or the second-order PDEs. In this section, PDEs in a 2D domain are used to classify the types of PDEs. A generic form of a PDE in a 2D space is given as,

$$A\frac{\partial^2 \phi}{\partial^2 x} + B\frac{\partial \phi}{\partial x}\frac{\partial \phi}{\partial y} + C\frac{\partial^2 \phi}{\partial^2 y} + D\frac{\partial \phi}{\partial x} + E\frac{\partial \phi}{\partial y} + F\phi + G = 0 \qquad (4.40)$$

where

$A, B, \dots$, and G are the constants in a PDE,
ϕ is the state variable of interest, and
(x, y) is an arbitrary position in the two-dimensional (2D) continuous domain.

As shown in Table 4.4, PDEs can be classified based on the discriminant $(B^2 - AC)$ of Eq. (4.40) (Hoffman and Frankel, 2001). Based on the discriminant, PDEs are classified into *elliptic*, *parabolic*, and *hyperbolic* PDEs.

The terminologies of PDE types were inspired by similar concepts for the classification of second-order algebraic equations in the form of

$$Ax^2 + Bxy + Cy^2 + Dx + Ey + F = 0 \qquad (4.41)$$

The type of a PDE determines its *characteristics*. The characteristics of a PDE are $(n - 1)$-dimensional hypersurfaces in an n-dimensional hyperspace that possess some special features. Generally, the prefix *hyper* is to describe a space with more than three dimensions, i.e., a combined domain (x, y, z, t) with *space* and *time* variables and curves and surfaces in such a domain.

Table 4.4 Types of Partial Differential Equations (PDEs) Based on Discriminant

Discriminant $(B^2 - AC)$	PDE Types		
	Elliptic	**Parabolic**	**Hyperbolic**
	$B^2 - AC < 0$	$B^2 - AC = 0$	$B^2 - AC > 0$

In a 2D domain, the characteristics of a PDE are paths in a solution domain along which information propagates. In other words, information propagates throughout the solution domain along the characteristics paths. If there is a dependent variable, the discontinuities of its derivative also propagate along the characteristics paths. If a PDE possesses real characteristics, then information propagates along these characteristics. If no real characteristics exist, then there are no preferred paths of information propagation. Consequently, the presence or absence of characteristics has a significant impact on the solution of a PDE (by both analytical and numerical methods). The classification of a PDE relates closely to the numerical approximation of the PDE.

4.6.1 PDEs for equilibrium problems

For an elliptic PDE that contains only second order of spatial derivatives, there are no preferred physical information propagation paths. Consequently, points are dependent on all other points and all points influence other points. This physical behavior should be accounted for in the numerical approximation of the PDE. An *equilibrium problem* is *a steady-state problem* in a *closed domain* $D(x, y)$. The solution $\phi(x, y)$ to such a problem is governed by an elliptic PDE subjected to BCs. BCs are specified on the boundary B of the domain D (Hoffman and Frankel, 2001).

An elliptic PDE has no real characteristics. Fig. 4.9 illustrates the closed solution domain $D(x, y)$ and its boundary B. Equilibrium problems exist in all fields of engineering and science. PDEs for equilibrium problems are analogous to boundary value problems in ordinary differential equations. A typical example of elliptic PDEs is the Laplace equation:

$$c_x \frac{\partial^2 \phi}{\partial^2 x} + c_y \frac{\partial^2 \phi}{\partial^2 x} + c_\phi = 0 \tag{4.42}$$

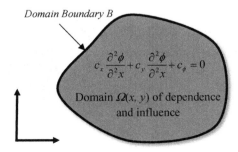

■ **FIGURE 4.9** Domain of dependence and influence of an elliptic partial differential equations.

where $\phi(x, y)$ is a state variable such as temperature in a heat transfer problem.

It is subjected to the BCs of,

$$\left.\begin{array}{c}\Gamma_1: \phi = \phi_0 \\[6pt] \Gamma_2: a\phi + b\dfrac{\partial \phi}{\partial n} = 0\end{array}\right\} \tag{4.43}$$

where Γ_1 and Γ_2 correspond to essential and natural boundaries and $\dfrac{\partial \phi}{\partial n}$ denotes the derivative normal to the boundary.

4.6.2 **PDEs for propagation problems**

For a parabolic PDE that contains only second-order spatial derivatives, the preferred physical information propagation paths are lines (or surfaces) of constant time. At each time step, the points are dependent on all of the other points and all points influence other points. A *propagation problem is* an initial-value problem in an open domain; one of independent variables, for example of time, does not have a closed range. The solution $f(x, t)$ in the domain of interest $D(x, t)$ is marched forward from the initial state, guided and modified by the given BCs. A propagation problem is governed by a parabolic or hyperbolic PDE. The majority of the propagation problems are unsteady problems. Taking an example, the following equation stands for an unsteady propagation problem,

$$\frac{\partial \phi}{\partial t} = \alpha \left(\frac{\partial^2 \phi}{\partial^2 x} + \frac{\partial^2 \phi}{\partial^2 y} \right) \tag{4.44}$$

where $\phi(x, y)$ is a state variable and its initial condition is given as,

$$\phi(x, y, t_0) = \phi_0(x, y) \text{ at } t = t_0$$

The solution to a parabolic or hyperbolic PDE is propagating from the initial property distribution at time t_0 and the solution is marched forward in the given time step.

4.6.3 **PDEs for eigenvalue problems**

An eigenvalue problem is a special type of problem where the solution exists only for special values (i.e., *eigenvalues*) of one system parameter. In solving an eigenvalue problem, the eigenvalues will be determined as well as the corresponding configurations of system. A typical PDE for an eigenvalue problem can be,

$$\frac{\partial^2 \phi}{\partial^2 x} + \frac{\partial^2 \phi}{\partial^2 y} + \lambda \phi = 0 \tag{4.45}$$

where $\phi(x, y)$ is a state variable and the BCs are given as,

$$\phi = \phi_0 \text{ on } \Gamma_1$$

$$\frac{\partial \phi}{\partial n} + \alpha\phi = 0 \text{ on } \Gamma_2$$

The numerical solution treats a continuous domain as a set of discrete nodes and elements. It calculates the values of the state variables on these discrete nodes. The calculation is repeated every time if any of the system parameters or BCs in the model are changed. In the rest of the chapter, a few examples of numerical simulation are introduced to solve PDEs for element modeling.

4.7 ELEMENT MODELING OF 2D EQUILIBRIUM PROBLEMS

An equilibrium problem is a steady problem. It is usually governed by an elliptic PDE. The solutions for state variables at any position are influenced with each other and are completely determined by the BCs. The Galerkin method is adopted to model element behaviors for a 2D equilibrium problem in Eq. (4.42) subjected to the BCs of Eq. (4.43).

To approximate the solution of PDE, the weak-form condition of the approximation is given as,

$$\overline{R} = \int w(x,y)\left(c_x\frac{\partial^2 \phi}{\partial^2 x} + c_y\frac{\partial^2 \phi}{\partial^2 y} + c_\phi - 0\right)dxdy = 0 \qquad (4.46)$$

where $w(x, y)$ is the selected test function and c_x, c_y, and c_ϕ are constants.

Eq. (4.43) can be converted into the following equation with the reduced derivatives as,

$$\overline{R} = \int \left\{ c_x\left\{\frac{\partial\left[w(x,y)\frac{\partial \phi}{\partial x}\right]}{\partial x} - \frac{\partial w(x,y)}{\partial x}\frac{\partial \phi}{\partial x}\right\} + c_y\left\{\frac{\partial\left[w(x,y)\frac{\partial \phi}{\partial y}\right]}{\partial y} - \frac{\partial w(x,y)}{\partial y}\frac{\partial \phi}{\partial y}\right\} + c_\phi \cdot w(x,y)\right\}dxdy$$

$$= c_x\oint \partial\left[w(x,y)\frac{\partial \phi}{\partial x}\right]dy - c_x\int\frac{\partial w(x,y)}{\partial x}\frac{\partial \phi}{\partial x}dxdy + c_y\oint \partial\left[w(x,y)\frac{\partial \phi}{\partial y}\right]dx - c_y\int\frac{\partial w(x,y)}{\partial y}\frac{\partial \phi}{\partial y}dxdy + \int w(x,y)\cdot c_\phi dxdy$$

$$\underbrace{\qquad}_{I} \quad \underbrace{\qquad}_{II} \quad \underbrace{\qquad}_{III} \quad \underbrace{\qquad}_{IV} \quad \underbrace{\qquad}_{V}$$

$$(4.47)$$

where portions I and III are the integral along the boundary of the 2D domain; portions II, IV, and V are the integrals in the continuous domain.

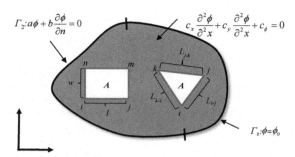

■ **FIGURE 4.10** Heat conduction as an example of elliptic partial differential equation.

In Fig. 4.10, a heat transfer problem is shown as an example for 2D equilibrium problems. For the given continuous domain, its boundaries include essential boundary Γ_1 and natural boundary Γ_2. Rectangle and triangular elements are shown as the examples for discretized elements and nodes in the domain.

To convert Eq. (4.47) into an element model, we concentrate on the portions II, IV, and V for the integrals in the continuous domain. Note that the portions of I and III are the integrals along the boundaries, which will be discussed in Chapter 5.

4.7.1 **Rectangle element**

For a rectangle element in Fig. 4.10, the interpolation of a state variable is given as Eq. (3.53) as,

$$\phi = [S]\{\boldsymbol{\varphi}\} = S_i(x,y)\cdot\phi_i + S_j(x,y)\cdot\phi_j + S_m(x,y)\cdot\phi_m + S_n(x,y)\cdot\phi_n \quad (4.48)$$

where

$\phi(x,y)$ is the state variable at an arbitrary position (x,y) in the element,
the shape functions for a rectangle element are defined in Eq. (3.53), and
ϕ_i, ϕ_j, ϕ_m, and ϕ_n are the values of the state variable at nodes.

Because the Galerkin method is used, the test functions are the shape functions in Eq. (4.48), i.e., $S_i(x,y)$, $S_j(x,y)$, $S_m(x,y)$, and $S_n(x,y)$. Substituting these shape functions into Eq. (4.47) can find the expressions of portions of II, IV, and V, respectively.

For portion II,

$$c_x\int\frac{\partial[S]^T}{\partial x}\frac{\partial\phi}{\partial x}dxdy = c_x\int\frac{\partial[S]^T}{\partial x}\frac{\partial[[S]\cdot[\boldsymbol{\varphi}]]}{\partial x}dxdy = [K_x]\cdot[\boldsymbol{\varphi}] \quad (4.49)$$

where

$$[K_x] = c_x \begin{bmatrix} \int \dfrac{\partial S_i}{\partial x} \cdot \dfrac{\partial S_i}{\partial x} dxdy & \int \dfrac{\partial S_i}{\partial x} \cdot \dfrac{\partial S_j}{\partial x} dxdy & \int \dfrac{\partial S_i}{\partial x} \cdot \dfrac{\partial S_m}{\partial x} dxdy & \int \dfrac{\partial S_i}{\partial x} \cdot \dfrac{\partial S_n}{\partial x} dxdy \\ \int \dfrac{\partial S_i}{\partial x} \cdot \dfrac{\partial S_j}{\partial x} dxdy & \int \dfrac{\partial S_j}{\partial x} \cdot \dfrac{\partial S_j}{\partial x} dxdy & \int \dfrac{\partial S_j}{\partial x} \cdot \dfrac{\partial S_m}{\partial x} dxdy & \int \dfrac{\partial S_j}{\partial x} \cdot \dfrac{\partial S_n}{\partial x} dxdy \\ \int \dfrac{\partial S_m}{\partial x} \cdot \dfrac{\partial S_i}{\partial x} dxdy & \int \dfrac{\partial S_m}{\partial x} \cdot \dfrac{\partial S_j}{\partial x} dxdy & \int \dfrac{\partial S_m}{\partial x} \cdot \dfrac{\partial S_m}{\partial x} dxdy & \int \dfrac{\partial S_m}{\partial x} \cdot \dfrac{\partial S_n}{\partial x} dxdy \\ \int \dfrac{\partial S_n}{\partial x} \cdot \dfrac{\partial S_i}{\partial x} dxdy & \int \dfrac{\partial S_n}{\partial x} \cdot \dfrac{\partial S_j}{\partial x} dxdy & \int \dfrac{\partial S_n}{\partial x} \cdot \dfrac{\partial S_m}{\partial x} dxdy & \int \dfrac{\partial S_n}{\partial x} \cdot \dfrac{\partial S_n}{\partial x} dxdy \end{bmatrix}$$

$$= \frac{c_x w}{6l} \begin{bmatrix} 2 & -2 & -1 & 1 \\ -2 & 2 & 1 & -1 \\ -1 & 1 & 2 & -2 \\ 1 & -1 & -2 & 2 \end{bmatrix}$$

and $[\boldsymbol{\varphi}]^T = \begin{bmatrix} \phi_i & \phi_j & \phi_m & \phi_n \end{bmatrix}$

Similarly, for portion IV,

$$c_y \int \frac{\partial [S]^T}{\partial y} \frac{\partial \phi}{\partial y} dxdy = c_y \int \frac{\partial [S]^T}{\partial y} \frac{\partial [[S] \cdot [\boldsymbol{\varphi}]]}{\partial y} dxdy = [K_y] \cdot [\boldsymbol{\varphi}] \qquad (4.50)$$

where

$$[K_y] = c_y \begin{bmatrix} \int \dfrac{\partial S_i}{\partial y} \cdot \dfrac{\partial S_i}{\partial y} dxdy & \int \dfrac{\partial S_i}{\partial y} \cdot \dfrac{\partial S_j}{\partial y} dxdy & \int \dfrac{\partial S_i}{\partial y} \cdot \dfrac{\partial S_m}{\partial y} dxdy & \int \dfrac{\partial S_i}{\partial y} \cdot \dfrac{\partial S_n}{\partial y} dxdy \\ \int \dfrac{\partial S_i}{\partial y} \cdot \dfrac{\partial S_j}{\partial y} dxdy & \int \dfrac{\partial S_j}{\partial y} \cdot \dfrac{\partial S_j}{\partial y} dxdy & \int \dfrac{\partial S_j}{\partial y} \cdot \dfrac{\partial S_m}{\partial y} dxdy & \int \dfrac{\partial S_j}{\partial y} \cdot \dfrac{\partial S_n}{\partial y} dxdy \\ \int \dfrac{\partial S_m}{\partial y} \cdot \dfrac{\partial S_i}{\partial y} dxdy & \int \dfrac{\partial S_m}{\partial y} \cdot \dfrac{\partial S_j}{\partial y} dxdy & \int \dfrac{\partial S_m}{\partial y} \cdot \dfrac{\partial S_m}{\partial y} dxdy & \int \dfrac{\partial S_m}{\partial y} \cdot \dfrac{\partial S_n}{\partial y} dxdy \\ \int \dfrac{\partial S_n}{\partial y} \cdot \dfrac{\partial S_i}{\partial y} dxdy & \int \dfrac{\partial S_n}{\partial y} \cdot \dfrac{\partial S_j}{\partial y} dxdy & \int \dfrac{\partial S_n}{\partial y} \cdot \dfrac{\partial S_m}{\partial y} dxdy & \int \dfrac{\partial S_n}{\partial y} \cdot \dfrac{\partial S_n}{\partial y} dxdy \end{bmatrix}$$

$$= \frac{c_y l}{6w} \begin{bmatrix} 2 & 1 & -1 & -2 \\ 1 & 2 & -2 & -1 \\ -1 & -2 & 2 & 1 \\ -2 & -1 & 1 & 2 \end{bmatrix}$$

For portion V,

$$[Q] = \int [S]^T c_\phi dxdy = c_\phi \begin{bmatrix} \int S_i dxdy \\ \int S_j dxdy \\ \int S_m dxdy \\ \int S_n dxdy \end{bmatrix} = \frac{c_\phi (lw)}{4} \begin{bmatrix} 1 \\ 1 \\ 1 \\ 1 \end{bmatrix} \qquad (4.51)$$

Substituting Eqs. (4.49)−(4.51) into Eq. (4.47) yields the model of a 2D rectangle element with no boundary edge (no portion I and III) as,

$$([K_x] + [K_y]) \cdot \{\varphi\} = [Q] \tag{4.52}$$

where $[K_x]$ and $[K_y]$ are determined in Eqs. (4.49) and (4.50), respectively, and the load vector is defined in Eq. (4.51).

4.7.2 Triangle element

For a triangle element in Fig. 4.10, the interpolation of state variable is given as Eq. (3.72) as,

$$\phi = [S]\{\varphi\} = S_i(x, y) \cdot \phi_i + S_j(x, y) \cdot \phi_j + S_k(x, y) \cdot \phi_k \tag{4.53}$$

where

$\phi(x, y)$ is the state variable at an arbitrary position (x, y) in the element,

the shape functions for a triangular element are defined in Eq. (3.72), and

ϕ_i, ϕ_j, and ϕ_k are the values of state variables at nodes.

Because the Galerkin method is used, the test functions are the constitutional functions in Eq. (4.53). There are three functions including, i.e., $S_i(x, y)$, $S_j(x, y)$, and $S_k(x, y)$. Substituting shape functions into Eq. (4.47), one can find the expressions of portions of II, IV, and V, respectively.

For portion II,

$$c_x \int \frac{\partial [S]^T}{\partial x} \frac{\partial \phi}{\partial x} dx dy = c_x \int \frac{\partial [S]^T}{\partial x} \frac{\partial [[S] \cdot [\phi]]}{\partial x} dx dy = [K_x] \cdot [\varphi] \tag{4.54}$$

where

$$[K_x] = c_x \begin{bmatrix} \int \frac{\partial S_i}{\partial x} \cdot \frac{\partial S_i}{\partial x} dx dy & \int \frac{\partial S_i}{\partial x} \cdot \frac{\partial S_j}{\partial x} dx dy & \int \frac{\partial S_i}{\partial x} \cdot \frac{\partial S_k}{\partial x} dx dy \\ \int \frac{\partial S_i}{\partial x} \cdot \frac{\partial S_j}{\partial x} dx dy & \int \frac{\partial S_j}{\partial x} \cdot \frac{\partial S_j}{\partial x} dx dy & \int \frac{\partial S_j}{\partial x} \cdot \frac{\partial S_k}{\partial x} dx dy \\ \int \frac{\partial S_k}{\partial x} \cdot \frac{\partial S_i}{\partial x} dx dy & \int \frac{\partial S_k}{\partial x} \cdot \frac{\partial S_j}{\partial x} dx dy & \int \frac{\partial S_k}{\partial x} \cdot \frac{\partial S_k}{\partial x} dx dy \end{bmatrix}$$

$$= \frac{c_x}{4A} \begin{bmatrix} \beta_i^2 & \beta_i \beta_j & \beta_j \beta_k \\ \beta_j \beta_i & \beta_j^2 & \beta_j \beta_k \\ \beta_k \beta_i & \beta_k \beta_j & \beta_k^2 \end{bmatrix}$$

and $[\varphi]^T = \begin{bmatrix} \phi_i & \phi_j & \phi_k \end{bmatrix}$

Similarly, for portion IV,

$$c_y \int \frac{\partial [S]^T}{\partial y} \frac{\partial \phi}{\partial y} dxdy = c_y \int \frac{\partial [S]^T}{\partial y} \frac{\partial [[S] \cdot [\varphi]]}{\partial y} dxdy = [K_y] \cdot [\varphi] \qquad (4.55)$$

where

$$[K_y] = c_y \begin{bmatrix} \int \frac{\partial S_i}{\partial y} \cdot \frac{\partial S_i}{\partial y} dxdy & \int \frac{\partial S_i}{\partial y} \cdot \frac{\partial S_j}{\partial y} dxdy & \int \frac{\partial S_i}{\partial y} \cdot \frac{\partial S_k}{\partial y} dxdy \\ \int \frac{\partial S_i}{\partial y} \cdot \frac{\partial S_j}{\partial y} dxdy & \int \frac{\partial S_j}{\partial y} \cdot \frac{\partial S_j}{\partial y} dxdy & \int \frac{\partial S_j}{\partial y} \cdot \frac{\partial S_k}{\partial y} dxdy \\ \int \frac{\partial S_k}{\partial y} \cdot \frac{\partial S_i}{\partial y} dxdy & \int \frac{\partial S_k}{\partial y} \cdot \frac{\partial S_j}{\partial y} dxdy & \int \frac{\partial S_k}{\partial x} \cdot \frac{\partial S_k}{\partial y} dxdy \end{bmatrix}$$

$$= \frac{c_y}{4A} \begin{bmatrix} \delta_i^2 & \delta_i \delta_j & \delta_i \delta_k \\ \delta_j \delta_i & \delta_j^2 & \delta_j \delta_k \\ \delta_k \delta_i & \delta_k \delta_j & \delta_k^2 \end{bmatrix}$$

For portion V,

$$[Q] = \int [S]^T c_\phi dxdy = c_\phi \begin{bmatrix} \int S_i dxdy \\ \int S_j dxdy \\ \int S_k dxdy \end{bmatrix} = \frac{c_\phi A}{3} \begin{bmatrix} 1 \\ 1 \\ 1 \end{bmatrix} \qquad (4.56)$$

Substituting Eqs. (4.54)−(4.56) into Eq. (4.47) yields the model of a 2D triangular element with no boundary edge (no portion I and III),

$$([K_x] + [K_y]) \cdot \{\varphi\} = [Q] \qquad (4.57)$$

where $[K_x]$ and $[K_y]$ are determined in Eqs. (4.51) and (4.55), respectively, and the load vector is defined in Eq. (4.56).

4.8 ELEMENT MODELING OF 2D PROPAGATE PROBLEMS

A propagate problem is an unsteady problem; it is governed by a parabolic or hyperbolic PDE. The solution in a continuous domain varies with respect to time and the solution also depends on initial conditions. The following propagate problem is discussed here to illustrate the procedure of modeling the element level (Thompson, 2005),

$$c_x \frac{\partial^2 \phi}{\partial^2 x} + c_y \frac{\partial^2 \phi}{\partial^2 x} + c_\phi = \alpha \frac{\partial \phi}{\partial t} \qquad (4.58)$$

where

$\phi(x, y)$ is the state variable at an arbitrary position (x, y) in element,

c_x, c_y, c_ϕ, α are the constants related to the propagation, and the solution is subjected to the following initial condition.

$$\phi(x, y, t_0) = \phi_0(x, y) \text{ at } t = t_0$$

The weak form solution to Eq. (4.58) can be expressed by,

$$\overline{R} = \int \left\{ c_x \left\{ \frac{\partial \left[w(x,y) \frac{\partial \phi}{\partial x} \right]}{\partial x} - \frac{\partial w(x,y)}{\partial x} \frac{\partial \phi}{\partial x} \right\} + c_y \left\{ \frac{\partial \left[w(x,y) \frac{\partial \phi}{\partial y} \right]}{\partial y} - \frac{\partial w(x,y)}{\partial y} \frac{\partial \phi}{\partial y} \right\} + c_\phi \cdot w(x,y) - \alpha \cdot w(x,y) \frac{\partial \phi}{\partial t} \right\} dxdy$$

$$= c_x \oint \partial \left[w(x,y) \frac{\partial \phi}{\partial x} \right] dy - c_x \int \frac{\partial w(x,y)}{\partial x} \frac{\partial \phi}{\partial x} dxdy + c_y \oint \partial \left[w(x,y) \frac{\partial \phi}{\partial y} \right] dx - c_y \int \frac{\partial w(x,y)}{\partial y} \frac{\partial \phi}{\partial y} dxdy + \int w(x,y) \cdot c_\phi dxdy$$

$$\underbrace{\hphantom{XXXX}}_{I} \quad \underbrace{\hphantom{XXXX}}_{II} \quad \underbrace{\hphantom{XXXX}}_{III} \quad \underbrace{\hphantom{XXXX}}_{IV} \quad \underbrace{\hphantom{XXXX}}_{V}$$

$$- \int \alpha w(x,y) \cdot \frac{\partial \phi}{\partial t} dxdy$$

$$\underbrace{\hphantom{XXXXXX}}_{VI}$$

$$(4.59)$$

We have discussed portions I−V in Section 4.6 and only portion VI is the new one. To evaluate the portion of integral VI, let us treat the derivative of $\phi(x, y)$ with respect to time as a dependent state variable. It can be found by the interpolation using the shape functions based on the nodal values of the same derivative,

$$\dot{\phi}(x,y) = \frac{\partial \dot{\phi}(x,y)}{\partial t} = [S]\{\dot{\varphi}\} \tag{4.60}$$

For a rectangle element, portion VI in Eq. (4.59) is evaluated as,

$$\alpha \int [S]^T \cdot [S][\dot{\varphi}]dxdy = [C][\dot{\varphi}] \tag{4.61}$$

where

$$[C] = \alpha \begin{bmatrix} \int S_i^2 dxdy & \int S_i S_j dxdy & \int S_i S_m dxdy & \int S_i S_n dxdy \\ \int S_j S_i dxdy & \int S_j^2 dxdy & \int S_i S_m dxdy & \int S_j S_n dxdy \\ \int S_m S_i dxdy & \int S_m S_j dxdy & \int S_m^2 dxdy & \int S_m S_n dxdy \\ \int S_n S_i dxdy & \int S_n S_j dxdy & \int S_n S_m dxdy & \int S_n^2 dxdy \end{bmatrix}$$

$$= \frac{(lw)\alpha}{18} \begin{bmatrix} 2 & 1 & 1 & 2 \\ 1 & 2 & 2 & 1 \\ 1 & 2 & 2 & 1 \\ 2 & 1 & 1 & 2 \end{bmatrix}$$

$[\dot{\varphi}]$ is the vector for the derivative of state variable with respect to time.

Combining Eq. (4.52) and Eq. (4.61), one gets the element model of Eq. (4.58) as,

$$([K_x] + [K_y]) \cdot \{\varphi\} + [C]\{\dot{\varphi}\} = [Q] \tag{4.62}$$

where $[K_x]$, $[K_y]$, and $[Q]$ are given in Eq. (4.52) and $[C]$ is given in Eq. (4.61).

If a linear triangle element is concerned, $[K_x]$, $[K_y]$, and $[Q]$ are given in Eq. (4.57) and $[C]$ can be defined by a similar procedure as,

$$[C] = \alpha \begin{bmatrix} \int S_i^2 dxdy & \int S_i S_j dxdy & \int S_i S_k dxdy \\ \int S_j S_i dxdy & \int S_j^2 dxdy & \int S_k S_j dxdy \\ \int S_k S_i dxdy & \int S_k S_j dxdy & \int S_k^2 dxdy \end{bmatrix} = \frac{A\alpha}{12} \begin{bmatrix} 2 & 1 & 1 \\ 1 & 2 & 1 \\ 1 & 1 & 2 \end{bmatrix} \tag{4.63}$$

Note that $\{\varphi\}$ and $\{\dot{\varphi}\}$ are dependent with each other. One has to find the complete solution using an iterative procedure from the given initial conditions.

For example, if the forward difference method is used in the iteration, Eq. (4.62) should be written as,

$$[C]\{\dot{\varphi}\} = [Q] - ([K_x] + [K_y]) \cdot \{\varphi\} \tag{4.64}$$

Eq. (4.64) shows that $\{\dot{\varphi}\}$ at time step t can be evaluated based on $\{\varphi\}$. Once $\{\dot{\varphi}\}$ at time t is obtained, $\{\varphi\}$ at the next time step $t + \Delta t$ can be found as,

$$\{\varphi\}_{t+\Delta t} = \{\varphi\}_t + \{\dot{\varphi}\}_t \cdot \Delta t \tag{4.65}$$

Once $\{\varphi\}_{t+\Delta t}$ is known, $\{\dot{\varphi}\}$ can be obtained from two time steps as,

$$\{\dot{\varphi}\}_t = \frac{1}{\Delta t}(\{\varphi\}_{t+\Delta t} - \{\varphi\}_t) \tag{4.66}$$

Substituting Eq. (4.66) back to Eq. (4.64) gets,

$$\frac{[C]}{\Delta t}(\{\varphi\}_{t+\Delta t} - \{\varphi\}_t) = [Q] - ([K_x] + [K_y]) \cdot \{\varphi\} \tag{4.67}$$

Rearranging Eq. (4.67) yields an explicit representation of the element model in finite difference approximation without a derivative term $\{\dot{\varphi}\}$,

$$\frac{[C]}{\Delta t}\{\varphi\}_{t+\Delta t} = (\{\varphi\}_{t+\Delta t} - \{\varphi\}_t) = [Q]_t + \left(\frac{[C]}{\Delta t} - ([K_x] + [K_y])\right) \cdot \{\varphi\}_t \tag{4.68}$$

In Eq. (4.68), the time step Δt must be set appropriately to the convergence and stability. We will discuss this issue when the transient heat transfer problem is concerned in Chapter 9.

4.9 ELEMENT MODELING OF EIGENVALUE PROBLEMS

In an eigenvalue problem, the solution exists only for special values (i.e., *eigenvalues*) of a system parameter. The Helmholtz equation is disused here as an example of 2D eigenvalue problems. Fig. 4.11 describes the eigenvalue problem governed by the Helmholtz equation (Bickford, 1990). The governing equation is represented by,

$$\frac{\partial^2 \phi}{\partial^2 x} + \frac{\partial^2 \phi}{\partial^2 y} + \lambda \phi = 0 \tag{4.69}$$

where $\phi(x, y)$ is the state variable at an arbitrary position (x, y) in the element and it is subjected to the BCs of

$$\left.\begin{array}{l} \Gamma_1 : \phi = \phi_0 \\[2mm] \Gamma_2 : a\phi + b\dfrac{\partial \phi}{\partial n} = 0 \end{array}\right\} \tag{4.70}$$

The weak-form residual of Eq. (4.69) is derived as

$$\bar{R} = \int \left\{ \left\{ \frac{\partial \left[w(x,y)\frac{\partial \phi}{\partial x} \right]}{\partial x} - \frac{\partial w(x,y)}{\partial x}\frac{\partial \phi}{\partial x} \right\} + \left\{ \frac{\partial \left[w(x,y)\frac{\partial \phi}{\partial y} \right]}{\partial y} - \frac{\partial w(x,y)}{\partial y}\frac{\partial \phi}{\partial y} \right\} + w(x,y)\lambda\phi \right\} dxdy$$

$$= \underbrace{\oint \partial \left[w(x,y)\frac{\partial \phi}{\partial x} \right] dy}_{I} - \underbrace{\int \frac{\partial w(x,y)}{\partial x}\frac{\partial \phi}{\partial x}dxdy}_{II} + \underbrace{\oint \partial \left[w(x,y)\frac{\partial \phi}{\partial y} \right] dx}_{III} - \underbrace{\int \frac{\partial w(x,y)}{\partial y}\frac{\partial \phi}{\partial y}dxdy}_{IV} + \underbrace{\int w(x,y) \cdot \lambda\phi dxdy}_{V}$$

$$\tag{4.71}$$

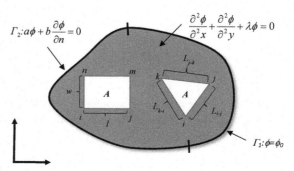

FIGURE 4.11 Eigenvalue problem example.

where $w(x, y)$ is a selected test function; portions I and III are the integral along the boundary of 2D domain, portions II, IV, and V are three integrals over the domain. The integrals along the boundary (portions I and III) will be addressed in Chapter 5 when the BCs of an FEA are discussed. In this section, the integrals over the domain are discussed as a part of an element model.

4.9.1 **Rectangle element**

Following the same procedure in Sections 4.6 and 4.7, the shape functions are used for the interpolation of state variable $\phi = [S]\{\varphi\}$. Shape functions are also used as the test functions in determining the expression of residuals caused by the approximation. The integrals in portions II, IV, and V are obtained as below, respectively.

For portion II,

$$\int \frac{\partial[S]^T}{\partial x} \frac{\partial \phi}{\partial x} dxdy = \int \frac{\partial[S]^T}{\partial x} \frac{\partial[[S]\cdot[\varphi]]}{\partial x} dxdy = [K_x]\cdot[\varphi] \qquad (4.72)$$

where

$$[K_x] = \begin{bmatrix} \int \frac{\partial S_i}{\partial x}\cdot\frac{\partial S_i}{\partial x} dxdy & \int \frac{\partial S_i}{\partial x}\cdot\frac{\partial S_j}{\partial x} dxdy & \int \frac{\partial S_i}{\partial x}\cdot\frac{\partial S_m}{\partial x} dxdy & \int \frac{\partial S_i}{\partial x}\cdot\frac{\partial S_n}{\partial x} dxdy \\ \int \frac{\partial S_i}{\partial x}\cdot\frac{\partial S_j}{\partial x} dxdy & \int \frac{\partial S_j}{\partial x}\cdot\frac{\partial S_j}{\partial x} dxdy & \int \frac{\partial S_j}{\partial x}\cdot\frac{\partial S_m}{\partial x} dxdy & \int \frac{\partial S_j}{\partial x}\cdot\frac{\partial S_n}{\partial x} dxdy \\ \int \frac{\partial S_m}{\partial x}\cdot\frac{\partial S_i}{\partial x} dxdy & \int \frac{\partial S_m}{\partial x}\cdot\frac{\partial S_j}{\partial x} dxdy & \int \frac{\partial S_m}{\partial x}\cdot\frac{\partial S_m}{\partial x} dxdy & \int \frac{\partial S_m}{\partial x}\cdot\frac{\partial S_n}{\partial x} dxdy \\ \int \frac{\partial S_n}{\partial x}\cdot\frac{\partial S_i}{\partial x} dxdy & \int \frac{\partial S_n}{\partial x}\cdot\frac{\partial S_j}{\partial x} dxdy & \int \frac{\partial S_n}{\partial x}\cdot\frac{\partial S_m}{\partial x} dxdy & \int \frac{\partial S_n}{\partial x}\cdot\frac{\partial S_n}{\partial x} dxdy \end{bmatrix}$$

$$= \frac{w}{6l} \begin{bmatrix} 2 & -2 & -1 & 1 \\ -2 & 2 & 1 & -1 \\ -1 & 1 & 2 & -2 \\ 1 & -1 & -2 & 2 \end{bmatrix}$$

and $[\varphi]^T = \begin{bmatrix} \phi_i & \phi_j & \phi_m & \phi_n \end{bmatrix}$

Similarly, for portion IV,

$$\int \frac{\partial[S]^T}{\partial y} \frac{\partial \phi}{\partial y} dxdy = \int \frac{\partial[S]^T}{\partial y} \frac{\partial[[S]\cdot[\varphi]]}{\partial y} dxdy = [K_y]\cdot[\varphi] \qquad (4.73)$$

where

$$
[K_y] = \begin{bmatrix}
\int \dfrac{\partial S_i}{\partial y} \cdot \dfrac{\partial S_i}{\partial y} dxdy & \int \dfrac{\partial S_i}{\partial y} \cdot \dfrac{\partial S_j}{\partial y} dxdy & \int \dfrac{\partial S_i}{\partial y} \cdot \dfrac{\partial S_m}{\partial y} dxdy & \int \dfrac{\partial S_i}{\partial y} \cdot \dfrac{\partial S_n}{\partial y} dxdy \\
\int \dfrac{\partial S_i}{\partial y} \cdot \dfrac{\partial S_j}{\partial y} dxdy & \int \dfrac{\partial S_j}{\partial y} \cdot \dfrac{\partial S_j}{\partial y} dxdy & \int \dfrac{\partial S_j}{\partial y} \cdot \dfrac{\partial S_m}{\partial y} dxdy & \int \dfrac{\partial S_j}{\partial y} \cdot \dfrac{\partial S_n}{\partial y} dxdy \\
\int \dfrac{\partial S_m}{\partial y} \cdot \dfrac{\partial S_i}{\partial y} dxdy & \int \dfrac{\partial S_m}{\partial y} \cdot \dfrac{\partial S_j}{\partial y} dxdy & \int \dfrac{\partial S_m}{\partial y} \cdot \dfrac{\partial S_m}{\partial y} dxdy & \int \dfrac{\partial S_m}{\partial y} \cdot \dfrac{\partial S_n}{\partial y} dxdy \\
\int \dfrac{\partial S_n}{\partial y} \cdot \dfrac{\partial S_i}{\partial y} dxdy & \int \dfrac{\partial S_n}{\partial y} \cdot \dfrac{\partial S_j}{\partial y} dxdy & \int \dfrac{\partial S_n}{\partial y} \cdot \dfrac{\partial S_m}{\partial y} dxdy & \int \dfrac{\partial S_n}{\partial y} \cdot \dfrac{\partial S_n}{\partial y} dxdy
\end{bmatrix}
$$

$$
= \frac{l}{6w} \begin{bmatrix}
2 & 1 & -1 & -2 \\
1 & 2 & -2 & -1 \\
-1 & -2 & 2 & 1 \\
-2 & -1 & 1 & 2
\end{bmatrix}
$$

For portion V,

$$
\int [S]^T \lambda [S][\varphi] dxdy = \lambda [M][\varphi] \tag{4.74}
$$

where

$$
[M] = \begin{bmatrix}
\int S_i^2 dxdy & \int S_i S_j dxdy & \int S_i S_m dxdy & \int S_i S_n dxdy \\
\int S_j S_i dxdy & \int S_j^2 dxdy & \int S_i S_m dxdy & \int S_j S_n dxdy \\
\int S_m S_i dxdy & \int S_m S_j dxdy & \int S_m^2 dxdy & \int S_m S_n dxdy \\
\int S_n S_i dxdy & \int S_n S_j dxdy & \int S_n S_m dxdy & \int S_n^2 dxdy
\end{bmatrix}
$$

$$
= \frac{(lw)}{18} \begin{bmatrix}
2 & 1 & 1 & 2 \\
1 & 2 & 2 & 1 \\
1 & 2 & 2 & 1 \\
2 & 1 & 1 & 2
\end{bmatrix}
$$

Substituting Eqs. (4.72)–(4.74) into Eq. (4.71) yields the element model as,

$$
\{([K_x] + [K_y]) - \lambda[M]\} \cdot \{\varphi\} = 0 \tag{4.75}
$$

where $[K_x]$, $[K_y]$, and $[M]$ are determined in Eqs. (4.72)–(4.74), respectively.

4.9.2 Triangle element

The similar procedure can be applied to model a 2D linear triangle element in an eigenvalue problem, and the resulted element model is expressed as,

$$
\{([K_x] + [K_y]) - \lambda \cdot [M]\}\{\varphi\} = 0 \tag{4.76}
$$

where

$$[K_x] = \begin{bmatrix} \int \frac{\partial S_i}{\partial x} \cdot \frac{\partial S_i}{\partial x} dxdy & \int \frac{\partial S_i}{\partial x} \cdot \frac{\partial S_j}{\partial x} dxdy & \int \frac{\partial S_i}{\partial x} \cdot \frac{\partial S_k}{\partial x} dxdy \\ \int \frac{\partial S_i}{\partial x} \cdot \frac{\partial S_j}{\partial x} dxdy & \int \frac{\partial S_j}{\partial x} \cdot \frac{\partial S_j}{\partial x} dxdy & \int \frac{\partial S_j}{\partial x} \cdot \frac{\partial S_k}{\partial x} dxdy \\ \int \frac{\partial S_k}{\partial x} \cdot \frac{\partial S_i}{\partial x} dxdy & \int \frac{\partial S_k}{\partial x} \cdot \frac{\partial S_j}{\partial x} dxdy & \int \frac{\partial S_k}{\partial x} \cdot \frac{\partial S_k}{\partial x} dxdy \end{bmatrix}$$

$$= \frac{1}{4A} \begin{bmatrix} \beta_i^2 & \beta_i\beta_j & \beta_i\beta_k \\ \beta_j\beta_i & \beta_j^2 & \beta_j\beta_k \\ \beta_k\beta_i & \beta_k\beta_j & \beta_k^2 \end{bmatrix}$$

$$[K_y] = \begin{bmatrix} \int \frac{\partial S_i}{\partial y} \cdot \frac{\partial S_i}{\partial y} dxdy & \int \frac{\partial S_i}{\partial y} \cdot \frac{\partial S_j}{\partial y} dxdy & \int \frac{\partial S_i}{\partial y} \cdot \frac{\partial S_k}{\partial y} dxdy \\ \int \frac{\partial S_i}{\partial y} \cdot \frac{\partial S_j}{\partial y} dxdy & \int \frac{\partial S_j}{\partial y} \cdot \frac{\partial S_j}{\partial y} dxdy & \int \frac{\partial S_j}{\partial y} \cdot \frac{\partial S_k}{\partial y} dxdy \\ \int \frac{\partial S_k}{\partial y} \cdot \frac{\partial S_i}{\partial y} dxdy & \int \frac{\partial S_k}{\partial y} \cdot \frac{\partial S_j}{\partial y} dxdy & \int \frac{\partial S_k}{\partial x} \cdot \frac{\partial S_k}{\partial y} dxdy \end{bmatrix}$$

$$= \frac{1}{4A} \begin{bmatrix} \delta_i^2 & \delta_i\delta_j & \delta_i\delta_k \\ \delta_j\delta_i & \delta_j^2 & \delta_j\delta_k \\ \delta_k\delta_i & \delta_k\delta_j & \delta_k^2 \end{bmatrix}$$

$$[M] = \begin{bmatrix} \int S_i^2 dxdy & \int S_i S_j dxdy & \int S_i S_k dxdy \\ \int S_j S_i dxdy & \int S_j^2 dxdy & \int S_k S_j dxdy \\ \int S_k S_i dxdy & \int S_k S_j dxdy & \int S_k^2 dxdy \end{bmatrix} = \frac{A}{12} \begin{bmatrix} 2 & 1 & 1 \\ 1 & 2 & 1 \\ 1 & 1 & 2 \end{bmatrix}$$

Note that for a triangle element, one has an integral formula as,

$$\int S_i^a \cdot S_j^b \cdot S_k^c dA = \frac{a!b!c!}{(a+b+c+2)!}(2A)$$

4.10 **SUMMARY**

Element models are the basic building blocks of a system model in FEA. To develop an element model, the governing equation of the element behavior must be known, and a boundary value problem can be defined as the mathematic model of elements. Different techniques have been introduced to solve mathematic models numerically. The direct formulation is applicable to some simple problems where the governing equations can be converted to the relations of state variables and loads easily. The method based on the minimal potential energy is widely applied to model solid mechanics problems. The weighted residual method has a number of variations including collocation method, subdomain method, least-square method, and Galerkin method. Due to the flexibility, the weighted residual method is widely used in computer implementation of FEA, and its applications have been illustrated in developing the models of elements for equilibrium problems, propagate problems, and eigenvalue problems.

REFERENCES

Bickford, W.B., 1990. A First Course in the Finite Element Method. Rickard D. Irwin, Inc., ISBN 0-256-07973-0

Chandrupatla, T.R., Belegundu, A.D., 2012. Introduction to Finite Elements in Engineering, fourth ed. Pearson. ISBN-10: 0-13-216274-1.

Hoffman, J.D., Frankel, S., 2001. Numerical Methods for Engineers and Scientists, second ed. CRC Press, Boca Raton, FL, USA. ISBN-10: 0824704436.

Thompson, E.G., 2005. Introduction to the Finite Element Method — Theory, Programming, and Applications. John Wiley & Sons, Inc., ISBN 0471-45253-X

■ PROBLEMS

4.1. Use the direct method to develop element stiffness matrix $[K]$ for an axial member in Fig. 4.12. It is assumed that the element has the elastic modulus of E, the cross-section of A, and the length of L. The stiffness matrix $[K]$ is to represent the relation of force and displacement in an element as

$$[K]\begin{Bmatrix} U_i \\ U_j \end{Bmatrix} = \begin{Bmatrix} F_i \\ F_j \end{Bmatrix}$$

The governing equation for an axial member under a tensile load can be derived from the constative equation of

$$\varepsilon = \frac{\sigma}{E} = \frac{F}{EA}$$

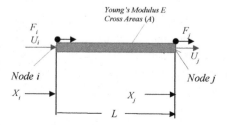

■ **FIGURE 4.12** Description of axial member.

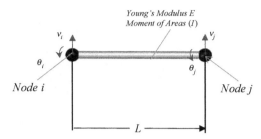

Young's Modulus E
Moment of Areas (I)

Node i   *Node j*

L

■ **FIGURE 4.13** Representation of a beam member.

4.2. Use the direct method to develop element stiffness matrix $[K]$ for a cantilever beam (see Fig. 4.13) with elastic modulus E, cross-section A, moment of area I, and length L. The stiffness matrix describes the relation of applied loads (Q_i, M_i, Q_j, M_j) with the deflections of (v_i, θ_i, v_j, θ_j) as

$$\begin{Bmatrix} Q_i \\ M_i \\ Q_j \\ M_j \end{Bmatrix} = [K] \begin{Bmatrix} v_i \\ \theta_i \\ v_j \\ \theta_j \end{Bmatrix} = \begin{bmatrix} k_{11} & k_{12} & k_{13} & k_{14} \\ k_{21} & k_{22} & k_{23} & k_{24} \\ k_{31} & k_{32} & k_{33} & k_{34} \\ k_{41} & k_{42} & k_{43} & k_{44} \end{bmatrix} \begin{Bmatrix} v_i \\ \theta_i \\ v_j \\ \theta_j \end{Bmatrix}$$

Hint: For an elastic deformation, the total displacement of a structure can be found by adding small displacements from individual loads as shown in Fig. 4.14. Therefore, the vector of displacements (v_i, θ_i, v_j, θ_j) can be treated as assembly of four separated cases as

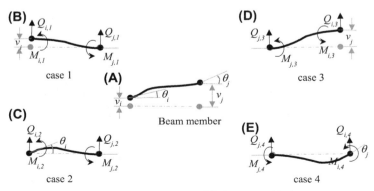

(B) $Q_{i,1}$ $Q_{j,1}$ $M_{i,1}$ $M_{j,1}$
case 1

(A) θ_j θ_i v_i v_j
Beam member

(C) $Q_{i,2}$ θ_i $Q_{j,2}$ $M_{i,2}$ $M_{j,2}$
case 2

(D) $Q_{i,3}$ $Q_{j,3}$ v_j $M_{j,3}$ $M_{i,3}$
case 3

(E) $Q_{j,4}$ $Q_{i,4}$ θ_j $M_{j,4}$ $M_{i,4}$
case 4

■ **FIGURE 4.14** Decomposition of displacements and loads.

4.3. Use two different weighted residual methods to find an approximated solution for the following boundary value problem:

$$\left. \begin{array}{l} \dfrac{d^2y}{dx^2} - y = x^2 \quad x \in [0, 1] \\[2mm] \text{subjected to} : \; y|_{x=0} = 0, \quad y|_{x=1} = 0 \end{array} \right\}$$

Plot the obtained approximated solutions to examine the difference.

4.4. Use the collocation method to develop an approximated solution for the following boundary value problem,

$$\left. \begin{array}{l} \dfrac{d^2\phi}{dx^2} - \phi = x \quad x \in [0, 1] \\[2mm] \text{subjected to} : \; \phi|_{x=0} = 0, \quad \phi|_{x=1} = 0 \end{array} \right\}$$

4.5. Use two different weighted residual methods to find an approximated solution for the following boundary value problem:

$$\left. \begin{array}{l} \dfrac{d^2y}{dx^2} + \dfrac{dy}{dx} - y - 1 = 0 \quad x \in [0, 1] \\[2mm] \text{subjected to} : \; y|_{x=0} = 0, \quad y|_{x=1} = 0 \end{array} \right\}$$

Plot the obtained approximated solutions to examine the difference.

System Analysis and Modeling

5.1 INTRODUCTION

In Chapter 2, different types of complexities in engineering problems have been discussed. To deal with the complexity in the formulation of a finite element analysis (FEA) model, the decomposition is performed at five aspects: *uncertainties*, *geometric shapes*, *physical behaviors*, *time factors*, and *boundary conditions*. The decomposition aims to represent a continuous domain by a set of discretized elements with associated properties, where the models and solutions for these elements are available in an FEA code. However, the couplings of element models and solutions are not tackled in the decomposition. When a system model is developed, the constraints on a continuous domain put aside in the course of the decomposition must be restored, so that element models can be assembled correctly into a system model.

Fig. 5.1 gives an overview on the relation of the decomposition and assembling of a system model. In the *decomposition*, a design problem is converted into a conceptual model with specified objects, physical behaviors, materials properties, load types, geometry, and state variables. To develop an FEA model, users provide all of the required inputs related to these aspects via graphic user interface. An FEA code then converts given inputs into the attributes in the selected *model template*. A model template specifies analysis types, element types, the types of boundary, and loading conditions. It also defines some simulation variables that are used in the algorithms of solving processes. The user sets the values for the variables or parameters related to the attributes in the model.

5.2 TYPES OF ANALYSIS MODELS

Depending on the nature of an engineering problem, analysis types can be different in the formulation of an FEA model. The selection of analysis type is represented by a model template including some basic attributes such as *number of objects* (single or multiple objects), *materials properties* (linear or

Finite Element Analysis Applications. http://dx.doi.org/10.1016/B978-0-12-809952-0.00005-4

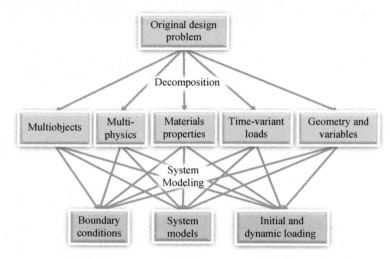

■ **FIGURE 5.1** Users' inputs for model types, boundary conditions and loads.

nonlinear, homogeneous or heterogeneous), *time dependence* (static, semi-static, or dynamic), and *simulation variables* (what-if or design study).

5.2.1 **Dimensions of analysis model**

The geometric object in an FEA model is often very complex. In many cases, computer aided design (CAD) models of objects have to be simplified to make the complexity of an FEA model manageable. A three-dimensional (3D) model is not always the best choice because the dimension of elements affects the complexity of an FEA model greatly. Strictly speaking, any physical object is a continuous domain in a 3D space. A commercial FEA code usually sets 3D elements as default for an object to be modeled. However, the amount of computation can be increased exponentially with an increase of the number of elements and the degrees of freedom (DOF) of nodes. The higher dimension of elements has, the higher computation of the FEA code is needed to obtain a solution. Therefore, a user should consider the feasibility of simplifying a 3D model into 1D or 2D model. Alternatively, element types can be mixed in an FEA model to balance the required accuracy at the regions of interest and the computation over the entire domain.

Despite the reduced computation by using low-dimensional elements, it may be risky to uncover some important design factors when one- or two-dimensional elements are used to represent an object in a 3D space. A designer should make his or her own engineering judgment to decide if a 3D model can be simplified or a simulation on a portion of 3D model is sufficient to represent the physical phenomenon of a continuous domain. Table 5.1 illustrates some common scenarios where 1D and 2D elements can be adopted in an analysis model.

Table 5.1 Exemplified Models by 1D or 2D Elements

Objects	Illustration	Simplification
Truss or beam structures		
Shell structures		
Axisymmetric structures		
Structure with plane stress or strain		

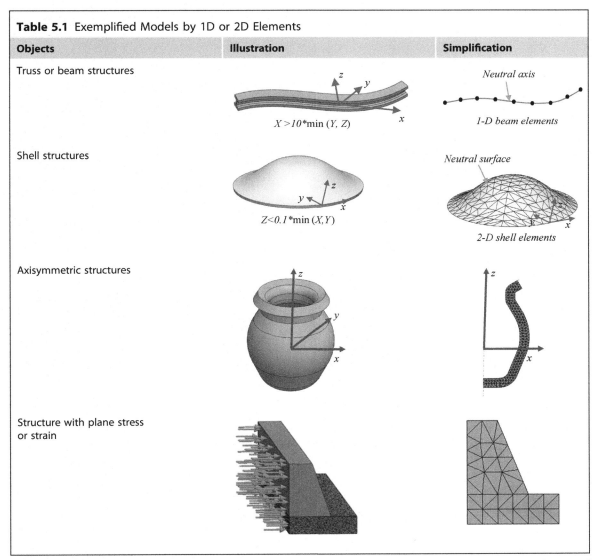

Continued

Table 5.1 Exemplified Models by 1D or 2D Elements *continued*

Objects	Illustration	Simplification
Structures with symmetric plane(s)		
Structure with circular patterns		

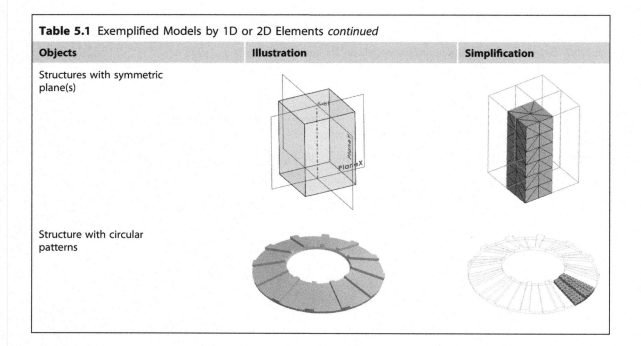

5.2.1.1 Models with 1D truss/beam elements

If an object has a neutral axis (x), and the length along this axis is much larger than those dimensions on other two axes (y, z); this object can be modeled by 1D truss or beam elements. Typically, a ratio of the dimensions over 10 is considered as large. 1D elements have been widely applied to model structural members in the construction. 1D elements can be *truss members*, *beam members*, or *frame members*. A truss member carries axial loads only, a beam member carries transverse loads, and a frame member carries both axial and transverse loads.

5.2.1.2 Models with shell elements

If an object has a neutral surface (x, y), and the dimensions over this neutral surface are much larger than the thickness of object (z); this object can be modeled by 2D shell elements. The loads on shell elements can be pressure or torque perpendicular to the neutral surface. Note that shell elements do not have the rigidity to carry moments. Some aircraft components such as fuselages and wings can be modeled by shell elements.

5.2.1.3 Models with axisymmetric elements

If an object is axisymmetric and the loads are applied symmetrically, this object should be modeled by 2D elements over a cross-section passing the revolving axis. For an axisymmetric model in solid mechanics, there is no strain perpendicular to the section of interest in the FEA model. Many pressure vessels and containers can be simplified by using axisymmetric elements.

5.2.1.4 Models with plane stress/strain elements

If a thin object only involves in the plane stress and there is no stress off the plane, this object should be modeled by 2D elements with plane stress. If an object has an extremely large length in contrast to those of other two axes, the displacement along this axis is ignorable in comparison with the displacements on other two axes; the object should be modeled by 2D elements with plane strain. For example, in modeling a dam structure, plane strain elements are usually applicable.

5.2.1.5 Models for objects with symmetric plane(s) or circular patterns

If an object has the shapes and loads, which are symmetrically about one or two planes, it cannot be modeled by axisymmetric elements. However, the behaviors of entire object can be represented by a model with a portion of the object. Only additional constraints applied on the partial model are roller conditions on cutting planes, i.e., the displacement perpendicular to the symmetric or cutting planes is not allowed.

5.2.2 Single- physics or multi- physics

Design problems may involve multiphysics behaviors. For a multiphysics problem, an FEA user has to decide if multiphysics behaviors can be decomposed in several decoupled analysis models and run the simulation in individual disciplines, respectively. This depends on how strong the coupling of multiphysics behaviors is. If the coupling is strong, the governing equations in multiple disciplines have to be solved simultaneously. Many industrial applications have coupled multiphysics behaviors. For example, the temperature change by the heat transfer induces mechanical force, an electric field interacts a magnetic field, and the material properties are affected by temperature. Design axioms by Suh (2005) can be applied to justify if a multiphysics model can be decoupled. Based on design axioms, the couplings in multiphysics can be classified into *weak-coupling* and *strong-coupling*. In Chapter 11, the multiphysics simulation will be discussed in details.

5.2.3 **Models with linear or nonlinear elements**

The selection of linear and nonlinear elements affects the complexity of an FEA model. In contrast to a linear element, a nonlinear element has more nodes to represent elemental behaviors. Generally, using nonlinear elements increases the accuracy of an FEA simulation result. In some cases, an analysis model does not have another choice except for using nonlinear elements, for example, when the modeled system has the physical behavior with nonlinear characteristics. In solid mechanics, the nonlinearity implies the nonlinear response of a mechanical structure when it experiences external loads. If the relations of loads $\{F\}$ and displacement $\{U\}$ are defined by the stiffness matrix $[K]$, the nonlinearity means that some entities in $[K]$ are not constant and are the functions of the displacements $\{U\}$. The stiffness may be changed because the shape changes. In other cases, an analysis model needs nonlinear elements when the object tends to have *large strains*, *large distortions*, *stress stiffening*, and *spin softening*, which are beyond the elastic regions of materials.

5.2.4 **Steady, semi- steady, and transient models**

An FEA model may be time dependent or time independent. In some cases, the states and responses of a model at its equilibrium conditions are the primary concern of analysis. The time factor can be ignored in the FEA model, and loads are applied in the model without a time-dependent change. Such an analysis model is called as a *steady analysis*. For example, if a designer cares about the safety factor or the deformation of a machine element subjected to static loads, steady FEA models will be appropriate.

In other cases, the time factor may be an important factor, and models have to take into consideration the time dependency of design variables. Corresponding FEA models are referred as *transient* models. Common examples where time-dependent models must be used are (1) fatigue analysis, (2) modal analysis of mechanical structures, and (3) transient heat transfer processes.

5.2.5 **What-if models and design studies**

FEA-based simulations are used to serve different design purposes. Two basic design activities are *design analysis* and *design synthesis*. In design analysis, all the design variables and system parameters are given; an FEA model is to find the system state or response to external loads. It is a type of task for design analysis, which is commonly referred as *what—if simulation*. In design synthesis, some design variables are to be determined based on the specified design criteria. In other words, design variables are given as changeable ranges rather than values, FEA models are applied to evaluate the responses of a system; although design variables can vary in a design space. FEA results for different scenarios are compared to optimize the solution based on the specified design criteria. *Design study* is a simulation-based design synthesis. The same FEA model is executed repeatedly under different settings of design variables to find the best system outcomes against design goals.

5.3 **BOUNDARY CONDITIONS**

Other than governing equations of system behaviors, *boundary conditions* (BCs) refer to additional constraints on state variables of an analysis model. BCs represent the interaction of model with its application environment. BCs are essential to solve a system model. Without BCs, the number of unknowns in a system model is larger than the number of governing equations, and there will not be unique system solutions. Additional constraints must be defined to describe the restraints by the environment or the interfaces of multiple objects.

5.3.1 **BCs on displacements of single object**

BCs of a single object are used to describe its relations to the application environment. BCs are defined in terms of state variables of nodes on the boundaries of an object. If a node in an element type has multiple DOF, the BCs on such a node can be confined fully or partially. Taking an example of a 1D beam element in Table 5.2, each node has six displacements along six directions, i.e., the translations u, v, w, and the rotations δ_x, δ_y, δ_z along x, y, and z, respectively. Theoretically, the number of possible BCs on such a node is,

$$C_6^0 + C_6^1 + C_6^2 + C_6^3 + C_6^4 + C_6^5 + C_6^6 = 64 \qquad (5.1)$$

These BCs can be classified according to the number of constrained DOF. Table 5.2 gives such a classification and examples are included for different types of BCs. In some cases, a node is set with a constant displacement along an axis and the fixed DOF should be offset by the given value of the displacement along DOF. An FEA code cannot find a solution to a system model when the BCs are insufficient. On the other hand, the solution exists when more BCs than minimum ones are defined. In such a case, the system model becomes an overconstrained model.

Table 5.2 Types of Boundary Conditions on a Node of Beam Member

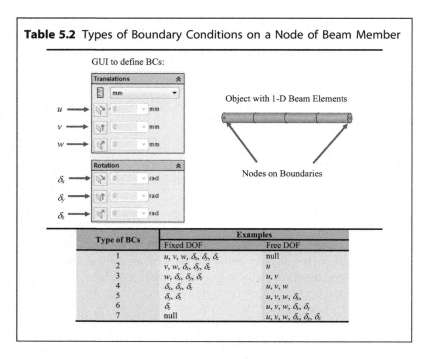

GUI to define BCs:

Object with 1-D Beam Elements

Nodes on Boundaries

Type of BCs	Examples	
	Fixed DOF	Free DOF
1	$u, v, w, \delta_x, \delta_y, \delta_z$	null
2	$v, w, \delta_x, \delta_y, \delta_z$	u
3	$w, \delta_x, \delta_y, \delta_z$	u, v
4	$\delta_x, \delta_y, \delta_z$	u, v, w
5	δ_y, δ_z	$u, v, w, \delta_x,$
6	δ_z	$u, v, w, \delta_x, \delta_y$
7	null	$u, v, w, \delta_x, \delta_y, \delta_z$

Example 5.1

The object and external loads in Fig. 5.2 are symmetric about the high-lighted plane. The half of object is modeled to calculate the stress distri-bution in a static analysis. Determine BCs on the symmetric plane of the model.

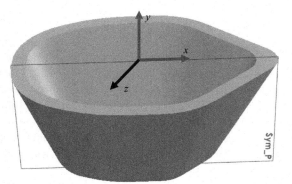

■ **FIGURE 5.2** Boundary conditions on symmetric plane.

Solution

Due to the symmetric relation, any node on the symmetric plane (*Sym_P* in Fig. 5.2) does not have a translational displacement along *z*-axis. It does not allow a rotational displacement along *x*-or *y*-axis, which leads to a situation where the node moves out of the symmetric plane. Therefore, the BCs on this plane are $w = 0$ and $\delta_x = \delta_y = 0$.

■

5.3.2 **Constraints in assembly**

If an FEA model is an assembly from a group of parts or components, the interfaces of each pair of two contacted objects must be considered and represented as contact conditions. Contact conditions are special types of BCs in an FEA model. If the assembly is treated as a structure, its objects are assumed to have *bonded* relations with each other. A bonded contact condition represents that the corresponding nodes on two contact surfaces from two objects have the same displacements. Thus, the nodes on two objects are bonded with no relative displacement. If a structure consists of multiple objects, a commercial FEA code sets "bonded" relations in default at the identified contact surfaces of objects.

If contact surfaces are not allowed to be penetrated, and they are allowed to be separated with each other when a deformation occurs, the appropriate contact condition will be "*no penetration*". Due to the possibility of a relative motion, a pair of contact surfaces with "no penetration" is associated with the friction coefficient to reflect external forces on contact surfaces. This brings a great deal of complexity in modeling and computation if such friction forces are of interests. To accommodate different contact conditions, an FEA code usually includes other types of BCs for contact conditions including "*virtual wall*", "*allow penetration*", and "*shrink fit*". Table 5.3 provides a list of contact conditions, explanations, and examples.

A special attention has to been paid when a CAD model of assembly is directly imported from other design platforms or tools. It is not uncommon that an analyst needs to spend an extreme amount of time to identify all the pairs of contact surfaces and determine appropriate types of contact conditions accordingly. When the number of parts in an assembly is large, the number of contact surfaces can also be large. It poses the challenge for an FEA package to identify all contact surface sets correctly. In particular, when there is a contact pair between a large surface and a point, a line or a tiny surface region.

Table 5.3 Types of Contact Conditions and Examples

Contact Condition	Explanation	Example
Bonded	Bonded entities behave as if they were welded. Displacements of the nodes on two entities are the same.	
No penetration	Selected components or bodies do not penetrate each other regardless of their initial contact condition. A body does not allow self-intersection by deflection. Surface to surface contact is set as no penetration.	
Allow penetration	Allow penetration overrides other existing contacts. Using this option is to skip *the inference check* for computation reduction. Use it when one can ensure the applied loads do not cause interference.	
Virtual wall	Virtual wall defines contact between the entities and a virtual wall defined by a target plane, which can be rigid or flexible. The friction coefficient at the target plane can be defined.	
Shrink fit	Shrink fit select faces from two components that are initially interfering with each other; the simulation ensures two interfering surface are compatible in the solutions.	

Example 5.2

A design engineer is developing a new type of biomedical materials. He needs to characterize the material properties via experiments and simulations. Fig. 5.3 shows the test setup for the measurement of material properties of bars. These bars have different cross-sections. Build an FEA model, which can be used to validate the relation of the tested deflection with cross-section and Young's modulus.

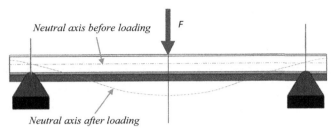

■ **FIGURE 5.3** Determination of boundary conditions of test setup.

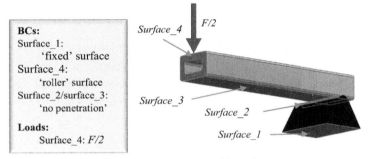

BCs:
Surface_1:
    'fixed' surface
Surface_4:
    'roller' surface
Surface_2/surface_3:
    'no penetration'

Loads:
    Surface_4: $F/2$

Surface_4 $F/2$
Surface_3
Surface_2
Surface_1

■ **FIGURE 5.4** Boundary conditions and loads of test setup.

Solution

Because the experiment platform is setup symmetrically about the middle plane, only half of the specimens should be modeled in the numerical simulation. Accordingly, the external load is applied in the middle plane. The half of the total test load will be applied in the model.

As shown in Fig. 5.4, due to symmetry, any node on the middle plane is not allowed to have any displacement along neutral axis. Therefore, one BC on the symmetric plan is the "*roller*" condition. The specimen is supported at two ends. The bottom surface of the support is defined as "*fixed geometry*" to immobile its motion. Between the support and specimen, the contact surface makes contacts but allows a relative sliding when the specimen is deformed. Therefore, the last BC is defined as "*no penetration*" for the top surface of support and the bottom surface of specimen.

5.3.3 **Equivalent nodal loads**

In this section, 2D field problems in Fig. 5.5 are taken as an example to illustrate how BCs affect a system model. A 2D domain is denoted as Ω; the boundaries of the domain may have two types of BCs, which are represented by Γ_1 and Γ_2, respectively.

A state variable of interest is denoted as ϕ. The first type of BCs is called an *essential BC* or Dirichlet BC. An essential BC corresponds to the given value of the state variable ϕ, which is given on Γ_1, and the second type of BCs is called as a *natural BC* or *Neumann BC*. A natural BC corresponds to the constraints on the derivative of state variable ϕ on Γ_2. The general representation of the second type of BCs on Γ_2 is,

$$k_x \frac{\partial \phi}{\partial x} \cos \theta + k_y \frac{\partial \phi}{\partial y} \sin \theta = -h\phi + q_0 \tag{5.2}$$

where k_x, k_y, and h are the coefficients related to material properties of the solid domain, and q_0 is a constant determined by the environmental condition. If the material is homogeneous, we have $k_x = k_y$ and Eq. (5.2) can be simplified as,

$$k_x \frac{\partial \phi}{\partial n} = -h\phi + q_0 \tag{5.3}$$

where $\frac{\partial \phi}{\partial n}$ is the derivative normal n to the boundary.

Note that ϕ in Eq. (5.2) is the state variable along boundary Γ_2; however, it is unknown and to be determined from the system model.

In a special case when the symmetric plane of a solid domain is modeled as boundary surface, there is no change of state variables on the normal

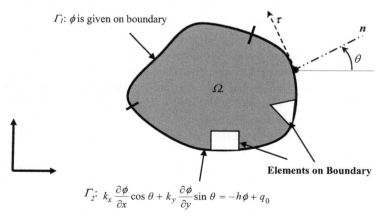

$$\Gamma_2 \colon k_x \frac{\partial \phi}{\partial x} \cos \theta + k_y \frac{\partial \phi}{\partial y} \sin \theta = -h\phi + q_0$$

■ **FIGURE 5.5** Types of boundary conditions in 2D domain.

direction of Γ_2 and $h = 0$, and $q_0 = 0$. Substituting these conditions in Eq. (5.3) gives,

$$k_x \frac{\partial \phi}{\partial x} \cos \theta + k_y \frac{\partial \phi}{\partial y} \sin \theta = 0 \tag{5.4}$$

Rewriting the item in Eq. (4.44) about the integral over the boundary Γ_2 as,

$$\{I_{bc}\} = \int_{\Gamma_2} [S]^T \left(k_x \frac{\partial \phi}{\partial x} \cos \theta + k_y \frac{\partial \phi}{\partial y} \sin \theta \right) d\Gamma \tag{5.5}$$

where $\{I_{bc}\}$ is the integral along boundary Γ_2 in a counter clockwise direction.

Substituting Eq. (5.2) into Eq. (5.5) gets,

$$\{I_{bc}\} = \int_{\Gamma_2} [S]^T (-h\phi + q_0) d\Gamma \tag{5.6}$$

Using the shape functions in an element for the interpolation of state variable ϕ, i.e.,

$$\phi = [S]^T \cdot [\varphi] \tag{5.7}$$

where $[\varphi]$ is the vector of field variables on nodes of element.

Substituting Eq. (5.7) into Eq. (5.6) gives,

$$\{I_{bc}\} = -h \int_{\Gamma_2} [S]^T \cdot [S] \cdot d\Gamma \cdot \{\varphi\} + q_0 \cdot \int_{\Gamma_2} [S]^T d\Gamma \tag{5.8}$$

In the following, we look into the cases when a rectangle element or triangle element has an edge on boundary.

5.3.3.1 Rectangle elements

Fig. 5.6 shows a rectangle element in a discretized domain. An local coordinate system (LCS) is used to describe the coordinates of four nodes, the element behaviors are represented by nodal values of state variables. Shape functions are used to determine the value of the state variable at any point within the element based on the given nodal values.

As shown in Eq. (4.54), the interpolation of the state variable ϕ in a rectangle element is,

$$\phi = [S] \cdot \{\varphi\} = [\, S_i \quad S_j \quad S_m \quad S_n \,] \cdot \begin{Bmatrix} \phi_i \\ \phi_j \\ \phi_m \\ \phi_n \end{Bmatrix} \tag{5.9}$$

where

$$S_i = \left(1 - \frac{x}{l}\right)\left(1 - \frac{y}{w}\right); S_j = \frac{x}{l}\left(1 - \frac{y}{w}\right); S_m = \frac{x}{l}\frac{y}{w}; S_n = \left(1 - \frac{x}{l}\right)\frac{y}{w}$$

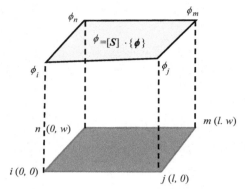

■ **FIGURE 5.6** Description of rectangle element.

If *edge i−j* is a part of Γ_2, y is set as 0 in Eq. (5.11); substituting it into Eq. (5.10) gives,

$$\{I_{bc}\}_{i-j} = -h \int_0^l \begin{bmatrix} \left(1-\dfrac{x}{l}\right) \\ \dfrac{x}{l} \\ 0 \\ 0 \end{bmatrix} \cdot \left[\left(1-\dfrac{x}{l}\right) \quad \dfrac{x}{l} \quad 0 \quad 0\right] dx \cdot \begin{Bmatrix} \phi_i \\ \phi_j \\ \phi_m \\ \phi_n \end{Bmatrix} + q_0 \int_0^l \begin{bmatrix} \left(1-\dfrac{x}{l}\right) \\ \dfrac{x}{l} \\ 0 \\ 0 \end{bmatrix} \cdot dx$$

$$= -h \begin{bmatrix} \int_0^l \left(1-\dfrac{x}{l}\right)^2 dx & \int_0^l \left(1-\dfrac{x}{l}\right)\dfrac{x}{l}dx & 0 & 0 \\ \int_0^l \left(1-\dfrac{x}{l}\right)\dfrac{x}{l}dx & \int_0^l \left(\dfrac{x}{l}\right)^2 dx & 0 & 0 \\ 0 & 0 & 0 & 0 \\ 0 & 0 & 0 & 0 \end{bmatrix} \begin{Bmatrix} \phi_i \\ \phi_j \\ \phi_m \\ \phi_n \end{Bmatrix} + q_0 \begin{bmatrix} \int_0^l \left(1-\dfrac{x}{l}\right)dx \\ \int_0^l \left(\dfrac{x}{l}\right)dx \\ 0 \\ 0 \end{bmatrix}$$

$$= -\frac{hl}{6}\begin{bmatrix} 2 & 1 & 0 & 0 \\ 1 & 2 & 0 & 0 \\ 0 & 0 & 0 & 0 \\ 0 & 0 & 0 & 0 \end{bmatrix}\begin{Bmatrix} \phi_i \\ \phi_j \\ \phi_m \\ \phi_n \end{Bmatrix} + \frac{q_0 l}{2}\begin{bmatrix} 1 \\ 1 \\ 0 \\ 0 \end{bmatrix}$$

(5.10)

where $\{I_{bc}\}_{i-j}$ is an integral of the second type of BCs along *edge i−j*.

Similarly, if *edge j—m*, *edge m—n*, and *edge n—i* are a part of Γ_2, respectively, the corresponding boundary conditions can be found as,

$$\{I_{bc}\}_{j-m} = -\frac{hw}{6}\begin{bmatrix} 0 & 0 & 0 & 0 \\ 0 & 2 & 1 & 0 \\ 0 & 1 & 2 & 0 \\ 0 & 0 & 0 & 0 \end{bmatrix}\begin{Bmatrix} \phi_i \\ \phi_j \\ \phi_m \\ \phi_n \end{Bmatrix} + \frac{q_0 w}{2}\begin{bmatrix} 0 \\ 1 \\ 1 \\ 0 \end{bmatrix} \tag{5.11}$$

$$\{I_{bc}\}_{m-n} = -\frac{hl}{6}\begin{bmatrix} 0 & 0 & 0 & 0 \\ 0 & 0 & 0 & 0 \\ 0 & 0 & 2 & 1 \\ 0 & 0 & 1 & 2 \end{bmatrix}\begin{Bmatrix} \phi_i \\ \phi_j \\ \phi_m \\ \phi_n \end{Bmatrix} + \frac{q_0 l}{2}\begin{bmatrix} 0 \\ 0 \\ 1 \\ 1 \end{bmatrix} \tag{5.12}$$

$$\{I_{bc}\}_{n-i} = -\frac{hw}{6}\begin{bmatrix} 2 & 0 & 0 & 1 \\ 0 & 0 & 0 & 0 \\ 0 & 0 & 0 & 0 \\ 1 & 0 & 0 & 2 \end{bmatrix}\begin{Bmatrix} \phi_i \\ \phi_j \\ \phi_m \\ \phi_n \end{Bmatrix} + \frac{q_0 w}{2}\begin{bmatrix} 1 \\ 0 \\ 0 \\ 1 \end{bmatrix} \tag{5.13}$$

where $\{I_{bc}\}_{j-m}$, $\{I_{bc}\}_{m-n}$, and $\{I_{bc}\}_{n-i}$ are the integrals of the second type of BCs along *edge j—m*, *edge m—n*, and *edge n—i*, respectively.

Example 5.3

A 2D solid domain in Fig. 5.7 is under an equilibrium state of the heat transfer. It is divided into four rectangle elements evenly. Element 3 involves the heat convection on two edges, and the nodes of the element are specified in the figure. Defines the boundary conditions of the element.

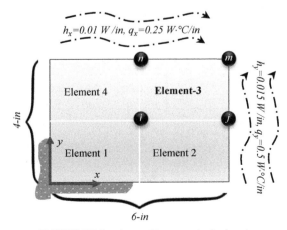

$h_x = 0.01\ W/in$, $q_x = 0.25\ W\cdot°C/in$

$h_y = 0.015\ W/in$, $q_y = 0.5\ W\cdot°C/in$

Element 4

Element-3

Element 1

Element 2

4-in

6-in

■ **FIGURE 5.7** Boundary conditions on rectangle element.

Solution

For a rectangle element, Eqs. (5.10)–(5.13) have shown that a loading condition will affect both state variables on nodes and the vector of nodal loads of the element.

The given rectangle element has two edges (j–m and m–n) with specified loads. For edge j–m, $h_{j-m} = h_y = 0.015$ W/in, $w = 4/2 = 2$ in, and $q_{m-n} = q_y = 0.5$ W°C/in. Applying Eq. (5.11) gets,

$$\{I_{bc}\}_{j-m} = -\frac{h_{j-m}w}{6}\begin{bmatrix} 0 & 0 & 0 & 0 \\ 0 & 2 & 1 & 0 \\ 0 & 1 & 2 & 0 \\ 0 & 0 & 0 & 0 \end{bmatrix}\begin{Bmatrix} \phi_i \\ \phi_j \\ \phi_m \\ \phi_n \end{Bmatrix} + \frac{q_{j-m}w}{2}\begin{bmatrix} 0 \\ 1 \\ 1 \\ 0 \end{bmatrix}$$

$$= -\begin{bmatrix} 0 & 0 & 0 & 0 \\ 0 & 0.01 & 0.005 & 0 \\ 0 & 0.005 & 0.01 & 0 \\ 0 & 0 & 0 & 0 \end{bmatrix}\begin{Bmatrix} \phi_i \\ \phi_j \\ \phi_m \\ \phi_n \end{Bmatrix} + \begin{bmatrix} 0 \\ 0.5 \\ 0.5 \\ 0 \end{bmatrix}$$

For edge m–n, $h_{m-n} = h_y = 0.01$ W/in, $l = 6/2 = 3$ in, $q_{m-n} = q_x = 0.25$ W°C/in. Applying Eq. (5.12) gets,

$$\{I_{bc}\}_{m-n} = -\frac{h_{m-n}l}{6}\begin{bmatrix} 0 & 0 & 0 & 0 \\ 0 & 0 & 0 & 0 \\ 0 & 0 & 2 & 1 \\ 0 & 0 & 1 & 2 \end{bmatrix}\begin{Bmatrix} \phi_i \\ \phi_j \\ \phi_m \\ \phi_n \end{Bmatrix} + \frac{q_{m-n}l}{2}\begin{bmatrix} 0 \\ 0 \\ 1 \\ 1 \end{bmatrix}$$

$$= -\begin{bmatrix} 0 & 0 & 0 & 0 \\ 0 & 0 & 0 & 0 \\ 0 & 0 & 0.01 & 0.005 \\ 0 & 0 & 0.005 & 0.01 \end{bmatrix}\begin{Bmatrix} \phi_i \\ \phi_j \\ \phi_m \\ \phi_n \end{Bmatrix} + \begin{bmatrix} 0 \\ 0 \\ 0.75 \\ 0.75 \end{bmatrix}$$

Therefore, the boundary conditions for element 3 are defined as,

$$\{I_{bc}\}_{j-m} + \{I_{bc}\}_{m-n} = -\begin{bmatrix} 0 & 0 & 0 & 0 \\ 0 & 0.01 & 0.005 & 0 \\ 0 & 0.005 & 0.02 & 0.005 \\ 0 & 0 & 0.005 & 0.01 \end{bmatrix}\begin{Bmatrix} \phi_i \\ \phi_j \\ \phi_m \\ \phi_n \end{Bmatrix}$$

$$+ \begin{bmatrix} 0 \\ 0.5 \\ 1.25 \\ 0.75 \end{bmatrix}$$

5.3.3.2 Triangle elements

As shown in Fig. 5.8, the natural coordinate system (CS) of a triangle element is used to simplify the deriving process. A linear triangle element consists of nodes i, j, and k. The lengths of *edge i—j*, *edge j—k*, and *edge i—k* are L_{ij}, L_{jk}, and L_{ik}, respectively. The element behaviors are represented by the values of state variables on three nodes. Shape functions are used to determine the value of state variable at any point of element based on values of state variable on three nodes.

The interpolation of state variable ϕ in a triangle element in natural CS is,

$$
\phi = [S] \cdot \{\varphi\} = [S_i \quad S_j \quad S_k] \cdot \begin{Bmatrix} \phi_i \\ \phi_j \\ \phi_k \end{Bmatrix} \tag{5.14}
$$

where

$$
S_i = \xi; S_j = \eta; S_k = 1 - \xi - \eta
$$

If *edge i—j* is a part of Γ_2, then let $S_k = 1 - \xi - \eta = 0$ or $\eta = 1 - \xi$ in Eq. (5.14); substituting it into Eq. (5.8) gives,

$$
\{I_{bc}\}_{i-j} = -h \int_1^0 \begin{bmatrix} \xi \\ 1-\xi \\ 0 \end{bmatrix} \cdot [\xi \quad 1-\xi \quad 0] d(-L_{ij} \cdot \xi) \cdot \begin{Bmatrix} \phi_i \\ \phi_j \\ \phi_k \end{Bmatrix} + q_0 \int_1^0 \begin{bmatrix} \xi \\ 1-\xi \\ 0 \end{bmatrix} \cdot d(-L_{ij} \cdot \xi)
$$

$$
= h \cdot L_{ij} \begin{bmatrix} \int_1^0 \xi^2 d\xi & \int_1^0 \xi(1-\xi)d\xi & 0 \\ \int_1^0 \xi(1-\xi)d\xi & \int_1^0 \xi^2 d\xi & 0 \\ 0 & 0 & 0 \end{bmatrix} \begin{Bmatrix} \phi_i \\ \phi_j \\ \phi_k \end{Bmatrix} - q_0 L_{ij} \begin{bmatrix} \int_1^0 \xi d\xi \\ \int_1^0 (1-\xi)d\xi \\ 0 \end{bmatrix}
$$

$$
= -\frac{hL_{ij}}{6} \begin{bmatrix} 2 & 1 & 0 \\ 1 & 2 & 0 \\ 0 & 0 & 0 \end{bmatrix} \begin{Bmatrix} \phi_i \\ \phi_j \\ \phi_k \end{Bmatrix} + \frac{q_0 L_{ij}}{2} \begin{bmatrix} 1 \\ 1 \\ 0 \end{bmatrix} \tag{5.15}
$$

where $\{I_{bc}\}_{i-j}$ is an integral of the second type of BCs along *edge i—j*.

Similarly, if *edge j—k*, and *edge k—i* are a part of Γ_2, respectively, the corresponding boundary conditions can be found as,

$$
\{I_{bc}\}_{j-k} = -\frac{hL_{jk}}{6} \begin{bmatrix} 0 & 0 & 0 \\ 0 & 2 & 1 \\ 0 & 1 & 2 \end{bmatrix} \begin{Bmatrix} \phi_i \\ \phi_j \\ \phi_k \end{Bmatrix} + \frac{q_0 L_{jk}}{2} \begin{bmatrix} 0 \\ 1 \\ 1 \end{bmatrix} \tag{5.16}
$$

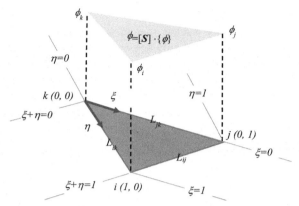

■ FIGURE 5.8 Description of triangular element.

$$\{I_{bc}\}_{i-k} = = -\frac{hL_{ik}}{6}\begin{bmatrix} 2 & 0 & 1 \\ 0 & 0 & 0 \\ 1 & 0 & 2 \end{bmatrix}\begin{Bmatrix} \phi_i \\ \phi_j \\ \phi_k \end{Bmatrix} + \frac{q_0 L_{ik}}{2}\begin{bmatrix} 1 \\ 0 \\ 1 \end{bmatrix} \qquad (5.17)$$

where $\{I_{bc}\}_{j-k}$ and $\{I_{bc}\}_{i-k}$ are the integrals of the second type of BCs along *edge j—k* and *edge i—k*, respectively.

5.4 EQUIVALENT LOADS

The Interaction of an object with its application environment is reflected by (1) the restraints on the boundaries of object, (2) the external loads applied on the body and/or surface, and (3) the time-dependent loads. Section 5.2 has discussed the restraints of a system model by boundary conditions. In this section, external loads in an FEA model are discussed.

Loads applied in a continuous domain can be quantified easily in an FEA model. For example, it is relatively easy to represent the gravity load in solid mechanics or the rate of heat generation in a thermal transfer problem for FEA. In contrast, for the loads on a surface, because the continuous domain is discretized as a set of elements and nodes, the loads have to be converted into equivalent loads on nodes. The principle to determine an equivalent load is to ensure that if an actual load is replaced by an equivalent load, the main physical behaviors of object is not affected. Based on applications, different engineering approaches are used to determine equivalent loads. In the following, we introduce the methods to determine equivalent loads for beam elements and 2D elements in static analysis.

5.4.1 **Equivalent loads of beam elements**

In this section, we discuss how to transfer a distributed load on a beam element into equivalent loads on nodes. Note that after the discretization, only nodal loads are accepted in an FEA model with beam elements. Therefore, any distributed load has to be transferred into the loads on nodes.

The criterion of finding a load equivalence is that the behaviors of element must be the same if the original load is replaced by equivalent loads. Because the behaviors of a beam element are represented by the displacements on two nodes; we can define the equivalent loads for the distributed loads based on the constraints on nodal displacements. For a beam element shown in Fig. 5.9, the governing equation of the beam's behaviors is,

$$EI\frac{d^4y}{dx^4} = \frac{dV(x)}{dx} = w(x) \tag{5.18}$$

where

E is Young's modulus,
I is the second moment of area (moment of inertia),
$V(x)$ is the functional of shear force over the cross-section at x, and
$w(x)$ is the distributed pressure load.

If the pressure is evenly distributed as illustrated in Fig. 5.9, i.e., $w(x) = -w_0$. Sequentially, we take four times of the integrals for Eq. (5.18), and we find,

$$\left.\begin{array}{l} EI\dfrac{d^3y}{dx^3} = V(x) = -w_0x + C_1 \\[3mm] EI\dfrac{d^2y}{dx^2} = M(x) = -\dfrac{w_0x^2}{2} + C_1x + C_2 \\[3mm] EI\dfrac{dy}{dx} = F(x) = -\dfrac{w_0x^3}{6} + \dfrac{C_1x^2}{2} + C_2x + C_3 \\[3mm] EIy = -\dfrac{w_0x^4}{24} + \dfrac{C_1x^3}{6} + \dfrac{C_2x^2}{2} + C_3x + C_4 \end{array}\right\} \tag{5.19}$$

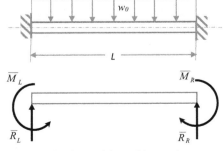

■ **FIGURE 5.9** Reactional forces of beam under pressure.

Because two ends of the beam element are fixed in Fig. 5.7, we have,

$$
\left.\begin{array}{r}
y|_{x=0} = 0 \\[6pt]
\dfrac{dy}{dx}\Big|_{x=0} = 0 \\[6pt]
y|_{x=l} = 0 \\[6pt]
\dfrac{dy}{dx}\Big|_{x=l} = 0
\end{array}\right\} \tag{5.20}
$$

Substituting Eq. (5.20) into Eq. (5.19), all of constants C_1, C_2, C_3, and C_4 can be solved,

$$
C_3 = C_4 = 0; C_1 = \frac{w_0 l}{2}; C_2 = -\frac{w_0 l^2}{12} \tag{5.21}
$$

Substituting the constants in Eq. (5.21) back to Eq. (5.19) yields the functions of displacement, slope, moment, and shear force as,

$$
\left.\begin{array}{l}
EI\dfrac{d^3y}{dx^3} = V(x) = -w_0 x + \dfrac{w_0 l}{2} \\[10pt]
EI\dfrac{d^2y}{dx^2} = M(x) = -\dfrac{w_0 x^2}{2} + \dfrac{w_0 l}{2}x - \dfrac{w_0 l^2}{12} \\[10pt]
EI\dfrac{dy}{dx} = F(x) = -\dfrac{w_0 x^3}{6} + \dfrac{w_0 l x^2}{4} - \dfrac{w_0 l^2}{12}x \\[10pt]
EIy = -\dfrac{w_0 x^4}{24} + \dfrac{w_0 l x^3}{12} - \dfrac{w_0 l^2 x^2}{24}
\end{array}\right\} \tag{5.22}
$$

Using Eq. (5.22) can find the reactional forces and moments at two ends as,

$$
\left.\begin{array}{l}
V_L = V(x)|_{x=0} = \dfrac{w_0 l}{2} \\[10pt]
V_R = V(x)|_{x=l} = -\dfrac{w_0 l}{2} \\[10pt]
M_L = M(x)|_{x=0} = -\dfrac{w_0 l^2}{12} \\[10pt]
M_R = M(x)|_{x=l} = -\dfrac{w_0 l^2}{12}
\end{array}\right\} \tag{5.23}
$$

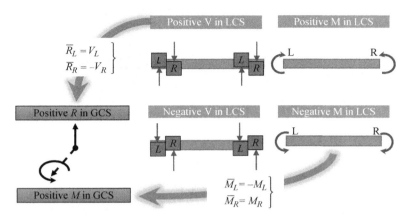

■ **FIGURE 5.10** Rules to convert shear force V and moment M from local coordinate system (LCS) to global coordinate system (GCS).

The equivalent nodal loads in Eq. (5.23) are in an LCS; they have to be transferred to the vector of loads in a system model in global coordinate system (GCS). Fig. 5.10 shows the rules of force and moment transformations from LCS to GCS. As a result, the vector $\boldsymbol{F}_e$ for equivalent loads at left and right nodes are,

$$
\left.\begin{aligned}
\overline{R}_L &= V_L = \frac{wl}{2} \\[4pt]
\overline{M}_L &= -M_L = \frac{wl^2}{12} \\[4pt]
\overline{R}_R &= -V_R = \frac{wl}{2} \\[4pt]
\overline{M}_R &= M_R = -\frac{wl^2}{12}
\end{aligned}\right\}
\Rightarrow \{F_e\} = -
\left\{\begin{aligned}
\overline{R}_L \\
\overline{M}_L \\
\overline{R}_R \\
\overline{M}_R
\end{aligned}\right\}
=
\left\{\begin{aligned}
-\frac{wl}{2} \\[4pt]
-\frac{wl^2}{12} \\[4pt]
-\frac{wl}{2} \\[4pt]
\frac{wl^2}{12}
\end{aligned}\right\}
\tag{5.24}
$$

where $\overline{R}_L, \overline{R}_R$ and $\overline{M}_L, \overline{M}_R$ are the reactional shear forces and moments at left and right nodes, respectively.

Using the different distribution shear function $w(x)$ in Eq. (5.18), the same procedure of integrals can be applied and the equivalent nodal loads for other types of transverse loads in a beam element can be defined. Table 5.4 summarizes the resulted equivalent nodal loads for different loading scenarios.

Table 5.4 Equivalent Loads for Common Transverse Loads on Beam

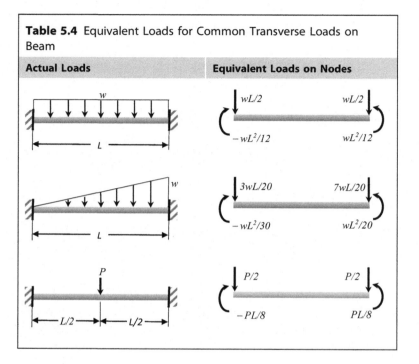

Actual Loads	Equivalent Loads on Nodes

Example 5.4

Fig. 5.11 shows the loading conditions on a cantilever beam. Determine the vectors of equivalent loads on nodes when two beam members are used to represent the cantilever.

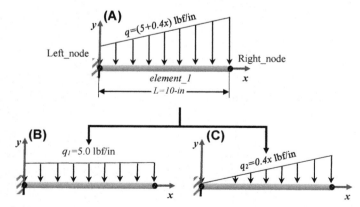

■ **FIGURE 5.11** Example of beam subjected to (A) combined loads, (B) evenly distributed load, (C) ramp load.

Solution

The cantilever is modeled as one beam element, which consists of two nodes. The load is decomposed into an evenly distributed load (Fig. 5.11B) and a ramp load (Fig. 5.11C). According to Table 5.4, the vector of load for this element can be found as,

$$\{F_e\} = -\begin{Bmatrix} V_L \\ M_L \\ V_R \\ M_R \end{Bmatrix} = \begin{Bmatrix} -\dfrac{wL}{2} \\ -\dfrac{wL^2}{12} \\ \dfrac{wL}{2} \\ \dfrac{wL^2}{12} \end{Bmatrix} + \begin{Bmatrix} -\dfrac{3wL}{20} \\ -\dfrac{wL^2}{30} \\ \dfrac{7wL}{20} \\ \dfrac{wL^2}{20} \end{Bmatrix}$$

$$= \begin{Bmatrix} -\dfrac{(5)(10)}{2} \\ -\dfrac{(5)(10)^2}{12} \\ -\dfrac{(5)(10)}{2} \\ \dfrac{(5)(10)^2}{12} \end{Bmatrix} + \begin{Bmatrix} -\dfrac{3(4)(10)}{20} \\ -\dfrac{(4)(10)^2}{30} \\ \dfrac{7(4)(10)}{20} \\ \dfrac{(4)(10)^2}{20} \end{Bmatrix} = \begin{Bmatrix} -31 \\ -55 \\ -39 \\ 61.66 \end{Bmatrix}$$

■

5.4.2 Equivalent loads of triangle elements

Subjected to external loads, the system response of a solid object is the deformation. The work done by external loads is converted into the strain energy associated with the deformation. Therefore, the equivalent loads can be determined by considering the equivalence of energy. Fig. 5.12 shows an example of triangular element subjected to an external load; it is uniformly distributed pressure p on edge $i-j$. The magnitude of pressure can be decomposed into p_x and p_y along two axes in GCS.

The work by the pressure on edge $i-j$ can be quantified as,

$$W^{(e)} = \int_{L_{i-j}} (u \cdot p_x + v \cdot p_y)(t \cdot dl) \tag{5.25}$$

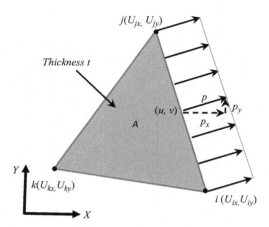

■ **FIGURE 5.12** Example of equivalent loads in triangular element.

where

t is the thickness of the triangular element,
L_{ij} is the length of edge $i-j$, and
p_x and p_y are the projection of the pressure p on x- and y-axis, respectively.
$W^{(e)}$ is denoted as an equivalent load for p.

Using the shape functions for the interpolation in the element, the displacements (u, v) at any position (x, y) can be obtained by,

$$\left. \begin{array}{l} u = S_i \cdot U_{ix} + S_j \cdot U_{jx} + S_k \cdot U_{kx} \\ v = S_i \cdot U_{iy} + S_j \cdot U_{jy} + S_k \cdot U_{ky} \end{array} \right\} \tag{5.26}$$

Substituting Eq. (5.26) into Eq. (5.25) yields,

$$W^{(e)} = \int_{L_{ij}} ((S_i \cdot U_{ix} + S_j \cdot U_{jx} + S_k \cdot U_{kx}) \cdot p_x + (S_i \cdot U_{iy} + S_j \cdot U_{jy} + S_k \cdot U_{ky}) \cdot p_y)(t \cdot dl) \tag{5.27}$$

Based on the minimal potential energy principle, the vector of the equivalent loads can be found by taking the derivative of $W^{(e)}$ with respect to $[U]$, i.e.,

$$\left\{F_{i-j}^{(e)}\right\} = \frac{\partial W^{(e)}}{\partial U} = \left\{\begin{array}{c} \dfrac{\partial W^{(e)}}{\partial U_{ix}} \\[2mm] \dfrac{\partial W^{(e)}}{\partial U_{iy}} \\[2mm] \dfrac{\partial W^{(e)}}{\partial U_{jx}} \\[2mm] \dfrac{\partial W^{(e)}}{\partial U_{jy}} \\[2mm] \dfrac{\partial W^{(e)}}{\partial U_{kx}} \\[2mm] \dfrac{\partial W^{(e)}}{\partial U_{ky}} \end{array}\right\} = \left\{\begin{array}{c} p_x \cdot \int_{L_{ij}} S_i(t \cdot dl) \\[2mm] p_y \cdot \int_{L_{ij}} S_i(t \cdot dl) \\[2mm] p_x \cdot \int_{L_{ij}} S_j(t \cdot dl) \\[2mm] p_y \cdot \int_{L_{ij}} S_j(t \cdot dl) \\[2mm] p_x \cdot \int_{L_{ij}} S_k(t \cdot dl) \\[2mm] p_z \cdot \int_{L_{ij}} S_k(t \cdot dl) \end{array}\right\} = \frac{t \cdot L_{ij}}{2}\left\{\begin{array}{c} p_x \\ p_y \\ p_x \\ p_y \\ 0 \\ 0 \end{array}\right\}$$

(5.28)

where $F_{i-j}^{(e)}$ is the vector of the equivalent loads on nodes when the pressure p is applied on edge $i-j$. Similarly, when there is an external load on edge $j-k$ or $k-i$, the corresponding vectors $F_{j-k}^{(e)}$ and $F_{k-i}^{(e)}$ of equivalent loads can be determined as,

$$\left\{F_{j-k}^{(e)}\right\} = \frac{t \cdot L_{jk}}{2}\left\{\begin{array}{c} 0 \\ 0 \\ p_x \\ p_y \\ p_x \\ p_y \end{array}\right\} \text{ and } \left\{F_{k-i}^{(e)}\right\} = \frac{t \cdot L_{ki}}{2}\left\{\begin{array}{c} p_x \\ p_y \\ 0 \\ 0 \\ p_x \\ p_y \end{array}\right\}$$

(5.29)

Example 5.6

A 0.05-in thick element has the pressure loads on edge $j-k$ and $k-i$ as shown in Fig. 5.13. Determine the vector of the equivalent nodal loads for this element.

Solution

The triangle element has a pressure load on two edges ($j-k$ and $k-i$). The corresponding parameters are (1) $t = 0.05$ in, $p_{1x} = -p_1*\cos(40.60) = -75.93$ lbf/in^2, $p_{1y} = -p_1*\sin(40.60) = -65.08$ lbf/in^2, and $L_{jk} = 2.305$ in

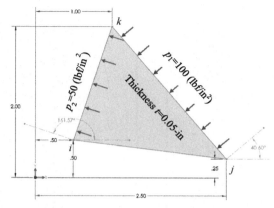

■ FIGURE 5.13 A triangle element under pressure.

for edge j–k, and (2) $p_{2x} = p_2*\cos(161.57) = -47.44$ lbf/in^2, $p_{2y} = p_2*\sin(161.57) = 15.81$ lbf/in^2, and $L_{ki} = 1.581$-in for edge k–i. Therefore, the vector of the nodal loads for this element is,

$$\left\{ F^{(e)} \right\} = \left\{ F^{(e)}_{j-k} \right\} + \left\{ F^{(e)}_{k-i} \right\}$$

$$= \frac{t \cdot L_{jk}}{2} \left\{ \begin{array}{c} 0 \\ 0 \\ p_{1x} \\ p_{1y} \\ p_{1x} \\ p_{1y} \end{array} \right\} + \frac{t \cdot L_{ki}}{2} \left\{ \begin{array}{c} p_{2x} \\ p_{2y} \\ 0 \\ 0 \\ p_{2x} \\ p_{2y} \end{array} \right\} = \left\{ \begin{array}{c} -1.875 \\ 0.625 \\ -4.38 \\ -3.75 \\ -6.25 \\ -3.125 \end{array} \right\} \text{ (lbf)}$$

5.4.3 Equivalent loads of rectangle elements

The similar procedure can be applied to define equivalent loads for a rectangle element. Fig. 5.14 shows a rectangle element with a uniformly distributed pressure p on edge i–j. The magnitude of pressure p can be decomposed into p_x and p_y along two axes in the GCS.

Using the shape functions for the interpolation in the element, the displacements (u, v) at any position (x, y) can be found as,

$$\left. \begin{array}{l} u = S_i \cdot U_{ix} + S_j \cdot U_{jx} + S_m \cdot U_{mx} + S_n \cdot U_{nx} \\ v = S_i \cdot U_{iy} + S_j \cdot U_{jy} + S_m \cdot U_{my} + S_n \cdot U_{ny} \end{array} \right\} \tag{5.30}$$

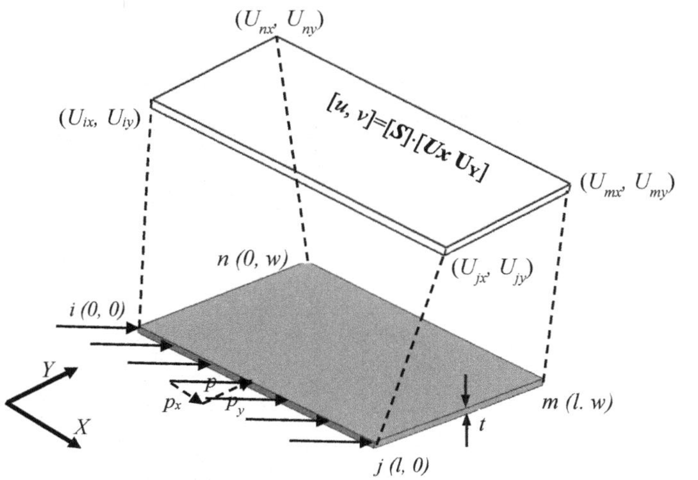

■ **FIGURE 5.14** Example of equivalent loads in rectangle element.

Substituting Eq. (5.30) into (5.25) yields,

$$W^{(e)} = \int_{L_{ij}} \begin{pmatrix} (S_i \cdot U_{ix} + S_j \cdot U_{jx} + S_m \cdot U_{mx} + S_n \cdot U_{nx}) \cdot p_x \\ + (S_i \cdot U_{iy} + S_j \cdot U_{jy} + S_m \cdot U_{my} + S_n \cdot U_{ny}) \cdot p_y \end{pmatrix} (t \cdot dl) \quad (5.31)$$

Based on the minimal potential energy principle, the vector of equivalent loads can be found by taking the derivative of $W^{(e)}$ with respect to $[U]$ as,

$$\left\{ F_{i-j}^{(e)} \right\} = \frac{\partial W^{(e)}}{\partial U} = \begin{Bmatrix} \dfrac{\partial W^{(e)}}{\partial U_{ix}} \\ \dfrac{\partial W^{(e)}}{\partial U_{iy}} \\ \dfrac{\partial W^{(e)}}{\partial U_{jx}} \\ \dfrac{\partial W^{(e)}}{\partial U_{jy}} \\ \dfrac{\partial W^{(e)}}{\partial U_{mx}} \\ \dfrac{\partial W^{(e)}}{\partial U_{my}} \\ \dfrac{\partial W^{(e)}}{\partial U_{nx}} \\ \dfrac{\partial W^{(e)}}{\partial U_{ny}} \end{Bmatrix} = \begin{Bmatrix} p_x \cdot \int_{L_{ij}} S_i(t \cdot dl) \\ p_y \cdot \int_{L_{ij}} S_i(t \cdot dl) \\ p_x \cdot \int_{L_{ij}} S_j(t \cdot dl) \\ p_y \cdot \int_{L_{ij}} S_j(t \cdot dl) \\ p_x \cdot \int_{L_{ij}} S_m(t \cdot dl) \\ p_y \cdot \int_{L_{ij}} S_m(t \cdot dl) \\ p_x \cdot \int_{L_{ij}} S_n(t \cdot dl) \\ p_y \cdot \int_{L_{ij}} S_n(t \cdot dl) \end{Bmatrix} = \frac{t \cdot l}{2} \begin{Bmatrix} p_x \\ p_y \\ p_x \\ p_y \\ 0 \\ 0 \\ 0 \\ 0 \end{Bmatrix}$$

$$(5.32)$$

where $F_{i-j}^{(e)}$ is the vector of equivalent loads on nodes when the pressure is on edge i—j. Similarly, when there is an external load on edge j—m, m—n, or n—i, the corresponding vectors $F_{j-m}^{(e)}$, $F_{m-n}^{(e)}$, and $F_{k-i}^{(e)}$ of equivalent loads are,

$$\left\{F_{j-m}^{(e)}\right\} = \frac{t \cdot w}{2}\begin{Bmatrix} 0 \\ 0 \\ p_x \\ p_y \\ p_x \\ p_y \\ 0 \\ 0 \end{Bmatrix}, \left\{F_{m-n}^{(e)}\right\} = \frac{t \cdot l}{2}\begin{Bmatrix} 0 \\ 0 \\ 0 \\ 0 \\ p_x \\ p_y \\ p_x \\ p_y \end{Bmatrix}, \text{ and } \left\{F_{n-i}^{(e)}\right\} = \frac{t \cdot l}{2}\begin{Bmatrix} p_x \\ p_y \\ 0 \\ 0 \\ 0 \\ 0 \\ p_x \\ p_y \end{Bmatrix}$$

$$(5.33)$$

5.5 ASSEMBLING SYSTEM MODELS

A system model is generally represented by a set of linear equations. The size of a system of linear equations depends on the number of nodes as well as the DOF of nodes. In addition, the complexity of a system model depends on a number of factors including (1) the size of linear equations set, (2) the time-dependence, and (3) the connections of elements. In this section, the general procedure for the development of a system model is introduced, and a number of system models from simple element types are used as the examples for the illustration purpose.

5.5.1 System model from 1D elements

Fig. 5.15A shows an example of a tapered cylinder, which is subjected to an axial load at one end and the fixed displacement at the other end. Fig. 5.15B shows a decomposed model with four axial members. We look into the procedure of creating a system model from the submodels of elements.

Fig. 5.16 gives the representation of an FEA model for the tapered cylinder where the nodes and elements in the discretized model are listed. As we discussed in Section 4.4, an axial member can be handled as an equivalent spring element, and the model for a spring system can be used for the tapered cylinder, in which the equivalent stiffness coefficient of an axial member is EA/L (see Table 4.2).

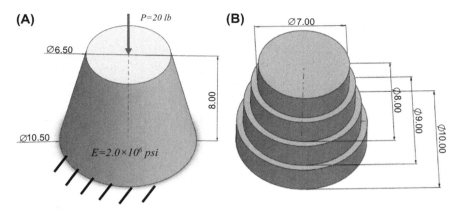

■ FIGURE 5.15 System model with axial members (A) object under compression (B) discretized model with four element.

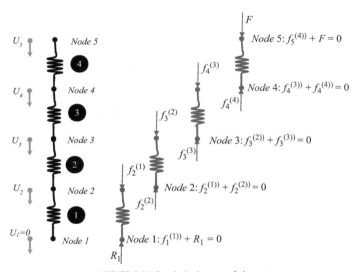

■ FIGURE 5.16 Free body diagrams of elements.

The correspondences of elements and nodes, as well as system parameters and nodal forces (internal and external) of elements, are all given in Table 5.5.

Based on free body diagrams of element in Fig. 5.16, the equations for the force equilibrium of four elements are written as,

Element **1**

$$k^{(1)}\left(U_1 - U_2\right) = R_1$$
$$k^{(1)}\left(U_2 - U_1\right) = f_2^{(1)}$$

(5.34)

Element **2**

$$k^{(2)}\left(U_2 - U_3\right) = f_2^{(2)}$$
$$k^{(2)}\left(U_3 - U_2\right) = f_3^{(2)}$$

(5.35)

Element **3**

$$k^{(3)}\left(U_3 - U_4\right) = f_3^{(3)}$$
$$k^{(3)}\left(U_4 - U_3\right) = f_4^{(3)}$$

(5.36)

Element **4**

$$k^{(4)}\left(U_4 - U_5\right) = f_4^4$$
$$k^{(4)}\left(U_5 - U_4\right) = F$$

(5.37)

Adding Eqs. (5.34)–(5.37) by merging two equations on the same node gets the system model as,

$$
\begin{aligned}
k^{(1)}U_1 \quad\quad -k^{(1)}U_2 &&&& &= -R_1 \\
-k^{(1)}U_1 \quad +\left(k^{(1)}+k^{(2)}\right)U_2 \quad -k^{(2)}U_3 &&&& &= f_2^{(1)}+f_2^{(2)} = 0 \\
-k^{(2)}U_2 \quad +\left(k^{(2)}+k^{(3)}\right)U_3 \quad -k^{(3)}U_4 &&& &= f_2^{(1)}+f_2^{(2)} = 0 \\
-k^{(3)}U_3 \quad +\left(k^{(3)}+k^{(4)}\right)U_4 \quad -k^{(4)}U_5 &&&= f_2^{(1)}+f_2^{(2)} = 0 \\
-k^{(4)}U_4 \quad +k^{(4)}U_5 &&&= F
\end{aligned}
$$

(5.38)

Table 5.5 System Parameters and Load Conditions of Elements

Element	Nodes i, j	Diameter (in)	Equivalent A_e (in^2) $\frac{\pi}{8}\left(D_i^2 + D_j^2\right)$	Equivalent k (lb/in) $\frac{EA_{e,i}}{l_i}$	Loads (lb)
1	1	10.5	78.73	7.9e7	$-R_1$
	2	9.5			$f_2^{(1)}$
2	2	9.5	63.81	6.4e7	$f_2^{(2)}$
	3	8.5			$f_3^{(2)}$
3	3	8.5	50.46	5.0e7	$f_3^{(3)}$
	4	7.5			$f_4^{(3)}$
4	4	7.5	38.68	3.9e7	$f_4^{(4)}$
	5	6.5			$F = 20$ lb

Eq. (5.38) can be rewritten as the matrix form as,

$$
\begin{bmatrix}
k^{(1)} & -k^{(1)} & 0 & 0 & 0 \\
-k^{(1)} & k^{(1)}+k^{(2)} & -k^{(2)} & 0 & 0 \\
0 & -k^{(2)} & k^{(2)}+k^{(3)} & -k^{(3)} & 0 \\
0 & 0 & -k^{(3)} & k^{(3)}+k^{(4)} & -k^{(4)} \\
0 & 0 & 0 & -k^{(4)} & k^{(4)}
\end{bmatrix}
\begin{Bmatrix}
U_1 \\ U_2 \\ U_3 \\ U_4 \\ U_5
\end{Bmatrix}
=
\begin{Bmatrix}
R_f \\ 0 \\ 0 \\ 0 \\ F
\end{Bmatrix}
\qquad (5.39)
$$

$U_1{=}0$

Furthermore, because the base of cylinder is fixed, we have $U_1 = 0$ as the BC. This BC must be applied in the system model to clean up the vector of unknowns. It can be done by removing the first column and row in Eq. (5.39). The system model after applying the BC becomes a set of four linear equations about four unknowns (U_2, U_3, U_4, U_5).

$$
\begin{bmatrix}
k^{(1)}+k^{(2)} & -k^{(2)}U_3 & 0 & 0 \\
-k^{(2)}U_3 & k^{(2)}+k^{(3)} & -k^{(3)}U_3 & 0 \\
0 & -k^{(3)}U_3 & k^{(3)}+k^{(4)} & k^{(4)} \\
0 & 0 & -k^{(4)} & k^{(4)}
\end{bmatrix}
\begin{Bmatrix}
U_2 \\ U_3 \\ U_4 \\ U_5
\end{Bmatrix}
=
\begin{Bmatrix}
0 \\ 0 \\ 0 \\ F
\end{Bmatrix}
\qquad (5.40)
$$

5.5.2 **2D element**

The system model for the modal analysis of a square vibrating membrane is used as an example for the development of system model for a 2D object.

Example 5.7

Fig. 5.17 shows a square vibrating membrane with the size of L by L. It has fixed boundary edges, and its behaviors are governed by the Helmholtz equation introduced in Chapter 4.8 as,

$$
\left.
\begin{aligned}
\frac{\partial^2 \phi}{\partial x^2} &+ \frac{\partial^2 \phi}{\partial y^2} + \lambda\phi = 0 \\
&\text{subjected to :} \\
\phi &= 0 \text{ on } \Gamma_1
\end{aligned}
\right\}
\qquad (5.41)
$$

where ϕ is the function of displacement in the transverse direction at any position.

Use the decomposition in Fig. 5.17 to determine the system model for the eigenvalues. Eigenvalues correspond to the natural frequencies of the vibrating membrane.

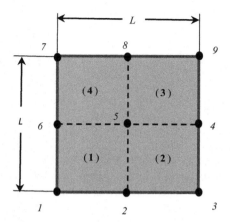

■ FIGURE 5.17 Decomposition of vibrating membrane.

Solution

The procedure with the following steps is performed to develop a system model for the modal analysis of the vibrating membrane.

Step 1

Decompose the continuous domain into elements and nodes. The result of decomposition is presented in Table 5.6 where the correspondences of elements and nodes are listed.

Step 2

Develop element models. The element solution to the Helmholtz equation has been developed in Section 4.8 of Chapter 4, and it is rewritten as,

$$\{([K_x] + [K_y]) - \lambda[M]\} \cdot \{\varphi\} = 0$$

Table 5.6 Elements and Nodes for Vibrating Membrane

Element	i	j	m	n	l	w
(1)	1	2	5	6	L/2	L/2
(2)	2	3	4	5		
(3)	5	4	9	8		
(4)	6	5	8	7		

where.

$$[K_x] = \frac{w}{6l}\begin{bmatrix} 2 & -2 & -1 & 1 \\ -2 & 2 & 1 & -1 \\ -1 & 1 & 2 & -2 \\ 1 & -1 & -2 & 2 \end{bmatrix}, \; [K_y]$$

$$= \frac{l}{6w}\begin{bmatrix} 2 & 1 & -1 & -2 \\ 1 & 2 & -2 & -1 \\ -1 & -2 & 2 & 1 \\ -2 & -1 & 1 & 2 \end{bmatrix}, \; [M] = \frac{(lw)}{18}\begin{bmatrix} 2 & 1 & 1 & 2 \\ 1 & 2 & 2 & 1 \\ 1 & 2 & 2 & 1 \\ 2 & 1 & 1 & 2 \end{bmatrix}$$

and $[\varphi]^T = \begin{bmatrix} \phi_i & \phi_j & \phi_m & \phi_n \end{bmatrix}$ are nodal values of displacement.. Note that $l = w = L/2$ for each element in the decomposition in Fig. 5.17. The element models can be established to define the corresponding matrices $[K]^{(e)}$ and $[M]^{(e)}$:

$$[K]^{(1)} = \begin{bmatrix} 2/3 & -1/6 & -1/3 & -1/6 \\ -1/6 & 2/3 & -1/6 & -1/3 \\ -1/3 & -1/6 & 2/3 & -1/6 \\ -1/6 & -1/3 & -1/6 & 2/3 \end{bmatrix} \begin{matrix} 1 \\ 2 \\ 5 \\ 6 \end{matrix}$$

$$[M]^{(1)} = \frac{L^2}{72}\begin{bmatrix} 2 & 1 & 1 & 2 \\ 1 & 2 & 2 & 1 \\ 1 & 2 & 2 & 1 \\ 2 & 1 & 1 & 2 \end{bmatrix} \begin{matrix} 1 \\ 2 \\ 5 \\ 6 \end{matrix}$$

$$[K]^{(2)} = \begin{bmatrix} 2/3 & -1/6 & -1/3 & -1/6 \\ -1/6 & 2/3 & -1/6 & -1/3 \\ -1/3 & -1/6 & 2/3 & -1/6 \\ -1/6 & -1/3 & -1/6 & 2/3 \end{bmatrix} \begin{matrix} 2 \\ 3 \\ 4 \\ 5 \end{matrix}$$

$$[M]^{(2)} = \frac{L^2}{72}\begin{bmatrix} 2 & 1 & 1 & 2 \\ 1 & 2 & 2 & 1 \\ 1 & 2 & 2 & 1 \\ 2 & 1 & 1 & 2 \end{bmatrix} \begin{matrix} 2 \\ 3 \\ 4 \\ 5 \end{matrix}$$

$$[K]^{(3)} = \begin{bmatrix} 2/3 & -1/6 & -1/3 & -1/6 \\ -1/6 & 2/3 & -1/6 & -1/3 \\ -1/3 & -1/6 & 2/3 & -1/6 \\ -1/6 & -1/3 & -1/6 & 2/3 \end{bmatrix} \begin{matrix} 5 \\ 4 \\ 9 \\ 8 \end{matrix}$$

$$[M]^{(3)} = \frac{L^2}{72}\begin{bmatrix} 2 & 1 & 1 & 2 \\ 1 & 2 & 2 & 1 \\ 1 & 2 & 2 & 1 \\ 2 & 1 & 1 & 2 \end{bmatrix} \begin{matrix} 5 \\ 4 \\ 9 \\ 8 \end{matrix}$$

$$[K]^{(4)} = \begin{bmatrix} 2/3 & -1/6 & -1/3 & -1/6 \\ -1/6 & 2/3 & -1/6 & -1/3 \\ -1/3 & -1/6 & 2/3 & -1/6 \\ -1/6 & -1/3 & -1/6 & 2/3 \end{bmatrix} \begin{matrix} 6 \\ 5 \\ 8 \\ 7 \end{matrix}$$

$$[M]^{(4)} = \frac{L^2}{72}\begin{bmatrix} 2 & 1 & 1 & 2 \\ 1 & 2 & 2 & 1 \\ 1 & 2 & 2 & 1 \\ 2 & 1 & 1 & 2 \end{bmatrix} \begin{matrix} 6 \\ 5 \\ 8 \\ 7 \end{matrix}$$

Step 3

Assembly element submodels into a system model. The system model is a collection of submodels of elements. A system model describes the relations of system unknowns and given boundary conditions of displacements and loads. For the modal analysis, the relations are represented by two metrics $[K]^{(G)}$ and $[M]^{(G)}$ in GCS.

To determine the system model, one has to figure out the sizes of global metrics at first. The size of a global matrix relates to the number of nodes as well as the DOF of state variable(s). In this example, the continuous domain is decomposed into four elements with nine nodes, and each node has 1 degree of freedom for its transverse displacement. Therefore, the size of a global matrix is 9. To assemble element models into a system model, one has to move the entities of $[K]^{(e)}$ or $[M]^{(e)}$ into the positions with the same indices of the nodes in the column and row.

$$[K]^{(G)} =$$

	1	2	3	4	5	6	7	8	9	
	2/3	-1/6	0	0	1/3	-1/6	0	0	0	1
	-1/6	4/3	-1/6	-1/3	-1/2	-1/3	0	0	0	2
	0	-1/6	2/3	-1/6	-1/6	0	0	0	0	3
	0	-1/3	-1/6	4/3	-1/3	0	0	-1/3	-1/6	4
	-1/3	-1/3	-1/3	-1/3	8/3	-1/3	-1/3	-1/3	-1/3	5
	-1/6	-1/3	0	0	-1/3	4/3	-1/6	-1/3	0	6
	0	0	0	0	-1/3	-1/6	2/3	-1/6	0	7
	0	0	0	-1/3	-1/3	-1/3	-1/6	4/3	-1/6	8
	0	0	0	-1/6	-1/3	0	0	-1/6	2/3	9

$$[M]^{(G)} = \frac{L^2}{72}$$

	1	2	3	4	5	6	7	8	9	
	2	1	0	0	3	2	0	0	0	1
	1	4	1	3	4	1	0	0	0	2
	0	1	2	2	1	0	0	0	0	3
	0	1	2	4	2	0	0	1	2	4
	1	2	1	2	8	3	1	4	1	5
	2	1	0	0	2	4	2	1	0	6
	0	0	0	0	1	2	2	1	0	7
	0	0	0	1	4	1	1	4	1	8
	0	0	0	2	1	0	0	1	2	9

Step 4

Apply BCs to simplify the system model. In this example, four edges are fixed, and all of nodes except node 5 are on edges. After remove the displacements from the system model, both of $[K]^{(G)}$ and $[M]^{(G)}$ includes one element only, i.e.,

$$\overline{K}^{(G)} = \left[\frac{8}{3}\right], \quad \overline{M}^{(G)} = \frac{L^2}{72} \cdot [8] = \left[\frac{L^2}{9}\right]$$

The first natural frequency of the vibration membrane can be estimated as,

$$\omega = \sqrt{\lambda} = \sqrt{\frac{8/3}{L^2/9}} = \frac{2\sqrt{6}}{L}$$

5.6 SUMMARY

In this paper, developing a system model from element models is considered, and some critical tasks in system modeling are discussed. These tasks are associated with the activities in the decomposition, the decomposed complexity in an original design problem must be represented appropriately in the system model: (1) the guides for the selection of analysis types are given based on physical behaviors and characteristics of continuous domains; (2) the methods to formulate boundary conditions are introduced; (3) equivalent nodal loads are defined based on the principle that physical behaviors of the continuous domain are not affected by replacement; (4) the procedure of assembling a system model from element models are illustrated via examples.

REFERENCE

Suh, N.P., 2005. Complexity: Theory and Applications. Oxford University Press, New York. ISBN-13: 978-0195178760.

FURTHER READING

Ansys Inc., 2013. ANSYS Mechanical APDL Coupled-field Analysis Guide. http://148.204.81.206/Ansys/150/ANSYS%20Mechanical%20APDL%20Coupled-Field%20Analysis%20Guide.pdf.

■ PROBLEMS

5.1. As shown in Fig. 5.18, an axial member has a uniformly distributed load q (lb/in): (1) define the interpolation for the displacement $U(X)$ at any position based on given U_i and U_j; (2) use the interpolation formula to estimate the work done by q on the axial member; (3) find the equivalent nodal loads on node i and j.

5.2. Explain the conditions in which the mirror symmetry can be applied to reduce the size of finite element model for an engineering problem. Prepare your answers with the consideration of geometry,

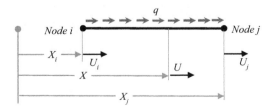

■ FIGURE 5.18 Axial member under shear load.

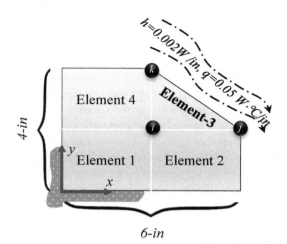

■ **FIGURE 5.19** Rectangle element with given boundary conditions.

materials, boundary conditions, and loads. Give an example of such an engineering problem.

5.3. A 2D solid domain in Fig. 5.19 is under an equilibrium state of heat transfer. It is divided into three rectangle elements and one triangle element. Element 3 involves the heat convection on the inclined edge, and the nodes of the element are specified in the figure. Defines the boundary conditions of the element.

5.4. Fig. 5.20 shows the loading conditions on a cantilever beam. Determine the vectors of equivalent loads on nodes if two beam members are used to represent the cantilever.

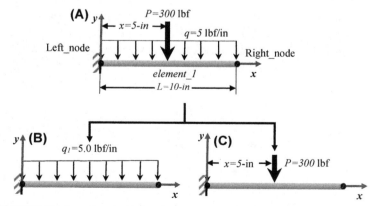

■ **FIGURE 5.20** Beam element with given loads (A) combined loads, (B) evenly distributed load, (C) ramp load.

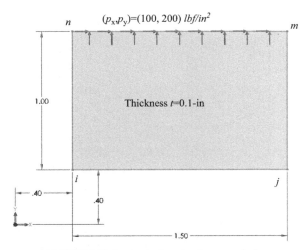

■ FIGURE 5.21 A rectangle element under pressure loads.

5.5. A 0.1-in thick rectangle element has the pressure loads on edge $j-m$ and $m-n$ as shown in Fig. 5.21, determine the vector of the equivalent nodal loads for this element.

5.6. Consider to use symmetric plane to develop a system model of the truss structure in Fig. 5.22. Define the load and displacement conditions for this half model; show your results by defining (1) a vector

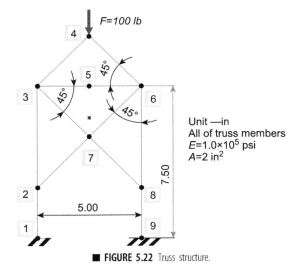

■ FIGURE 5.22 Truss structure.

(A)   **(B)**

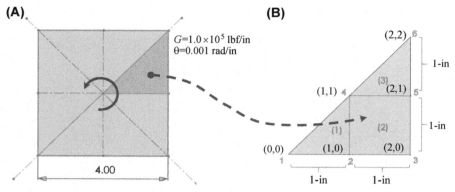

$G=1.0\times10^5$ lbf/in
$\theta=0.001$ rad/in

■ **FIGURE 5.23** Torsional member: (A) a square shaft with torsional load, (B) discretized elements and nodes.

of state variables and (2) a vector of external loads on relevant nodes.

5.7. As shown in Fig. 5.23A, a shaft with a square cross-section is subjected to a torsional load (twist angle θ for a unit length is 0.001 rad/in). The governing equation for a twist number is

$$\frac{\partial^2 \phi}{\partial x^2} + \frac{\partial^2 \phi}{\partial y^2} + 2G\theta = 0$$

Due to the symmetric geometry, one-eighth of the cross-section is used to model the part. The decomposition is made in Fig. 5.23B where the FEA consists of one rectangle element, two triangle elements, and six nodes. (1) Develop the system model, and (2) apply BCs on system model.

Solutions to System Models

6.1 INTRODUCTION

Following the procedure of finite element analysis (FEA), a structure is divided into elements and nodes; the discretized nodes are used to represent the behaviors of the continual domain, element models are formulated according to governing differential equations, element models are then assembled into a system model, and finally, boundary conditions and loads are incorporated in a system model. Once a system model is obtained, the next is to solve the system model by computer programs. Because FEA models in general are very complex that require a great deal of computation. Therefore, we are especially interested in how to reduce the computation in the solving process to find system solution efficiently. In addition, the system model is used to calculate primary unknowns. We will also discuss how to perform the postprocessing to calculate dependent variables in the system solution. Let us review the notation and properties of vectors, tensors, and matrices before the solutions to a system of linear equations are discussed.

6.2 TENSORS AND VECTORS, AND MATRICES

Due to a large amount of the information for design variables, system parameters, and their relations, it is very tedious and error prone to use scalars to represent mathematic models and solutions. Concise notations and expressions of these entities are very helpful in developing mathematic models and computer algorithms to solve FEA problems. *Tensors*, *vectors*, and *matrices* are widely adopted in the introduction of mathematic fundamentals of the FEA theory.

In mathematics, a *tensor* refers to an array of numbers or functions that transforms one set of numbers from another based on the specified rules. In physics, a tensor is usually used to characterize the properties of a physical

Finite Element Analysis Applications. http://dx.doi.org/10.1016/B978-0-12-809952-0.00006-6

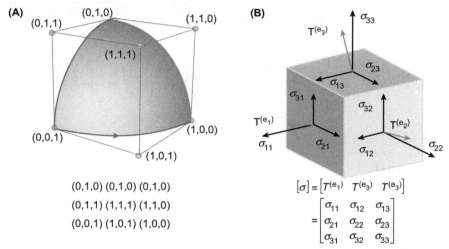

(A)

(0,1,0) (0,1,0) (0,1,0)

(0,1,1) (1,1,1) (1,1,0)

(0,0,1) (1,0,1) (1,0,0)

(B)

$$[\sigma] = \begin{bmatrix} T^{(e_1)} & T^{(e_3)} & T^{(e_3)} \end{bmatrix}$$

$$= \begin{bmatrix} \sigma_{11} & \sigma_{12} & \sigma_{13} \\ \sigma_{21} & \sigma_{22} & \sigma_{23} \\ \sigma_{31} & \sigma_{32} & \sigma_{33} \end{bmatrix}$$

■ **FIGURE 6.1** Two examples of tensors: (A) tensor for coordinates of vertices; (B) tensor for stress state.

system. Fig. 6.1A shows an example of tensor, which is used to represent the positions of a set of discrete points in space; Fig. 6.1B is a tensor to illustrate the stress state of the materials at a position of a solid object in solid mechanics.

A *scalar* corresponds to a tensor with a single number. In such a case, the single number is referred to as a tensor of order zero, or simply a scalar. A scalar is a tensor or an array of dimension zero (same as the order of the tensor). A *vector* is a tensor of the dimension one. Narrowly speaking, a matrix is a tensor of dimension two. However, in the computer implementation, a matrix can be in any dimensions and any sizes. Therefore, a tensor in any dimension can be represented by a matrix. The latter definition for matrices is adopted in this chapter.

A *matrix* is an array of numbers or mathematic functions arranged in rows along the horizontal direction and in columns along the vertical direction. The entities in the array are called the elements of matrix. From the chapter, a matrix is denoted as a *bolded capital letter* in square brackets [ ], for example [*K*]. If the number of the columns is reduced to "1", the corresponding matrix is reduced into a vector. It is denoted as a *bolded noncapital letter* in curly brackets { }, for example, {*u*}.

In the previous chapters, we have so many examples where matrices are used to represent physical terms. A few of examples of matrices expressed with given notations are shown as follows.

$$[I] = \begin{bmatrix} \int S_i^2\,dxdy & \int S_i S_j\,dxdy & \int S_i S_k\,dxdy \\ \int S_j S_i\,dxdy & \int S_j^2\,dxdy & \int S_k S_j\,dxdy \\ \int S_k S_i\,dxdy & \int S_k S_j\,dxdy & \int S_k^2\,dxdy \end{bmatrix} \text{ for an integration on matrix}$$

$$[T] = \begin{bmatrix} \cos\theta & -\sin\theta & X_i \\ \sin\theta & \cos\theta & Y_i \\ 0 & 0 & 1 \end{bmatrix} \text{ for the coordinate transformation}$$

$$\{p\} = \begin{Bmatrix} x \\ y \\ 1 \end{Bmatrix} \text{ for a position in a 2D space}$$

$$[S] = [\, S_i(x,y) \quad S_j(x,y) \quad S_k(x,y) \,] \text{ for a set of shape functions.}$$

A matrix should be specified by its size. Conventionally, the matrix size is defined by the number m of its rows and the number n of its columns, respectively. Mathematically, a matrix with the size of $m \times n$ with its elements can be written as,

$$[A] = \begin{bmatrix} a_{1,1} & a_{1,2} & \cdots & a_{1,j} & \cdots & a_{1,n} \\ a_{2,1} & a_{2,2} & \cdots & a_{2,j} & \cdots & a_{2,n} \\ \vdots & \vdots & \vdots & \vdots & \vdots & \vdots \\ a_{i,1} & a_{i,2} & \cdots & a_{i,j} & \cdots & a_{i,n} \\ \vdots & \vdots & \vdots & \vdots & \vdots & \vdots \\ a_{m,1} & a_{m,2} & \cdots & a_{m,j} & \cdots & a_{m,n} \end{bmatrix} \tag{6.1}$$

where $a_{i,j}$ represents the element of matrix at the i-th row and the j-th column.

The matrix $[A]$ also has its concise notation of $a_{i,j}$.

To make two matrices $[A]$ and $[B]$ be equal; these matrices have to be with the same size. Moreover, all of the elements at the corresponding positions must have the same value, i.e., $a_{i,j} = b_{i,j}$ for ($i = 1, 2,...m$; $j = 1, 2,...n$).

Example 6.1

Let $[A] = \begin{bmatrix} 1 & 2 \\ 0 & -5 \end{bmatrix}$ and $[B] = \begin{bmatrix} 6x+1 & 3-u \\ y-4 & v+7 \end{bmatrix}$

Find the values of $x, y, u,$ and v to satisfy the conditions of $[A] = [B]$.

Solution

Based on the condition of the equality of two matrices, all of the elements of two matrices at the corresponding positions must have the same value, i.e., $a_{i,j} = b_{i,j}$. Therefore, one can find the correspondences as,

$$\left.\begin{array}{llr}(1,1): & 1 = 6x + 1 \\ (1,2): & 2 = 3 - u \\ (2,1): & 0 = y - 4 \\ (2,2): & -5 = v + 7\end{array}\right\}$$

Solving the above set of linear equations leads to the solution of $x = 0$, $u = 1$, $y = 4$, and $v = -2$.

6.2.1 Matrix types

As shown in Fig. 6.2, matrices can be classified based on the characteristics of rows, columns, as well as elements.

Based on the characteristics of elements and the classification of matrices in Fig. 6.2, the examples of different types of matrices are provided as follows.

Null matrix:

$$[O] = \begin{bmatrix} 0 & 0 & 0 \\ 0 & 0 & 0 \\ 0 & 0 & 0 \end{bmatrix}$$

where the value of any element in the matrix is "0."

Identity matrix:

$$[I] = \begin{bmatrix} 1 & 0 & 0 \\ 0 & 1 & 0 \\ 0 & 0 & 1 \end{bmatrix}$$

where the value of any element in a diagonal position is "1" and the value of an element not on a diagonal position is "0."

Diagonal matrix:

$$[D] = \begin{bmatrix} d_{11} & 0 & 0 \\ 0 & d_{22} & 0 \\ 0 & 0 & d_{33} \end{bmatrix}$$

where the value of an element not on a diagonal position is "0".

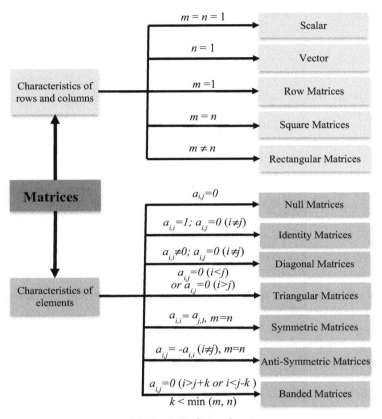

FIGURE 6.2 Classification of matrices.

Triangular matrix:

$$[L] = \begin{bmatrix} l_{11} & 0 & 0 \\ l_{21} & l_{22} & 0 \\ l_{31} & l_{32} & l_{33} \end{bmatrix} \text{ or } [U] = \begin{bmatrix} u_{11} & u_{12} & u_{13} \\ 0 & u_{22} & u_{23} \\ 0 & 0 & u_{33} \end{bmatrix}$$

where the elements with a nonzero value locate only at a lower or upper triangular area of the matrix.

Symmetric matrix:

$$[A] = \begin{bmatrix} a_{11} & 12 & 13 \\ 12 & a_{22} & 23 \\ 13 & 23 & a_{33} \end{bmatrix}$$

where the element values are symmetric about the diagonal line, i.e., $a_{i,j} = a_{j,i}$.

Antisymmetric matrix:

$$[A] = \begin{bmatrix} a_{11} & 12 & 13 \\ -12 & a_{22} & 23 \\ -13 & -23 & a_{33} \end{bmatrix}$$

where the element values are antisymmetric about the diagonal line, i.e., $a_{i,j} = -a_{j,i}$.

Banded matrix:

$$[K] = \begin{bmatrix} k_{1,1} & k_{1,2} & 0 & 0 & 0 & 0 \\ k_{2,1} & k_{2,2} & k_{2,3} & k_{2,4} & 0 & 0 \\ k_{3,1} & k_{3,2} & k_{3,3} & k_{3,4} & k_{3,5} & 0 \\ 0 & 0 & k_{4,3} & k_{4,4} & 0 & k_{4,6} \\ 0 & 0 & k_{5,3} & k_{5,4} & k_{5,5} & k_{5,6} \\ 0 & 0 & 0 & k_{6,4} & k_{6,5} & k_{6,6} \end{bmatrix}$$

where the elements with a nonzero value are located within a band around the diagonal line. In this example, $k = 2$ and the width of the band is 5.

Sometimes, it is convenient to divide a matrix into a number of submatrices with small sizes. The matrix to be divided is called as a partitioned matrix or supermatrix. An example of a partitioned matrix is,

$$[A] = \begin{bmatrix} a_{11} & a_{12} & a_{13} \\ a_{21} & a_{22} & a_{23} \\ \hline a_{31} & a_{32} & a_{33} \end{bmatrix} = \begin{bmatrix} [A_{11}] & [A_{12}] \\ [A_{21}] & [A_{22}] \end{bmatrix}$$

where the horizontal and vertical lines are dividing lines in the partitioning.

To make the above partition valid, the following conditions must be satisfied:

$$[A_{11}] = \begin{bmatrix} a_{11} & a_{12} \\ a_{21} & a_{22} \end{bmatrix}, \quad [A_{12}] = \begin{bmatrix} a_{13} \\ a_{23} \end{bmatrix}, \quad [A_{21}] = \begin{bmatrix} a_{31} & a_{32} \end{bmatrix}, \quad [A_{22}] = [a_{33}]$$

6.2.2 Matrix properties

A matrix has a number of properties, which can be utilized in the matrix algebra.

6.2.2.1 Transpose of matrix

The transpose of a matrix $[A]$ is denoted as $[A]^T$. The transpose matrix $[A]^T$ is obtained by interchanging the columns and rows of matrix $[A]$. If the size of $[A]$ is $m \times n$, then the size $[A]^T$ of is $n \times m$.

Example 6.2

Find the transport matrix for $[\boldsymbol{A}] = \begin{bmatrix} 1 & 3 & 2 \\ 4 & 6 & 5 \end{bmatrix}$

Solution

Applying the definition of transpose of matrix about $[\boldsymbol{A}]$ gets,

$$[\boldsymbol{A}]^T = \begin{bmatrix} 1 & 4 \\ 3 & 6 \\ 2 & 5 \end{bmatrix}$$

where the elements $a_{i,\,j}$ ($i = 1, 2$ and $j = 1, 2, 3$) of $[\boldsymbol{A}]$ are transposed to the elements $a_{j,\,i}$ ($j = 1, 2, 3$ and $i = 1, 2$) of $[\boldsymbol{A}]^T$. The transpose of a transpose of a matrix is equal to this matrix itself, that is,

$$\left(\left([\boldsymbol{A}] \right)^T \right)^T = [\boldsymbol{A}] \tag{6.2}$$

6.2.2.2 Trace of matrix

The trace of a matrix $[A]$ is denoted as $\mathrm{tr}[A]$ if the matrix $[A]$ is square. The trace is defined as the sum of the diagonal elements of matrix $[A]$.

Example 6.3

Find the trace of the matrix $[\boldsymbol{A}] = \begin{bmatrix} 5 & 1 & 1 \\ 3 & -4 & -2 \\ -2 & 1 & 6 \end{bmatrix}$.

Solution

Applying the definition of trace of matrix, the trace $\mathrm{tr}[\boldsymbol{A}]$ of $[\boldsymbol{A}]$ can be found by,

$$\mathrm{tr}[\boldsymbol{A}] = (5) + (-4) + (6) = 7.$$

6.2.2.3 Determinant of matrix

If a matrix $[A]$ is square, the determinant can be denoted as $\det[A]$. It is a scalar corresponding to a sum of the products:

$$\det[A] = (-1)^{i+j} a_{i,j} \det[\boldsymbol{M}_{i,j}] \tag{6.3}$$

where $a_{i,j}$ are the element at the i-th row and the j-th column, and

$[M_{i,j}]$ is derived from $[A]$ by eliminating the i-th row and j-th column of matrix $[A]$.

Let $[A]$ to be.

$$[A] = \begin{bmatrix} a_{1,1} & a_{1,2} & \cdots & a_{1,j} & \cdots & a_{1,n} \\ a_{2,1} & a_{2,2} & \cdots & a_{2,j} & \cdots & a_{2,n} \\ \vdots & \vdots & \vdots & \vdots & \vdots & \vdots \\ a_{i,1} & a_{i,2} & \cdots & a_{i,j} & \cdots & a_{i,n} \\ \vdots & \vdots & \vdots & \vdots & \vdots & \vdots \\ a_{n,1} & a_{n,2} & \cdots & a_{n,j} & \cdots & a_{n,n} \end{bmatrix}$$

If the i-th row is selected to calculate the determinant of $[A]$,

$$\det[A] = (-1)^{i+1}(a_{i,1})\det[M_{i,1}] + (-1)^{i+2}(a_{i,2})\det[M_{i,2}] + \cdots$$
$$+ (-1)^{i+n}(a_{i,n})\det[M_{i,n}] \tag{6.4}$$

Note that i can be any value from $1, 2,..., n$, and

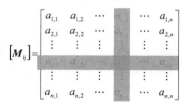

$$[M_{ij}] =$$

For a special case when the matrix $[A]$ has one element, $[A]$ becomes $[a_{11}]$, then

$$\det[A] = a_{11} \tag{6.5}$$

For the case when the size of matrix $[A]$ is 2×2, that is $[A] = \begin{bmatrix} a_{11} & a_{12} \\ a_{21} & a_{22} \end{bmatrix}$, then

$$\det[A] = a_{11}a_{22} - a_{12}a_{21} \tag{6.6}$$

For the case when the size of matrix $[A]$ is 3×3, that is $[A] = \begin{bmatrix} a_{11} & a_{12} & a_{13} \\ a_{21} & a_{22} & a_{23} \\ a_{31} & a_{32} & a_{33} \end{bmatrix}$, then

$$\det[A] = a_{11}(-1)^{1+1}\det[M_{1,1}] + a_{12}(-1)^{1+2}\det[M_{1,2}] + a_{13}(-1)^{1+3}\det[M_{1,3}]$$

$$= a_{11}\det\begin{bmatrix} a_{22} & a_{23} \\ a_{32} & a_{33} \end{bmatrix} - a_{12}\det\begin{bmatrix} a_{21} & a_{23} \\ a_{31} & a_{33} \end{bmatrix} + a_{13}\det\begin{bmatrix} a_{21} & a_{22} \\ a_{31} & a_{32} \end{bmatrix}$$

$$= a_{11}(a_{22}a_{33} - a_{32}a_{23}) - a_{12}(a_{21}a_{33} - a_{31}a_{23}) + a_{13}(a_{21}a_{32} - a_{31}a_{22}) \tag{6.7}$$

Example 6.4

Find the determinant of $[A] = \begin{bmatrix} 5 & 1 & 1 \\ 3 & -4 & -2 \\ -2 & 1 & 6 \end{bmatrix}$

Solution

Using Eq. (6.7) gives,

$$\det[A] = a_{11}(a_{22}a_{33} - a_{32}a_{23}) - a_{12}(a_{21}a_{33} - a_{31}a_{23}) + a_{13}(a_{21}a_{32} - a_{31}a_{22})$$
$$= (5)((-4)(6) - (1)(-2)) - (1)((3)(6) - (-2)(-2)) + (1)((3)(1) - (-2)(-4))$$
$$= -129$$

6.2.2.4 Differentiation of matrix

The differentiation $[\dot{A}]$ of a matrix is the matrix whose elements are the differentiations of corresponding elements of $[A]$. Note that the elements in matrix $[A]$ are the functions of t, the differentiation $[\dot{A}]$ of a matrix can be denoted by $\dfrac{d[A]}{dt}$ and its elements are $\dfrac{da_{i,j}}{dt}$ where $i = 1, 2,...m, j = 1, 2,...n$.

Example 6.5

Find the differentiation $[\dot{A}]$ with respect to x for $[A] = \begin{bmatrix} x^3 + 2 & e^x \\ 1 & x^2 - 2 \end{bmatrix}$.

Solution

Based on the definition of the differentiation of a matrix, one can find it taking a derivative for every element in the matrix as,

$$\frac{d[A]}{dt} = \begin{bmatrix} \dfrac{d(x^3 + 2)}{dx} & \dfrac{d(e^x)}{dx} \\ \dfrac{d(1)}{dx} & \dfrac{d(x^2 - 2)}{dx} \end{bmatrix} = \begin{bmatrix} 3x^2 & e^x \\ 0 & 2x \end{bmatrix}$$

6.2.2.5 Integration of matrix

The integration of a matrix $\int_a^b [A] dt$ is a matrix whose elements are the integrations of elements in matrix $[A]$. Note that the elements in the matrix $[A]$ are the functions of t, and the elements in the integration of a matrix $\int_a^b [A] dt$

are $\displaystyle\int_a^b a_{i,j} dt$ where $i = 1, 2,...m, j = 1, 2,...n$.

Example 6.6

Find the integration $\int_a^b [A] dt$ of a matrix $[A]$ over the range of $[0, 1]$ of x for

$$[A] = \begin{bmatrix} x^3 + 2 & 4 \\ x & -x^2 \end{bmatrix}.$$

Solution

Based on the definition of the integration $\int_a^b [A] dt$ of a matrix $[A]$, one can find the integration of a matrix by taking integral for every element in $[A]$ as,

$$\int_a^b [A] dx = \begin{bmatrix} \int_0^1 (x^3 + 2) dx & \int_0^1 (4) dx \\ \int_0^1 (x) dx & \int_0^1 (-x^2) dx \end{bmatrix} = \begin{bmatrix} 2\frac{1}{4} & 4 \\ \frac{1}{2} & -\frac{1}{3} \end{bmatrix}$$

6.2.3 Matrix operations

Similar to scalars, different operations or functions may be applied to matrices. However, some specific conditions must be satisfied for certain types of operations.

6.2.3.1 Addition or subtraction

Addition and subtraction of two or more matrices can be performed if two or more matrices with the same size, i.e., all of the matrices in an addition or subtraction must have the same numbers of rows or columns. Let $[C]$ be the result of the addition or subtraction of two matrices $[A]$ and $[B]$, addition and subtraction can be represented by,

$$[C] = [A] \pm [B] \quad or \quad c_{i,j} = a_{i,j} \pm b_{i,j} \tag{6.8}$$

where $[A]$, $[B]$, and $[C]$ have the same size of $m \times n$, and $a_{i,j}$, $b_{i,j}$, and $c_{i,j}$ are the elements of $[A]$, $[B]$, and $[C]$, respectively.

Example 6.7

Find the results of addition and subtraction for $[A] = \begin{bmatrix} 6 & 3 \\ 1 & 9 \end{bmatrix}$ and $[B] = \begin{bmatrix} 2 & 3 \\ 0 & 4 \end{bmatrix}.$

Solution

The addition and subtraction of [**A**] and [**B**] can be obtained, respectively, as,

$$[\boldsymbol{C}] = [\boldsymbol{A}] + [\boldsymbol{B}] = \begin{bmatrix} 6 & 3 \\ 1 & 9 \end{bmatrix} + \begin{bmatrix} 2 & 3 \\ 0 & 4 \end{bmatrix} = \begin{bmatrix} 8 & 6 \\ 1 & 13 \end{bmatrix}$$

and

$$[\boldsymbol{C}] = [\boldsymbol{A}] - [\boldsymbol{B}] = \begin{bmatrix} 6 & 3 \\ 1 & 9 \end{bmatrix} - \begin{bmatrix} 2 & 3 \\ 0 & 4 \end{bmatrix} = \begin{bmatrix} 4 & 0 \\ 1 & 5 \end{bmatrix}$$

Addition/subtraction of matrices are *associative*. Assume three matrices [**A**], [**B**], and [**C**] are added/subtracted sequentially; the result of compounded operations is not affected by the order of the operations to be performed, that is,

$$[\boldsymbol{A}] \pm [\boldsymbol{B}] \pm [\boldsymbol{C}] = ([\boldsymbol{A}] \pm [\boldsymbol{B}]) \pm [\boldsymbol{C}] = [\boldsymbol{A}] \pm ([\boldsymbol{B}] \pm [\boldsymbol{C}]) \qquad (6.9)$$

In Eq. (6.9), the addition/subtraction can be performed either on [**A**] and [**B**] or [**B**] and [**C**]. It does not affect the final result.

Example 6.8

Find the result of [**A**] + [**B**] + [**C**] when

$$[\boldsymbol{A}] = \begin{bmatrix} 2 & 3 \\ 1 & 0 \end{bmatrix}, \ [\boldsymbol{B}] = \begin{bmatrix} 1 & 0 \\ 1 & 6 \end{bmatrix} \text{ and } [\boldsymbol{C}] = \begin{bmatrix} 1 & 2 \\ 5 & 2 \end{bmatrix},$$

Solution

Using Eq. (6.9) yields,

$$[\boldsymbol{A}] + [\boldsymbol{B}] + [\boldsymbol{C}] = ([\boldsymbol{A}] + [\boldsymbol{B}]) + [\boldsymbol{C}] = \begin{bmatrix} 2+1 & 3+0 \\ 1+1 & 0+6 \end{bmatrix} + \begin{bmatrix} 1 & 2 \\ 5 & 2 \end{bmatrix}$$

$$= [\boldsymbol{A}] + ([\boldsymbol{B}] + [\boldsymbol{C}]) = \begin{bmatrix} 2 & 3 \\ 1 & 0 \end{bmatrix} + \begin{bmatrix} 1+1 & 0+2 \\ 1+5 & 6+2 \end{bmatrix}$$

$$= \begin{bmatrix} 4 & 5 \\ 7 & 8 \end{bmatrix}$$

Addition/subtraction of two matrices are also commutative, *i.e.,*

$$\left. \begin{array}{l} [\boldsymbol{A}] + [\boldsymbol{B}] = [\boldsymbol{B}] + [\boldsymbol{A}] \\ [\boldsymbol{A}] - [\boldsymbol{B}] = -[\boldsymbol{B}] + [\boldsymbol{A}] \end{array} \right\} \qquad (6.10)$$

Example 6.9

Verify the commutative properties for the subtraction of [*A*] and [*B*] when

$$[A] = \begin{bmatrix} 5 & 3 \\ 1 & 3 \end{bmatrix}, \; [B] = \begin{bmatrix} 2 & 5 \\ 7 & 0 \end{bmatrix}$$

Solution

Calculate the subtraction of [*A*] − [*B*] and the addition of −[*B*] + [*A*],

$$[A] - [B] = \begin{bmatrix} 5 & 3 \\ 1 & 3 \end{bmatrix} - \begin{bmatrix} 2 & 5 \\ 7 & 0 \end{bmatrix} = \begin{bmatrix} 3 & -2 \\ -6 & 3 \end{bmatrix}$$

$$-[B] + [A] = \begin{bmatrix} -2 & -5 \\ -7 & 0 \end{bmatrix} + \begin{bmatrix} 5 & 3 \\ 1 & 3 \end{bmatrix} = \begin{bmatrix} 3 & -2 \\ -6 & 3 \end{bmatrix}$$

Both of two operations reach the same result.

6.2.3.2 Transpose of combined matrices

Applying a transpose on a couple of matrices with an addition/subtraction gives,

$$([A] + [B])^T = [A]^T + [B]^T \tag{6.11}$$

The properties of transpose of matrix are very useful in FEA modeling; it is extremely useful in assembling element models into a system model. Because the sizes of the matrices in element models are different from that of a system model, and even two element models have different sizes. New matrices can be defined based on the information of elements about nodes, and the transpose of an element matrix is then used to add an element model into a system model.

Example 6.10

A structure consists of six nodes for two triangular elements and one rectangle element. Build a system model based on the given element models as follows.

$$[K]^{(1)} = \begin{array}{c} \\ \begin{array}{ccc} 1 & 2 & 4 \end{array} \\ \left[\begin{array}{ccc} 0.5 & -0.5 & 0 \\ -0.5 & 1 & -0.5 \\ 0 & -0.5 & 0.5 \end{array}\right] \begin{array}{c} 1 \\ 2 \\ 4 \end{array} \end{array} \quad [K]^{(2)} = \begin{array}{c} \\ \begin{array}{ccc} 4 & 5 & 6 \end{array} \\ \left[\begin{array}{ccc} 0.5 & -0.5 & 0 \\ -0.5 & 1 & -0.5 \\ 0 & -0.5 & 0.5 \end{array}\right] \begin{array}{c} 4 \\ 5 \\ 6 \end{array} \end{array} \quad [K]^{(3)} = \begin{array}{c} \\ \begin{array}{cccc} 2 & 3 & 5 & 4 \end{array} \\ \left[\begin{array}{cccc} 4 & -1 & -2 & -1 \\ -1 & 4 & -1 & -2 \\ -2 & -1 & 4 & -1 \\ -1 & -2 & -1 & 4 \end{array}\right] \begin{array}{c} 2 \\ 3 \\ 5 \\ 4 \end{array} \end{array}$$

Solution

Element (1) has three nodes; an assistive matrix is defined for the correlation of nodes of element (1) to a system model as

$$[A]^{(1)} = \begin{bmatrix} 1 & 0 & 0 & 0 & 0 & 0 \\ 0 & 1 & 0 & 0 & 0 & 0 \\ 0 & 0 & 0 & 1 & 0 & 0 \end{bmatrix}$$

where the number of rows of $[A]^{(1)}$ is the number of nodes (3), and the number of columns is the number of nodes (6) in the system model. At the i-th row of $[A]^{(1)}$, the element at the column with the index of the i-th node of element is set as "1". Using $[A]^{(1)}$ and its transpose, the model of element (1) can be assembled into the system model as,

$$[K]^{(G,1)} = \left([A]^{(1)}\right)^T [K]^{(1)}[A]^{(1)} = \begin{bmatrix} 0.5 & -0.5 & 0 & 0 & 0 & 0 \\ -0.5 & 1 & 0 & -0.5 & 0 & 0 \\ 0 & 0 & 0 & 0 & 0 & 0 \\ 0 & -0.5 & 0 & 0.5 & 0 & 0 \\ 0 & 0 & 0 & 0 & 0 & 0 \\ 0 & 0 & 0 & 0 & 0 & 0 \end{bmatrix}$$

Element (2) has three nodes; an assistive matrix is defined as

$$[A]^{(2)} = \begin{bmatrix} 0 & 0 & 0 & 1 & 0 & 0 \\ 0 & 0 & 0 & 0 & 1 & 0 \\ 0 & 0 & 0 & 0 & 0 & 1 \end{bmatrix}$$

Using $[A]^{(2)}$ and its transpose, the model of element (2) can be assembled into the system model as,

$$[K]^{(G,2)} = \left([A]^{(2)}\right)^T [K]^{(2)}[A]^{(2)} = \begin{bmatrix} 0 & 0 & 0 & 0 & 0 & 0 \\ 0 & 0 & 0 & 0 & 0 & 0 \\ 0 & 0 & 0 & 0 & 0 & 0 \\ 0 & 0 & 0 & 0.5 & -0.5 & 0 \\ 0 & 0 & 0 & -0.5 & 1 & -0.5 \\ 0 & 0 & 0 & 0 & -0.5 & 0.5 \end{bmatrix}$$

Element (3) has four nodes; the assistive matrix is defined for the correlation of nodes of element (3) to a system model as

$$[A]^{(3)} = \begin{bmatrix} 0 & 1 & 0 & 0 & 0 & 0 \\ 0 & 0 & 1 & 0 & 0 & 0 \\ 0 & 0 & 0 & 0 & 1 & 0 \\ 0 & 0 & 0 & 1 & 0 & 0 \end{bmatrix}$$

where the number of rows of $[A]^{(3)}$ is the number of nodes (4), and the number of columns is the number of nodes (6) in the system model. At the i-th row of $[A]^{(3)}$, the element at the column with the index of the i-th node of element is set as "1". Using $[A]^{(3)}$ and its transpose, the model of element (3) can be assembled into the system model as,

$$[K]^{(G,3)} = \left([A]^{(3)} \right)^T [K]^{(3)} [A]^{(3)} = \begin{bmatrix} 0 & 0 & 0 & 0 & 0 & 0 \\ 0 & 4 & -1 & -1 & -2 & 0 \\ 0 & -1 & 4 & -2 & -1 & 0 \\ 0 & -1 & -2 & 4 & -1 & 0 \\ 0 & -2 & -1 & -1 & 4 & 0 \\ 0 & 0 & 0 & 0 & 0 & 0 \end{bmatrix}$$

Finally, the system model can be generated by adding all of components from three elements as,

$$[K]^{(G)} = [K]^{(G,1)} + [K]^{(G,2)} + [K]^{(G,3)} = \begin{bmatrix} 0.5 & -0.5 & 0 & 0 & 0 & 0 \\ -0.5 & 5 & -1 & -1.5 & -2 & 0 \\ 0 & -1 & 4 & -2 & -1 & 0 \\ 0 & -1.5 & -2 & 5 & -1.5 & 0 \\ 0 & -2 & -1 & -1.5 & 5 & -0.5 \\ 0 & 0 & 0 & 0 & -0.5 & 0.5 \end{bmatrix}$$

6.2.3.3 Multiplication by scalar

If a matrix $[A]$ is multiplied by a scalar s, the result of multiplication is the matrix $[B]$ with the same size with the scaled elements by s as,

$$[\mathbf{B}] = s \cdot [\mathbf{A}] \quad \text{or} \quad b_{i,j} = s \cdot a_{i,j} \tag{6.12}$$

Example 6.11

Find the new matrix $[B]$ by multiplying $s = 0.5$ with $[A] = \begin{bmatrix} 2 & 3 \\ 1 & -2 \end{bmatrix}$.

Solution

Applying Eq. (6.12) yields,

$$[B] = s \cdot [A] = \begin{bmatrix} 1 & 1.5 \\ 0.5 & -1 \end{bmatrix}$$

6.2.3.4 Transpose of a matrix with a product of scalar

If a transpose operation is applied on a matrix with a product of scalar, it has,

$$(s \cdot [A])^T = s \cdot [A]^T \tag{6.12}$$

6.2.3.5 Product of matrices

For given two matrices $[A]$ and $[B]$, a product of matrices exists if and only if the number of columns of $[A]$ equals to the number of rows of $[B]$. A product of two matrices $[C]$ can be defined as,

$$[C] = [A] \cdot [B] \quad \text{or} \quad c_{i,j} = \sum_{k=1}^{k=n} a_{i,k} \cdot b_{k,j} \tag{6.13}$$

where the sizes of $[A]$ and $[B]$ are $m \times n$ and $n \times r$, respectively. The size of $[C]$ is determined by the number of rows of $[A]$ and the columns of $[B]$, that is, $m \times r$. The elements $c_{i,j}$ in $[C]$ are obtained by taking the dot product of the i-th row of $[A]$ and the j-th column of $[B]$.

To find an element at the i-th row and the j-th column of $[C]$, one has to single out the i-th row of $[A]$ and the j-th column of $[B]$, multiply the corresponding elements, and sum as the resulting products.

Example 6.12

Find the matrix $[C]$ for a product of $[A] = \begin{bmatrix} a_{11} & a_{12} \\ a_{21} & a_{22} \end{bmatrix}$ and $[B] = \begin{bmatrix} b_{11} & b_{12} & b_{13} \\ b_{21} & b_{22} & b_{23} \end{bmatrix}$.

Solution

Based on Eq. (6.13), a product of two matrices can be defined legally because the number of columns of $[A]$ equals to the number of rows of $[B]$. Furthermore, $[C]$ can be calculated as,

$$
\begin{aligned}
[C] &= [A] \cdot [B] \\
&= \begin{bmatrix} a_{11} & a_{12} \\ a_{21} & a_{22} \end{bmatrix} \cdot \begin{bmatrix} b_{11} & b_{12} & b_{13} \\ b_{21} & b_{22} & b_{23} \end{bmatrix} \\
&= \begin{bmatrix} a_{11}b_{11} + a_{12}b_{21} & a_{11}b_{12} + a_{12}b_{22} & a_{11}b_{13} + a_{12}b_{13} \\ a_{21}b_{11} + a_{22}b_{21} & a_{21}b_{12} + a_{22}b_{22} & a_{21}b_{13} + a_{22}b_{23} \end{bmatrix}
\end{aligned}
$$

Taking an example of an element $c_{2,2}$ in $[C]$, it is determined as the sum of a dot product of the second row of $[A]$ and the second column of $[B]$.

6.2.3.6 Transpose of a product of two matrices

A transpose operation on the product of matrices meets the following condition:

$$([A] \cdot [B])^T = [B]^T \cdot [A]^T \tag{6.14}$$

Example 6.13

Verify Eq. (6.14) for a transpose of a product of $[A]$ and $[B]$ when

$[A] = \begin{bmatrix} a_{11} & a_{12} \\ a_{21} & a_{22} \end{bmatrix}$ and $[B] = \begin{bmatrix} b_{11} & b_{12} & b_{13} \\ b_{21} & b_{22} & b_{23} \end{bmatrix}$.

Solution

The expressions on the left and right side of Eq. (6.14) are evaluated to see if the results are the same.

$$[A] \cdot [B] = \begin{bmatrix} a_{11}b_{11} + a_{12}b_{12} & a_{11}b_{12} + a_{12}b_{22} & a_{11}b_{13} + a_{12}b_{23} \\ a_{21}b_{11} + a_{22}b_{12} & a_{21}b_{12} + a_{22}b_{22} & a_{21}b_{13} + a_{22}b_{13} \end{bmatrix}$$

$$([A] \cdot [B])^T = \begin{bmatrix} a_{11}b_{11} + a_{12}b_{12} & a_{21}b_{11} + a_{22}b_{12} \\ a_{11}b_{12} + a_{12}b_{22} & a_{21}b_{12} + a_{22}b_{22} \\ a_{11}b_{13} + a_{12}b_{23} & a_{21}b_{13} + a_{22}b_{13} \end{bmatrix}$$

$$[A]^T = \begin{bmatrix} a_{11} & a_{21} \\ a_{12} & a_{22} \end{bmatrix}, [B]^T = \begin{bmatrix} b_{11} & b_{12} \\ b_{12} & b_{22} \\ b_{13} & b_{13} \end{bmatrix}$$

$$[B]^T \cdot [A]^T = \begin{bmatrix} b_{11} & b_{12} \\ b_{12} & b_{22} \\ b_{13} & b_{13} \end{bmatrix} \cdot \begin{bmatrix} a_{11} & a_{21} \\ a_{12} & a_{22} \end{bmatrix} = \begin{bmatrix} a_{11}b_{11} + a_{12}b_{12} & a_{21}b_{11} + a_{22}b_{12} \\ a_{11}b_{12} + a_{12}b_{22} & a_{21}b_{12} + a_{22}b_{22} \\ a_{11}b_{13} + a_{12}b_{23} & a_{21}b_{13} + a_{22}b_{13} \end{bmatrix}$$

Therefore, we prove the validity of Eq. (6.14) because the sizes and elements of two resulted matrices are all the same.

6.2.3.7 Partition of matrix

The manipulation on a matrix with a large size is computationally expansive. The computation can be reduced if the operation of the matrix occurs only to a part of the matrix. The part with the elements to be changed can be partitioned out. For example, a large-scale system model is usually a sparse matrix where the majority of elements are zeros and the elements in a narrow band in the diagonal direction are nonzero. The regions with all zero elements should be partitioned out to reduce the computation of matrix manipulation. That is, the required manipulation can be performed directly on submatrices. For two matrices $[A]$ and $[B]$, assume that they can be partitioned as,

$$[A] = \begin{bmatrix} [A_{11}] & [A_{12}] \\ [A_{21}] & [A_{22}] \end{bmatrix} \text{ and } [B] = \begin{bmatrix} [B_{11}] & [B_{12}] \\ [B_{21}] & [B_{22}] \end{bmatrix} \tag{6.15}$$

If the sizes of all submatrices of $[A]$ and $[B]$ are matched, the addition/subtraction of two matrices can be performed on submatrices as,

$$[A + B] = \begin{bmatrix} [A_{11} + B_{11}] & [A_{12} + B_{12}] \\ [A_{21} + B_{21}] & [B_{11} + B_{22}] \end{bmatrix} \tag{6.16}$$

And the matrix multiplication can be performed as,

$$[A \times B] = \begin{bmatrix} [A_{11} \times B_{11} + A_{12} \times B_{21}] & [A_{11} \times B_{12} + A_{12} \times B_{22}] \\ [A_{21} \times B_{11} + A_{22} \times B_{21}] & [A_{21} \times B_{12} + A_{22} \times B_{22}] \end{bmatrix} \tag{6.17}$$

6.3 SOLUTIONS TO A SYSTEM OF LINEAR EQUATIONS

For a static analysis, a system model is represented by a system of linear equations, which can be described as,

$$[A]\{x\} = \{b\} \tag{6.18}$$

where $\{b\}$ is the vector of loads, $\{x\}$ is the vector of state variables, and $[A]$ is the matrix for the relation of loads and state variables, such as the stiffness matrix in static analysis. Theoretically, the solution of $\{x\}$ can be found directly by,

$$\{x\} = [A]^{-1} \cdot \{b\} \tag{6.19}$$

where $[A]^{-1}$ is the inverse matrix of $[A]$.

However, obtaining $[A]^{-1}$ brings a great deal of computing challenges. An FEA model for a real-world problem often involves thousands and even several millions of unknown variables. Effective algorithms for solving system models become imperative.

Note that a system model can be either a linear or a nonlinear system model. In a *linear system model*, all of the elements in $[A]$ and $\{b\}$ are constant. In *a nonlinear system model*, the elements in $[A]$ or $\{b\}$ are the functions of state variables $\{x\}$. To solve a nonlinear system model, an iterative procedure is applied to solve $\{x\}$. At each iteration, the system model is treated as a linear system model for $\{x\}$. The iterative procedure is terminated until all of the constraints are satisfied. Therefore, only the solutions to a system of linear equations are taken into consideration in this chapter.

Algorithms to solve algebraic or eigenvalue system models are quite mature and reliable now. The solvers to system models can generally be classified into *direct* or *iterative* methods. A direct method obtains the solution in a definite number of steps, and the computation can be predicted based on the size of $[A]$. On the other hand, the computation of an iterative method for system solution depends on the type of model, it is difficult to estimate how many steps it takes to find a converged solution; but it usually takes less time to solve a system model in FEA practice.

6.3.1 Gaussian elimination

A set of algebraic equations with multiple variables differs from one algebraic equation with one variable; in a sense that variables are coupled in a group of equations. If these variables can be decoupled with each other in the set of equations, all of variables can be solved from their individual equations without coupling others. *The Gaussian elimination* is to decouple the variables for a set of linear equations. Assume that a system of linear equations in Eq. (6.19) is represented in its scalar form as,

$$\left.\begin{array}{l} a_{11}x_1 + a_{12}x_2 + \cdots + a_{1n}x_n = b_1 \\ a_{21}x_1 + a_{22}x_2 + \cdots + a_{2n}x_n = b_2 \\ \cdots \\ a_{n1}x_1 + a_{n2}x_2 + \cdots + a_{nn}x_n = b_n \end{array}\right\} \quad (6.20)$$

Assume $a_{11} \neq 0$ in the first equation, x_1 can be represented by $(x_2, x_3, \ldots, x_n)$ as,

$$x_1 = \frac{b_1}{a_{11}} - \left(\frac{a_{12}}{a_{11}}x_2 + \cdots + \frac{a_{1n}}{a_{11}}x_n \right) \quad (6.21)$$

Substituting Eq. (6.21) into the rest of $n - 1$ equations in (6.20) yields,

$$\left.\begin{array}{l} \left(a_{22} - \dfrac{a_{21}a_{12}}{a_{11}} \right)x_2 + \cdots + \left(a_{2n} - \dfrac{a_{21}a_{1n}}{a_{11}} \right)x_n = b_2 - \dfrac{a_{21}b_1}{a_{11}} = b_2^* \\[2mm] \left(a_{32} - \dfrac{a_{31}a_{12}}{a_{11}} \right)x_2 + \cdots + \left(a_{3n} - \dfrac{a_{31}a_{1n}}{a_{11}} \right)x_n = b_3 - \dfrac{a_{31}b_1}{a_{11}} = b_3^* \\[2mm] \cdots \\[1mm] \left(a_{n2} - \dfrac{a_{n1}a_{12}}{a_{11}} \right)x_2 + \cdots + \left(a_{nn} - \dfrac{a_{n1}a_{1n}}{a_{11}} \right)x_n = b_n - \dfrac{a_{n1}b_1}{a_{11}} = b_n^* \end{array}\right\} \quad (6.22)$$

In the above operation, a_{11} is chosen as a *pivot*, and x_1 has been eliminated from the $n - 1$ equations in Eq. (6.20). Next, a new pivot $a_{22}^* = a_{22} - \dfrac{a_{21}a_{12}}{a_{11}}$ can be chosen in the first equation of (6.22), and the same operation can be performed to eliminate x_2 from the rest of $n - 2$ equations. The same operation can be repeated until $x_1, x_2 \ldots$ are x_{n-1} are all eliminated from the last equation. At the end, the set of decoupled equations can be written as,

$$\left.\begin{array}{r}
a_{11}x_1 + a_{12}x_2 + \cdots + a_{1n}x_n = b_1 \\
a_{22}^*x_2 + \cdots + a_{2n}^*x_n = b_2^* \\
\cdots \\
a_{nn}^*x_n = b_n^*
\end{array}\right\} \tag{6.23}$$

The above procedure reduces a full $[A]$ into a triangular matrix $[A]^*$, and this procedure is referred as the *forward reduction*. After the forward reduction is completed, in Eq. (6.23), x_n can be directly found from the last equation as $x_n = b_n^*/a_{nn}^*$, and the other variables can be solved in a reverse order of the equations as,

$$x_i = \frac{1}{a_{ii}^*}\left(b_i^* - \sum_{j=i+1}^{n} a_{ij}x_j\right) \tag{6.24}$$

where $i = 1, 2, \ldots n$.

The procedure to use Eq. (6.24) to find all unknown variables is called the *back substitution*.

In the case where the modified coefficient at the pivot position $a_{ii}^* = 0$ or extremely small, the rest of the equations should be reordered so that an appropriate value of the pivot element can be utilized. For a robust implementation, the interchanging rows/columns is performed at each step to put the largest element on the diagonal. This practice keeps the zero off the diagonal and keeps small numbers off the diagonal—reduces computer round off error.

Example 6.14

Use the Gaussian elimination to find the solution of the following linear equations.

$$\left.\begin{array}{r}
x_1 + 2x_2 - 3x_3 = 0 \\
x_1 + 3x_2 + x_3 = 5 \\
2x_1 - x_2 - x_3 = 25
\end{array}\right\} \tag{6.25}$$

Solution

Eqs. (6.23) and (6.24) are applied in sequence for forward and back reduction.

1. Forward reduction

Step f-1:

$$\left.\begin{array}{r} x_1 + 2x_2 - 3x_3 = 0 \\ x_2 + 4x_3 = 5 \\ -5x_2 + 5x_3 = 25 \end{array}\right\}$$

Step f-2:

$$\left.\begin{array}{r} x_1 + 2x_2 - 3x_3 = 0 \\ x_2 + 4x_3 = 5 \\ 25x_3 = 50 \end{array}\right\}$$

2. Back substitution

Step b-1:

$$\left.\begin{array}{r} x_1 + 2x_2 - 3x_3 = 0 \\ x_2 + 4x_3 = 5 \\ x_3 = 2 \end{array}\right\}$$

Step b-2:

$$\left.\begin{array}{r} x_1 + 2x_2 - 3x_3 = 0 \\ x_2 = 5 - 4(2) = -3 \\ x_3 = 2 \end{array}\right\}$$

Step b-3:

$$\left.\begin{array}{r} x_1 = -2(-3) + 3(2) = 12 \\ x_2 = 5 - 4(2) = -3 \\ x_3 = 2 \end{array}\right\}$$

6.3.2 *Lower and upper* decomposition

It is found that a full matrix $[A]$ can always be reduced and converted into a triangular matrix $[A]^*$, so that the unknown variables in the equation set can be decoupled. In a *lower and upper (LU)* decomposition, it is assumed that a full matrix $[A]$ can be decomposed into a lower triangular matrix $[L]$ and an upper triangular matrix $[U]$ directly to satisfy $[A] = [L][U]$. Accordingly, the set of linear equations with $[A]$ can be found by two processes similar to forward reduction and back substitution in the Gaussian elimination.

$$\text{Let } [A] = \begin{bmatrix} a_{11} & a_{12} & a_{13} \\ a_{21} & a_{22} & a_{23} \\ a_{31} & a_{32} & a_{33} \end{bmatrix}, [L] = \begin{bmatrix} 1 & 0 & 0 \\ l_{21} & 1 & 0 \\ l_{31} & l_{32} & 1 \end{bmatrix} \text{ and}$$

$$[U] = \begin{bmatrix} u_{11} & u_{12} & u_{13} \\ 0 & u_{22} & u_{23} \\ 0 & 0 & u_{33} \end{bmatrix}$$

For a matrix $[A]$ with a size of 3 by 3, $[L]$ and $[U]$ have a total of nine unknown and they can be readily determined completely based on the condition of $[A] = [L][U]$,

$$[A] = \begin{bmatrix} a_{11} & a_{12} & a_{13} \\ a_{21} & a_{22} & a_{23} \\ a_{31} & a_{32} & a_{33} \end{bmatrix} = \begin{bmatrix} u_{11} & u_{12} & u_{13} \\ l_{21}u_{11} & l_{21}u_{12}+u_{22} & l_{21}u_{13}+u_{23} \\ l_{31}u_{11} & l_{31}u_{12}+l_{32}u_{22} & l_{31}u_{13}+l_{32}u_{23}+u_{33} \end{bmatrix} = [L][U]$$

The procedure in determining $[L]$ and $[U]$ from $[A]$ is as follows,

Step ①: $\quad u_{1i} = a_{1i} \quad (i=1,2,3)$

Step ②: $\quad l_{i1} = \dfrac{a_{i1}}{u_{11}} \quad (i=2,3)$

Step ③: $\quad u_{2i} = a_{2i} - l_{21}u_{1i} \quad (i=2,3)$

Step ④: $\quad l_{32} = \dfrac{a_{32} - l_{31}u_{12}}{u_{22}}$

Step ⑤: $\quad u_{33} = a_{33} - l_{31}u_{13} - l_{32}u_{23}$

After the LU decomposition is performed, the solution to the equation set $[A]\{x\} = \{b\}$ can be replaced by two subsolutions, i.e., $[L]\{y\} = \{b\}$ and $[U]\{x\} = \{y\}$.

The solution $\{y\}$ to $[L]\{y\} = \{b\}$ is found by a *forward substitution* as,

$$\left. \begin{aligned} y_1 &= b_1 \\ y_2 &= b_2 - l_{21}y_1 \\ y_3 &= b_3 - l_{31}y_1 - l_{32}y_2 \end{aligned} \right\}$$

The solution $\{x\}$ to $[U]\{x\} = \{y\}$ is found by a *backward substitution* as,

$$\left. \begin{aligned} x_1 &= \frac{1}{u_{11}}\left(y_1 - u_{12}x_2 - u_{13}x_3\right) \\ x_2 &= \frac{1}{u_{22}}\left(y_2 - u_{23}x_3\right) \\ x_3 &= \frac{y_3}{u_{33}} \end{aligned} \right\}$$

Example 6.15

Use the LU decomposition to obtain the solution of the following linear equations.

$$\left.\begin{array}{rcl} x_1 - 2x_2 + 3x_3 &=& 2 \\ 2x_1 - 5x_2 + 12x_3 &=& 10 \\ 2x_2 - 10x_3 &=& 4 \end{array}\right\}$$

Solution

Obtain the matrix $[A]$ from the given equation as,

$$[A] = \begin{bmatrix} 1 & -2 & 3 \\ 2 & -5 & 12 \\ 0 & 2 & -10 \end{bmatrix}$$

1. Let $[A] = [L][U]$, Eq. (6.26) is used to determine $[L]$ and $[U]$ as,

$$[L] = \begin{bmatrix} 1 & 0 & 0 \\ 2 & 1 & 0 \\ 0 & -2 & 1 \end{bmatrix} \text{ and } [U] = \begin{bmatrix} 1 & -2 & 3 \\ 0 & -1 & 6 \\ 0 & 0 & 2 \end{bmatrix}$$

2. Let $[L]\{y\} = \{b\}$ where $\{b\} = \{2,10,4\}^T$, a forward substitution can find $\{y\}$ as,

$$\left.\begin{array}{l} y_1 = 2 \\ y_2 = 10 - 4 = 6 \\ y_3 = 4 - 0 + 2(6) = 16 \end{array}\right\} \Downarrow$$

3. Let $[L]\{x\} = \{y\}$ where $\{y\} = \{2,6,16\}^T$, a backward substitution can find $\{x\}$ as,

$$\left.\begin{array}{l} x_1 = 2 + 2(42) - 3(8) = 62 \\ x_2 = \dfrac{6 - 6(8)}{(-1)} = 42 \\ x_3 = \dfrac{16}{2} = 8 \end{array}\right\} \Uparrow$$

6.3.3 **Iterative methods**

Theoretically, the Gaussian elimination method or the LU decomposition method will find an exact solution to a system of linear equations. However, they suffer a number of significant disadvantages in their implementations. For example,

1. Both of these two methods are open-loop solutions. An open-loop solution implies that the solution does not have the mechanism to
   compensate round-off errors by computers. It could be a fatal issue if
   computation errors are accumulated rapidly during a solving process.
2. The implementation of the Gaussian elimination or the LU decomposition algorithm can be very intriguing if all of the special cases are
   considered. For example, the possibility of a zero value at the pivoting
   position in the Gaussian elimination method.
3. These two methods are fairly generic but they lack the flexibility to
   reduce the computation for the system model with a sparse matrix.
   Moreover, the computation and storage needs will be exponentially
   increased with the size of a system model.

An alternative to overcome these issues is to use closed-loop iterations. The
accuracy of the approximated solution is checked as every iterative step,
and the iteration will not be terminated until the obtained solution meets
the specified requirement of accuracy.

To develop an iterative solving procedure, rewrite $[A]\{x\} = \{b\}$ in
Eq. (6.18) as $([L] + [D] + [U])\{x\} = \{b\}$, so that the original full $[A]$ is
decomposed into three submatrices, i.e., *lower*, *diagonal*, and *upper*
submatrices.

$$[L] = \begin{bmatrix} 0 & 0 & \cdots & 0 \\ a_{2,1} & \ddots & \cdots & \cdots \\ \cdots & \ddots & \ddots & \cdots \\ a_{n,1} & \cdots & a_{n,n-1} & 0 \end{bmatrix}, \quad [D] = \begin{bmatrix} a_{1,1} & 0 & \cdots & 0 \\ 0 & \ddots & \cdots & \cdots \\ \cdots & \ddots & \ddots & 0 \\ 0 & \cdots & 0 & a_{n,n} \end{bmatrix}, \quad [U] = \begin{bmatrix} 0 & a_{1,2} & \cdots & a_{1,n} \\ 0 & \ddots & \ddots & \cdots \\ \cdots & & \ddots & a_{n-1,n} \\ 0 & \cdots & 0 & 0 \end{bmatrix}$$

In such a way, the solution to the original system model can be converted
into,

$$\{x\} = [D]^{-1}(\{b\} - [L + U]\{x\}) \tag{6.29}$$

where

$$[D]^{-1} = \begin{bmatrix} a_{1,1}^{-1} & 0 & \cdots & 0 \\ 0 & \ddots & \cdots & \cdots \\ \cdots & \ddots & \ddots & 0 \\ 0 & \cdots & 0 & a_{n,n}^{-1} \end{bmatrix}, \quad [L + U] = \begin{bmatrix} 0 & a_{1,2} & \cdots & a_{1,n} \\ a_{2,1} & \ddots & \ddots & \cdots \\ \cdots & & \ddots & a_{n-1,n} \\ a_{n,1} & \cdots & a_{n,n-1} & 0 \end{bmatrix}$$

Eq. (6.29) can be rewritten into an iterative format where two $\{x\}$ vectors on
the left and right represent the results at two steps, i.e.,

$$\{x\}^{(k+1)} = [D]^{-1}(\{b\} - [L + U]\{x\}^{(k)}) \tag{6.30}$$

The iteration will continue until the following terminate condition is satisfied,

$$\left\| \{x\}^{(k+1)} - \{x\}^{(k)} \right\| < \varepsilon \tag{6.31}$$

where ε is specified accuracy of system solution.

The above method is called the *Jacobi' method*, and it is assumed that all diagonal entries in $[A]$ are nonzero, and the scalar version of Eq. (6.30) can be written as,

$$x_i^{(k+1)} = \frac{1}{a_{ii}} \left[b_i - \sum_{\substack{j=1 \\ j \neq i}}^{n} a_{ij} x_j^{(k)} \right], \quad (i = 1, 2 \cdots n) \tag{6.32}$$

Bear in mind that not all of the systems of linear equations can be solved by an iterative method. To ensure that the convergence $[A]$ should be strictly diagonally dominant by rows, i.e.,

$$|a_{ii}| > \sum_{j=1}^{n} |a_{ij}| \quad \text{for } j \neq i \text{ and } i = 1, 2, \cdots n \tag{6.33}$$

As a consequence,

$$\left\| [D]^{-1}([L + U]) \right\|_{\infty} = \max_{i=1,2,\cdots n} |a_{ij}/a_{ii}| < 1 \tag{6.34}$$

The condition in Eq. (6.34) ensures that the solution is gradually converged until the terminating condition is satisfied. Many system models of FEA modeling may not strictly meet the convergent conditions in Eq. (6.33), but they can still be solved by an iterative method effectively.

Example 6.16

Use the iterative method to solve $[A]\{u\} = \{b\}$ where

$$[A] = \begin{bmatrix} 2 & -1 & 0 & 0 & 0 \\ -1 & 2 & -1 & 0 & 0 \\ 0 & -1 & 2 & -1 & 0 \\ 0 & 0 & -1 & 2 & -1 \\ 0 & 0 & 0 & -1 & 2 \end{bmatrix} \text{ and } \{b\} = \begin{bmatrix} 5 \\ 2 \\ 2 \\ 2 \\ 5 \end{bmatrix}$$

Let $\{x\}^T = [1\ 1\ 1\ 1\ 1]$ and the terminate tolerance $\varepsilon = 0.0001$.

Using Eq. (6.29) to execute the iterative process, the process to obtain the final solutions is,

Step	$\{x_i\}^T$					Convergence
0	[1	1	1	1	1]	
1	[3.0000	2.0000	2.0000	2.0000	3.0000]	3.5000
10	[6.5762	8.6270	9.1523	8.6270	6.5762]	0.7750
20	[7.6621	10.4369	11.3242	10.4369	7.6621]	0.1839
30	[7.9198	10.8664	11.8396	10.8664	7.9198]	0.0436
40	[7.9810	10.9683	11.9619	10.9683	7.9810]	0.0104
50	[7.9955	10.9925	11.9910	10.9925	7.9955]	0.0025
60	[7.9989	10.9982	11.9979	10.9982	7.9989]	5.8325e-04
70	[7.9997	10.9996	11.9995	10.9996	7.9997]	1.3841e-04
73	[7.9998	10.9997	11.9997	10.9997	7.9998]	8.9898e-05

6.3.4 **Computation in matrix manipulation**

The amount of computation is one of the primary criteria in assessing the efficiency of a computer program. For a given task, the amount of computation varies with the selected algorithms or programs. Conventionally, the computation can be estimated based on the number of multiplications in the execution of an algorithm or program. The computation for some common operations is estimated as below.

Table 6.1 can be applied to evaluate the performance of computer algorithms. For example, if an inverse matrix method and the Gauss elimination are used to solve a system of linear equations, the Gauss elimination method will reduce the computation greatly in contrast to the inverse matrix method.

6.4 **EIGENVALUE PROBLEMS**

Time-dependent problems are mostly modeled as a system of ordinary differential equations. The integration has to be performed explicitly or implicitly in the time domain using eigenvalues and eigenvectors. Therefore, one primary task is to find eigenvalues and eigenvectors for the given system models.

Let a system model for an eigenvalue problem be

$$[A]\{x\} - \lambda[B]\{x\} = 0 \qquad (6.35)$$

Table 6.1 Computation of Some Exemplified Matrix Manipulation

Operation	Break Down Computation	Total Computation
$[C]_{n \times n} = [A]_{n \times n}[B]_{n \times n}$	■ A total $n \times n$ elements in $[C]$, and n multiplications for each element $c_{ij} = a_{ik}b_{kj}$ ($i, j, k = 1,2,...n$),	n^3
$\det\|[A]_{n \times n}\|$ $= \sum_{i_1,i_2\cdots i_n = 1}^{n} \varepsilon_{i_1,i_2\cdots i_n} a_{1,i_1} a_{2,i_2} \cdots a_{n,i_n}$	■ A total $n!$ permutations of $a_{1,i_1}, a_{2,i_2} \cdots a_{n,i_n}$, each permutation has n multiplications.	$n \cdot n!$
$[A]^{-1}_{n \times n} = \frac{[A]^C}{\det[A]}$ where $[A]^C = (-1)^{i+j}[m_{ij}]$ $m_{ij} = \det\big[[a_{i^* j^*}]\big]$ $i^* = 1, \cdots, i-1, i+1, \cdots n$ $j^* = 1, \cdots, j-1, j+1, \cdots n$	■ A total $n \cdot n!$ multiplications to calculate the determinant for $[A]$; ■ A total $n \times n$ elements in $[A]^C$; each element requires $(n-1) \cdot (n-1)!$ multiplications to calculate m_{ij}. ■ A total $n \times n$ elements for the division of $\frac{[A]^C}{\det[A]}$	$n^2 \cdot n! + n^2$
Cramer method to solve $[A]_{n \times n}\{x\} = \{b\}$ where $x_i = \frac{\det[A^i]}{\det[A]}$ $a^i_{jk} = \begin{cases} a_{jk} & k \neq i \\ b_j & k \neq i \end{cases}$	■ A total $n \cdot n!$ multiplications to calculate the determinant for $[A]$; ■ A total of n elements, each element needs to calculate the determinant for $[A^i]$ with $n \cdot n!$ multiplications; ■ n divisions to find all elements in $\{x\}$.	$n(n+1)! + n$
Gauss elimination to solve $[A]_{n \times n}\{x\} = \{b\}$: $\left.\begin{array}{l} a_{11}x_1 + a_{12}x_2 + \cdots + a_{1n}x_n = b_1 \\ a^*_{22}x_2 + \cdots + a^*_{2n}x_n = b^*_2 \\ \cdots \\ a^*_{nn}x_n = b^*_n \end{array}\right\}$	■ $k \times k$ divisions ($k = n, n-1, ...,1$) in the forward elimination ■ k divisions ($k = 1, 2,..., n$) in backward substitution	$\sum_{k=1}^{n} k^2 + \sum_{k=1}^{n} k$ $= \frac{n^3}{3} + n^2 - \frac{n}{3}$

An eigenvalue λ_i should satisfy the condition for a valid solution of the system model,

$$\det([A] - \lambda[B]) = 0 \tag{6.36}$$

Eq. (6.36) is also called as *the characteristic equation* of a system model.

Correspondingly, an eigenvector v_i from λ_i satisfies the following condition,

$$([A] - \lambda[B])v_i = 0 \tag{6.37}$$

If using a system model with a few of variables, it is possible to determine the eigenvalues and eigenvectors by solving a polynomial equation for the determinate of the matrix.

Example 6.17

Find the eigenvalues and eigenvectors for a system model:

$$\begin{bmatrix} 3 & -1 & 0 \\ -1 & 2 & -1 \\ 0 & -1 & 3 \end{bmatrix} \{x\} - \lambda\{x\} = 0$$

Solution

The characteristic equation of the given system model is,

$$\det \begin{bmatrix} 3-\lambda & -1 & 0 \\ -1 & 2-\lambda & -1 \\ 0 & -1 & 3-\lambda \end{bmatrix} = (3-\lambda)^2(2-\lambda) - 2(3-\lambda) = 0$$

Solving the above polynomial equation gets the solutions to the characteristic equation as $\lambda_1 = 1$, $\lambda_2 = 3$, and $\lambda_3 = 4$.

Substituting three eigenvalues in the system model of the problem gets the following unit eigenvectors, respectively.

$$\{\overline{x}_1\} = \frac{1}{\sqrt{6}} \begin{Bmatrix} 1 \\ 2 \\ 1 \end{Bmatrix}, \{\overline{x}_2\} = \frac{1}{\sqrt{2}} \begin{Bmatrix} 1 \\ 0 \\ -1 \end{Bmatrix}, \{\overline{x}_3\} = \frac{1}{\sqrt{3}} \begin{Bmatrix} 1 \\ -1 \\ 1 \end{Bmatrix}$$

The characteristic Eq. (6.36) is actually a polynomial equation about natural frequencies. When the size of a system model is large, solving a polynomial equation with a high order in a reasonable time frame becomes impractical. A generic iterative method should be employed to find eigenvalues and eigenvectors.

6.4.1 **Range of natural frequencies**

In a numerical simulation, we expect to know the range of natural frequencies. Here, *the Rayleigh quotient* and the corresponding method are introduced to estimate the range of natural frequencies (Phani and Adhikari, 2008).

Assume that the model of an eigenvalue problem as $[A][u] - \lambda[B] = 0$. An eigenvalue λ_i and the corresponding eigenvector u_i satisfies the condition of,

$$[A]u_i = \lambda_i[B]u_i \tag{6.38}$$

Reformatting Eq. (6.38) by multiplying u_i^T yields the expression of λ_i as,

$$\lambda_i = \frac{u_i^T[A]u_i}{u_i^T[B]u_i} \tag{6.39}$$

Note that Eq. (6.39) is valid for any one of eigenvalue and eigenvector in a system model. Therefore, it can be generalized as,

$$\rho(u) = \frac{u^T[A]u}{u^T[B]u} \tag{6.40}$$

$\rho(u)$ in Eq. (6.40) is well known as *the Rayleigh quotient*, and it has the properties that for any vector $u \neq 0$, $\rho(u)$ will satisfy,

$$\lambda_{min} \leq \rho(u) \leq \lambda_{max} \tag{6.41}$$

where λ_{min} and λ_{max} are the minimized and maximized eigenvalues of the system model.

Moreover, $\rho(u)$ is treated as a function of u, the value $\rho(u)$ is stationary when u is one of eigenvectors, and the corresponding ρ is an eigenvalue.

A *power method* can be used to find λ_{min} and λ_{max} of an ordinary algebraic eigenvalue problem $[A][u] - \lambda[u] = 0$ (Larson and Edwards, 2012). A power method is an iterative method. In its implementation, a nonzero vector u_0 is selected initially, this vector is iteratively updated for the next approximation of eigenvector u_i as,

$$u_{i+1} = [A]u_i \quad (i = 0, 1, 2, \cdots) \tag{6.42}$$

If u_i is evolved to be orthogonal to an eigenvector u_{max} corresponding to the eigenvalues of the largest magnitude λ_{max}, the process converges to satisfy the condition of $[A]u_{max} = (\lambda_{max}) u_{max}$. The application of the power method to find the maximized λ_{max} is called the *forward iteration*.

By defining a mirror problem of the original one as $[A]^{-1}[u] - \lambda^{-1}[u] = 0$, the similar iteration of Eq. (6.42) can be applied, and it will converge to the smallest λ_{min}, i.e., it satisfies the condition of $[A]^{-1}u_{min} = (1/\lambda_{min}) u_{min}$. The application of the power method to find the minimized λ_{min} is called the *inverse iteration*.

Example 6.18

Use the power method to find the smallest and largest eigenvalues and corresponding eigenvectors of $[A][u] - \lambda[u] = 0$ where

$$[A] = \begin{bmatrix} 5 & -1 & 0 \\ -1 & 3 & -1 \\ 0 & -1 & 4 \end{bmatrix}$$

Let $u_0 = [1, 0, 0]^T$, and the terminate accuracy of convergence be 0.001.

The processes of the forward and inverse iterations are shown in Tables 6.2 and 6.3, respectively.

Table 6.2 The Forward Iteration for the Largest Eigenvalue and Eigenvector

Initialization : $i = 0$, $\boldsymbol{u}_i = [1\,0\,0]^T$, $\rho_i = |\boldsymbol{u}_i|$

Iteration : $\boldsymbol{u}_{i+1} = [\boldsymbol{A}]\boldsymbol{u}_i$, $\rho_{i+1} = |\boldsymbol{u}_{i+1}|$

Update : $i \leftarrow i + 1$, $\rho_i \leftarrow \rho_{i+1}$; $\boldsymbol{u}_i \leftarrow \boldsymbol{u}_{i+1}/\rho_{i+1}$

Terminate condition : $\left| \dfrac{\rho_{i+1} - \rho_i}{\rho_{i+1}} \right| < \varepsilon$

Convergence : $\lambda_{max} = \rho_{i+1}$, $\boldsymbol{u}_{max} = \boldsymbol{u}_{i+1}/\rho_{i+1}$

Step	$\boldsymbol{u}_i^T$	ρ_i	Convergence
0	[1, 0, 0]	1	
1	[0.9806, −0.1961, 0]	5.099	0.8039
2	[0.9551, −0.2939, 0.0367]	5.3385	0.0449
3	[0.9349, −0.3455, 0.0813]	5.4226	0.0155
4	[0.9186, −0.3756, 0.1227]	5.4647	0.0077
5	[0.9050, −0.3949, 0.1578]	5.4900	0.0046
6	[0.8936, −0.4082, 0.1864]	5.5058	0.0029
7	[0.8841, −0.4179, 0.2092]	5.5157	0.0018
8	[0.8762, −0.4250, 0.2272]	5.5219	0.0011
9	**[0.8697, -0.4304, 0.2414]**	**5.5258**	**0.0007**

Table 6.3 The Inverse Iteration for the Smallest Eigenvalue and Eigenvector

Initialization : $i = 0$, $\boldsymbol{u}_i = [1\,0\,0]^T$, $\rho_i = |\boldsymbol{u}_i|$

Iteration : $\boldsymbol{u}_{i+1} = [\boldsymbol{A}]^{-1}\boldsymbol{u}_i$, $\rho_{i+1} = |\boldsymbol{u}_{i+1}|$

Update : $i \leftarrow i + 1$, $\rho_i \leftarrow \rho_{i+1}$; $\boldsymbol{u}_i \leftarrow \boldsymbol{u}_{i+1}/\rho_{i+1}$

Terminate condition : $\left| \dfrac{\rho_{i+1} - \rho_i}{\rho_{i+1}} \right| < \varepsilon$

Convergence : $\lambda_{min} = 1/\rho_{i+1}$, $\boldsymbol{u}_{min} = \boldsymbol{u}_{i+1}/\rho_{i+1}$

Step	$\boldsymbol{u}_i^T$	$1/\rho_i$	Convergence
0	[1, 0, 0]	1	
1	[0.9364, 0.3405, 0.0851]	4.3414	3.3414
2	[0.7106, 0.6643, 0.2317]	3.0852	0.2894
3	[0.4957, 0.8005, 0.3369]	2.3615	0.2346
4	[0.3824, 0.8368, 0.3919]	2.1689	0.0816
5	[0.3319, 0.8447, 0.4199]	2.1301	0.0178
6	[0.3100, 0.8458, 0.4342]	2.1226	0.0036
7	**[0.3006, 0.8454, 0.4416]**	**2.1210**	**0.0007**

6.4.2 **Significant natural frequencies from subspace iteration**

The power method in Section 6.4.1 could only find the smallest and the largest eigenvalues and their eigenvectors. In most of the real-world applications, a system can be adequately characterized by a set of lower responsive modes. Therefore, the lowest several eigenvalues and eigenvectors have the most significant impact on a system. The iteration method on the subspace of eigenvectors can be applied to meet this need (Bathe, 2012).

For a system model $[A][u] - \lambda[B][u] = 0$, assume that the sizes of $[A]$ or $[B]$ are $n \times n$, and m lowest eigenvalues and the corresponding eigenvectors are of interest. The iteration in the subspace of eigenvectors is described as follows:

1. Select m nonzero vectors to generate an $n \times m$ modal matrix $[X_i]$, where each of its columns corresponds to one vector, and set the iteration index $i = 0$.
2. Solve the linear equation set $[A][Y_i] = [B][X_i]$ to get $n \times m$ modal matrix $[Y_i]$.
3. Build a reduced eigenvalue problem with the size of $m \times m$ as

$$[A]_i^* u^* - \lambda[B]_i^* u^* = 0$$

where $[A]_i^* = Y_i^T[A]Y_i$, $[B]_i^* = Y_i^T[B]Y_i$
4. Solve the reduced eigenvalue problem for m eigenvalues and eigenvalues as,

$$\lambda_m^{(i)} = \text{diag}\left[\lambda_1^{(i)}, \lambda_2^{(i)}, \cdots \lambda_m^{(i)}\right]$$

$$[u^*] = \left[u_1^{(i)}, u_2^{(i)}, \cdots u_m^{(i)}\right]$$

5. Update the iteration as: $i \leftarrow i + 1$, $[X_i] \leftarrow [X_{i+1}] = \text{normc}([Y_i][u^*])$, where normc($\bullet$) is a function to make every vector (column) in a matrix as a unit vector).
6. Check the terminating condition as,

$$\max\left(\left|\frac{\lambda_1^{(i)} - \lambda_1^{(i-1)}}{\lambda_1^{(i)}}\right|, \left|\frac{\lambda_2^{(i)} - \lambda_2^{(i-1)}}{\lambda_2^{(i)}}\right|, \cdots \left|\frac{\lambda_m^{(i)} - \lambda_m^{(i-1)}}{\lambda_m^{(i)}}\right|\right) \leq \varepsilon$$

If it is satisfied, the iteration is terminated; otherwise, return to step 2 for the continuation.

Example 6.19

Find the lowest two natural frequencies of the following system model with the given accuracy ε of 0.001:

$$[M]\{\ddot{U}\} + [K]\{U\} = 0 \tag{6.43}$$

where

$$[M] = \begin{bmatrix} 20 & 5 & 0 \\ 5 & 10 & 5 \\ 0 & 5 & 10 \end{bmatrix}, \quad [K] = \begin{bmatrix} 2 & -1 & 0 \\ -1 & 2 & -1 \\ 0 & -1 & 1 \end{bmatrix}$$

Solution with subspace iterative method

We are interested in finding the first two natural frequencies of the system model. Therefore, we have to set the dimension of a subspace as 2. The iteration starts with an assumption of two initial vectors for $\{u\}_{sub}$, i.e.,

$$[u]_{sub} = \begin{bmatrix} u_1 & u_2 \end{bmatrix} = \begin{bmatrix} 1 & 0 \\ 0 & 1 \\ 0 & 0 \end{bmatrix}$$

Let $i = 1$ and $[X_1] = [u]_{sub}$ to initialize the iterative process.

Iteration 1

Solving the equation set $[K][Y_i] = [M][X_i]$ yields,

$$[Y_i] = \begin{bmatrix} 25 & 20 \\ 30 & 35 \\ 30 & 40 \end{bmatrix}$$

The reduced eigenvalue problem at this iteration becomes,

$$[K]^*[u]^* - \lambda [M]^*[u]^* = 0$$

where

$$[K]^* = [Y_i]^T[K][Y_i] = \begin{bmatrix} 650 & 575 \\ 575 & 650 \end{bmatrix}$$

$$[M]^* = [Y_i]^T[M][Y_i] = 1.0 \times 10^4 \begin{bmatrix} 4.7 & 5.1125 \\ 5.1125 & 5.725 \end{bmatrix}$$

The eigenvalues and eigenvectors of the above reduced problem can be found as $\lambda_1 = 0.1052$, $\lambda_2 = 0.0113$, and $[u]^* = \begin{bmatrix} 0.7455 & 0.0450 \\ -0.6665 & 0.9990 \end{bmatrix}$.

Next, $[X_i]$ can be updated as,

$$[X_i] = \text{normc}([Y_i][u]^*) = \text{normc}\left(\begin{bmatrix} 5.3075 & 21.1040 \\ -0.9625 & 36.3137 \\ -4.295 & 41.3087 \end{bmatrix}\right)$$

$$= \begin{bmatrix} 0.7697 & 0.3582 \\ -0.1396 & 0.6164 \\ -0.6229 & 0.7012 \end{bmatrix}$$

Iteration 2

Solving the equation set $[K][Y_i] = [M][X_i]$ with the updated $[X_i]$ yields,

$$[Y_i] = \begin{bmatrix} 7.1085 & 31.8025 \\ -0.4800 & 53.3581 \\ -7.4069 & 63.4522 \end{bmatrix}$$

The reduced eigenvalue problem at this iteration becomes,

$$[K]^*[u]^* - \lambda[M]^*[u]^* = 0$$

where

$$[K]^* = [Y_i]^T[K][Y_i] = 1.0 \times 10^3 \begin{bmatrix} 0.1561 & -0.0074 \\ -0.0074 & 1.5779 \end{bmatrix}$$

$$[M]^* = [Y_i]^T[M][Y_i] = 1.0 \times 10^5 \begin{bmatrix} 0.0156 & -0.0074 \\ -0.0074 & 1.3979 \end{bmatrix}$$

The eigenvalues and eigenvectors of the above reduced problem can be found as $\lambda_1 = 0.1052$, $\lambda_2 = 0.0113$, and $[u]^* = \begin{bmatrix} 1.000 & -0.0069 \\ 0.0054 & 1.0000 \end{bmatrix}$.

Next, $[X_i]$ can be updated as,

$$[X_i] = \text{normc}([Y_i][u]^*) = \text{normc}\left(\begin{bmatrix} 7.2799 & 31.7526 \\ -0.1923 & 53.3601 \\ -7.0647 & 63.5019 \end{bmatrix}\right)$$

$$= \begin{bmatrix} 0.7175 & 0.3575 \\ -0.0190 & 0.6008 \\ -0.6963 & 0.7150 \end{bmatrix}$$

Check the convergence: $\max\left(\left|\frac{\lambda_1^{i+1}-\lambda_1^{i+1}}{\lambda_1^{i+1}}\right|, \left|\frac{\lambda_1^{i+1}-\lambda_1^{i+1}}{\lambda_1^{i+1}}\right|\right) = 0.05 > \varepsilon$; therefore, the iteration has to be continued.

Iteration 3

Solving the equation set $[K][Y_i] = [M][X_i]$ with the updated $[X_i]$ yields,

$$[Y_i] = \begin{bmatrix} 7.1139 & 31.6789 \\ -0.0274 & 53.2034 \\ -7.0852 & 63.3574 \end{bmatrix}$$

The reduced eigenvalue problem at this iteration becomes,

$$[K]^*[u]^* - \lambda[M]^*[u]^* = 0$$

where

$$[K]^* = [Y_i]^T[K][Y_i] = 1.0 \times 10^3 \begin{bmatrix} 0.1514 & -0.0000 \\ -0.0000 & 1.5700 \end{bmatrix}$$

$$[M]^* = [Y_i]^T[M][Y_i] = 1.0 \times 10^5 \begin{bmatrix} 0.0151 & -0.0000 \\ -0.0000 & 1.3908 \end{bmatrix}$$

The eigenvalues and eigenvectors of the above reduced problem can be found as $\lambda_1 = 0.1001$, $\lambda_2 = 0.0113$, and $[u]^* = \begin{bmatrix} 1.000 & -0.000 \\ 0.000 & 1.0000 \end{bmatrix}$.

Next, $[X_i]$ can be updated as,

$$[X_i] = \text{normc}([Y_i][u]^*) = \text{normc}\left(\begin{bmatrix} 7.1143 & 31.6788 \\ -0.0267 & 53.2034 \\ -7.0844 & 63.3575 \end{bmatrix} \right)$$

$$= \begin{bmatrix} 0.7086 & 0.3576 \\ -0.0027 & 0.6006 \\ -0.7056 & 0.7152 \end{bmatrix}$$

Check the convergence: $\max\left(\left| \frac{\lambda_1^{i+1} - \lambda_1^{i+1}}{\lambda_1^{i+1}} \right|, \left| \frac{\lambda_1^{i+1} - \lambda_1^{i+1}}{\lambda_1^{i+1}} \right| \right) = 0.001 < \varepsilon.$

The convergence condition is satisfied.

In addition, the solution from above iterative method can be validated by defining the matrix $[A]$ as,

$$[A] = [M]^{-1}[K] = \begin{bmatrix} 0.16 & -0.16 & 0.06 \\ -0.24 & 0.44 & -0.24 \\ 0.12 & -0.32 & 0.22 \end{bmatrix}$$

Using the build-in function in the Matlab can find the eigenvalues of $[A]$ are $\lambda_1 = 0.0113$, $\lambda_2 = 0.1000$, $\lambda_3 = 0.7087$, and the corresponding eigenvectors are

$$[X] = [x_1 \quad x_2 \quad x_3] = \begin{bmatrix} \begin{Bmatrix} 0.3576 \\ 0.6005 \\ 0.7152 \end{Bmatrix} & \begin{Bmatrix} 0.7071 \\ 0 \\ 0.7071 \end{Bmatrix} & \begin{Bmatrix} -0.2865 \\ 0.7678 \\ -0.5731 \end{Bmatrix} \end{bmatrix}$$

6.4.3 **Transformation method**

The power method or subspace method are designed to find the range or a few of eigenvalues and eigenvectors. If one is interested in calculating all of the eigenvalues and eigenvectors, an iterative method based on the transformation can be applied (Sinayoko, 2008).

Let the eigenvalue problem to be as $[A]u - \lambda u = 0$, a matrix $[T]$ is defined so that u is the result of transformation from u^* as, $u = [T]u^*$. By using u^* and premultiplying $[T]^T$ in the original eigenvalue problem, one can find,

$$[T]^T[A][T]u^* - \lambda[T]^T[T]u^* = 0 \tag{6.44}$$

Eq. (6.44) shows that if $[T]$ can satisfy $[T]^T[T] = [I]$, i.e., $[T]$ is an orthogonal matrix, the corresponding $[T]^T[A][T] = \text{diag}[\lambda_1, \lambda_2, \dots \lambda_n]$. Therefore, an iterative procedure of the transformation can be applied on $[A]$ until it is turned into a diagonal matrix, and the corresponding modal matrix u^* is for all of eigenvectors.

The transformation is based on a plane rotation matrix $R(\theta)$ defined as,

$$R(\theta) = \begin{bmatrix} 1 & & & & & \\ & \ddots & & & & \\ & & \cos\theta & & -\sin\theta & \\ & & & \ddots & & \\ & & \sin\theta & & \cos\theta & \\ & & & & & \ddots \\ & & & & & & 1 \end{bmatrix} \begin{matrix} \\ \\ i \\ \\ j \\ \\ \end{matrix}$$

In matrix $R(\theta)$, all of the diagonal elements are 1 except the i-th and j-th, which are given as $\cos\theta$; all nondiagonal elements are 0 except the elements on (i, j) and (i, j), which are $-\sin\theta$ and $\sin\theta$, respectively. Applying the transformation $R(\theta)$ gets $[A]^* = [R(\theta)]^T[A][R(\theta)]$ where the following elements are revised:

$$\left.\begin{aligned}
a_{ik}^* &= a_{ki}^* = a_{ik}\cos\theta + a_{jk}\sin\theta \quad k \neq i, j \\
a_{jk}^* &= a_{kj}^* = -a_{ik}\sin\theta + a_{jk}\cos\theta \quad k \neq i, j \\
a_{ii}^* &= a_{ii}\cos^2\theta + a_{ij}\sin 2\theta + a_{jj}\sin^2\theta \\
a_{jj}^* &= a_{ii}\sin^2\theta - a_{ij}\sin 2\theta + a_{jj}\cos^2\theta \\
a_{ij}^* &= a_{ji}^* = a_{ij}\cos 2\theta + \frac{(a_{jj} - a_{ii})\sin 2\theta}{2}
\end{aligned}\right\} \tag{6.46}$$

The variable θ in each transformation is determined by the last relation in Eq. (6.46) that the nondiagonal element a_{ij} is set to be 0, i.e.,

$$\tan 2\theta = \frac{2a_{ij}}{a_{ii} - a_{jj}} \tag{6.47}$$

In the implementation, nondiagonal elements are examined in a certain order and the plan rotations are performed to eliminate the elements whose absolute values are above a preset threshold. The execution of plan rotations for nondiagonal elements in one round is called a sweep. The sweeps will be repeated until the absolute values of all nondiagonal elements of $[A]^*$ are below the specified threshold ε. Note that for a symmetric $[A]$ with the size of $n \times n$, only the sweep on upper triangle matrix is sufficient. The transformation method is implemented as follows:

1. initialize $[A]$ and set $[u]$ to be $[I]$
2. for $i = 1$ to n, do
       for $j = i + 1$, n, do
           calculate θ by eq. (6.xx) and use eq. (6.xx) to create $[A]^*$
           update $[u] \leftarrow \text{normc}([u][R(\theta)])$
           reset $[A] \leftarrow [A]^*$
       end
     end
3. check the maximized nonzero values of nondiagonal elements
     error $= \max (A(i, j)$ where $i \neq j$, and $i, j = 1, 2, \ldots n)$
4. if the error is larger than the specified threshold value ε, return step 2 for the next sweep; otherwise, the transformation is terminated. As the converged results, all of the diagonal elements of $[A]^*$ are eigenvalues, and $[u]$ is the modal matrix with all corresponding eigenvectors.

Example 6.20

Use the transformation method to find all of three eigenvalues and eigenvectors of the eigenvalue problem $[A]u - \lambda u = 0$, where

$$[A] = \begin{bmatrix} 3 & -1 & 0 \\ -1 & 5 & -1 \\ 0 & -1 & 6 \end{bmatrix}$$

Solution

Let the threshold value ε of nondiagonal elements be $\varepsilon = 0.0001$. The transformation method will be repeatedly executed in the order of $A(1, 2)$, $A(1, 3)$, and $A(2, 3)$, and the intermediate results of the execution are given in Table 6.4.

Table 6.4 Example of Transformation Method for All Eigenvalues and Vectors

Step	Element	[A]			θ	[u]			ε
0		3	−1	0		1	0	0	
		−1	5	−1		0	1	0	
		0	−1	6		0	0	1	
1	(1, 2)	2.5868	0.0000	−0.3827	3.9e-01	0.9239	−0.3827	0	
		0.0000	5.4142	−0.9239		0.3827	0.9239	0	
		−0.3827	−0.9239	6.000		0	0	1.0000	
2	(1, 3)	2.5434	−0.1017	0	1.1e-01	0.9183	−0.3827	−0.1017	
		−0.1017	5.4142	−0.9183		0.3804	0.9239	−0.0421	
		0	−0.9183	6.0424		0.1100	0	0.9939	
3	(2, 3)	2.5434	−0.0827	0.0591	6.2e-01	0.9183	−0.3704	0.1398	0.0827
		−0.0827	4.7578	−0.0000		0.3804	0.7271	−0.5715	
		0.0591	−0.0000	6.6988		0.1100	0.5780	0.8086	
4	(1, 2)	2.5403	0	0.0591	3.7e-02	0.9038	−0.4044	0.1398	
		0	4.7609	−0.0022		0.4072	0.7124	−0.5715	
		0.0591	−0.0022	6.6988		0.1315	0.5735	0.8086	
5	(1, 3)	2.5395	0.0000	0	−1.4e-02	0.9017	−0.4044	0.1527	
		0.0000	4.7609	−0.0022		0.4153	0.7124	−0.5657	
		0	−0.0022	6.6996		0.1200	0.5735	0.8104	
6	(2, 3)	2.5395	0.0000	0	1.1e-03	0.9017	−0.4042	0.1531	3.1296e-05
		0.0000	4.7609	0		0.4153	0.7118	−0.5665	
		0	0	6.6996		0.1200	0.5744	0.8097	

6.5 TRANSIENT PROBLEMS

In a transient problem, the time domain will also be discretized other than the discretization over a space domain. The solution to a transient problem consists of a number of subsolutions corresponding to each time step. At each step, the system model is updated based on the subsolutions to precedent steps. The simulation completes until the max number of time steps is reached. Although there is some special consideration on the size of time steps as well as the modification of system models, the solving methods to a system model at each step are similar to the solutions to equilibrium problems, i.e., the methods a system of linear equations are still applicable to find a subsolution at each time step. Other parameters related to a transient process will be discussed in Chapter 10 when the application of transient heat transfer problems is discussed.

6.6 **SUMMARY**

The fundamentals of mathematics used to represent and solving FEA models are reviewed. The concepts of matrices, vectors, and system of linear equations related to the FEA theory have been discussed thoroughly. The amount of computation is introduced to evaluate the performance of computer algorithms. System models for FEA are classified into the models for equilibrium problems, eigenvalue problems, and transient problems. Popular methods to solve these system models have been developed.

In commercial FEA codes, all of the computation for solving system models is automatically done by computers. However, user's basic understanding on computer algorithms help to provide the best settings for the solvers in obtaining system solutions effectively.

REFERENCES

Bathe, K.J., 2012. The subspace iteration method — revisited. Computers and Structures 126, 177—183.

Larson, R., Edwards, B.H., 2012. Elementary Linear Algebra, seventh ed. ISBN-13: 978-1133110873.

Phani, A.S., Adhikari, S., 2008. Rayleigh quotient and dissipative systems. ASME Journal of Applied Mechanics 75, 061005-1.

Sinayoko, S., 2008. Eigenvalue Problems I: Introduction and Jacobi Method. http://www.southampton.ac.uk/~feeg6002/lecturenotes/feeg6002_numerical_methods08.pdf.

▪ PROBLEMS

6.1. Let $[A] = \begin{bmatrix} 2 & 5 \\ 1 & -5 \end{bmatrix}$ and $[B] = \begin{bmatrix} x+1 & 3+u \\ y+2 & v+3 \end{bmatrix}$, find the values of x, y, u, and v to satisfy the conditions of $[A] = [B]$.

6.2. Use Eq. (6.7) to find the determinant of $[A] = \begin{bmatrix} 3 & 2 & 0 & 4 \\ 2 & 3 & 2 & 2 \\ -2 & 2 & 3 & -1 \\ 1 & 2 & -1 & 1 \end{bmatrix}$

6.3. Find the trace of the matrix $[B] = \begin{bmatrix} 7 & 2 & -8 \\ 3 & 2 & 1 \\ -2 & -5 & -3 \end{bmatrix}$

6.4. Define the differentiation $[\dot{A}]$ with respect to t for $[A] = \begin{bmatrix} e^t & \cos^2(\omega t) \\ \sin(\omega t) & e^t \end{bmatrix}$.

(A)

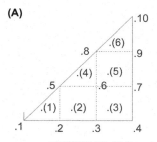

(B)

elements	i	j	k/m	n
(1)	1	2	5	
(2)	2	3	6	5
(3)	3	4	7	6
(4)	5	6	8	
(5)	6	7	9	8
(6)	8	9	10	

■ **FIGURE 6.3** (A) Elements and nodes; (B) relations of nodes and elements.

6.5. As shown in Fig. 6.3, a structure consists of 10 nodes for three triangular elements and three rectangle elements. Build a system model based on the given element models as follows.
The stiffness matrices $[K]^{(e)}$ for elements are found as.

Triangle Elements (1), (4), and (6)	Rectangle Elements (2), (3), and (4)
$[K]^{(e)} = \begin{bmatrix} 0.5 & -0.5 & 0 \\ -0.5 & 1 & -0.5 \\ 0 & -0.5 & 0.5 \end{bmatrix}$	$[K]^{(e)} = \begin{bmatrix} 4 & -1 & -2 & -1 \\ -1 & 4 & -1 & -2 \\ -2 & -1 & 4 & -1 \\ -1 & -2 & -1 & 4 \end{bmatrix}$

Assembly element models into a system model to obtain the stiffness matrix for the system.
(Hint: write a simple program for assembly based on the solution of Example 6.10)

6.6. Use the Gaussian elimination to find the solution of the following linear equations.

$$\left.\begin{array}{l} 3x_1 - x_2 + 2x_3 = 6 \\ x_1 + 5x_2 \quad\quad = 8 \\ 2x_1 - 4x_2 + x_3 = 0 \end{array}\right\}$$

6.7. Use the LU decomposition to find the solution of the following linear equations.

$$\left.\begin{array}{l} 2x_1 - x_2 + 6x_3 = 4 \\ x_2 + 2x_3 = 5 \\ 4x_1 - 3x_2 - x_3 = -8 \end{array}\right\}$$

6.8. Write and run your program to verify the solution provided in Example 6.16.

6.9. Write and run your program to verify the solution provided in Example 6.17.

6.10. Write and run your program to verify the solution provided in Example 6.18.

6.11. Write and run your program to verify the solution provided in Example 6.19.

6.12. Write and run your program to verify the solution provided in Example 6.20.

Computer Implementation

7.1 INTRODUCTION

On the one hand, finite element analysis (FEA) is the computer implementation of generic solutions to boundary value partial differential equations; on the other hand, users are required to have a good understanding of physical systems to model and solve design problems by FEA. Austrell et al. (2004) have identified a number of the limitations in traditional FEA course designs. Either students are required to write their own programs, which are very time-demanding, or the teaching is tied to one of commercial FEA packages. When the commercial code is used, some of the concept-based goals are hard to be achieved because students can solve design problems without a thorough understanding of underlying physical and numerical methods. It is important for students to know how to write their own finite element code for some basic problems, but it should be in more time-efficient ways than traditional programming.

In this chapter, object-oriented (OO) programming and Matlab are used to illustrate computer implementation of FEA modeling. Students will focus on the understanding of the functions, relations, and interactions of modules in computer programs, rather than the detailed implementation of methods and algorithms.

7.2 OVERVIEW OF COMPUTER PROGRAMMING

Even though the steps and activities involved in FEA modeling are straightforward and can be well formulated. When design problems become complicated and diversified, it is extremely trivial, error-prone, and impractical to define and solve an FEA model by hands. *Computer programming* is a process to convert problem formulation and the problem-solving method into *computing programs*. Using same programs, the numerical computation can be executed by computers to obtain the solutions to different design problems with the same formulation of mathematic models.

A computer program often refers to a sequence of the instructions that are used to fulfill specific tasks or solve given problems automatically.

Finite Element Analysis Applications. http://dx.doi.org/10.1016/B978-0-12-809952-0.00007-8

Therefore, the design of computer program requires a good understanding of the domain knowledge of design problems as well as computer languages and algorithms. The outcome of computer programming is program code; but computer programming also includes the activities of selecting language and operating systems, determining software architecture, developing algorithms, and verifications of algorithms.

7.2.1 **Commercial and open access FEA packages**

In regards to computer programming for FEA, a great deal of resources is available online. The software tools in the FEA field can be classified into two categories: commercial FEA packages and open access tools. A commercial FEA tool usually has better capabilities in terms of the versatility for a wide scope of design problems, the robustness and efficiency in obtaining design solutions, and the support in guiding users to define and run FEA simulations. However, a commercial FEA tool is often encapsulated; it prevents users to learn the algorithms behind the scenes. An open-access FEA tool is usually very sophisticated, and it only deals with limited types of design problems. However, one can access source code, modify and customize the tool to meet their own needs if he or she has the basic understanding of the FEA methodology. From this perspective, open-access FEA tools are most suitable for students to learn the computer implementation of the FEA theory. Tables 7.1 and 7.2 list some popular commercial tools and open-access tools, respectively.

7.2.2 **Computer programs**

Hands-on experience of computer programming helps students to understand the computer implementation of FEA modeling, thus to use available FEA tools adequately. The function of a computer program is analogue to that of a manufacturing process. A manufacturing process transfers the tangible materials in one state to another; although a computer program is to convert intangible input data into output data. As shown in Fig. 7.1, a program includes some methods or algorithms to process input data and transfer to output data. In data processing, some intermediate data must be stored to facilitate the executions of algorithms. Depending on the functions of a program, in some cases, one program has to be performed repeatedly to optimize some design variables or design criteria.

To define and solve an FEA model by a computer program, one must clarify what types of data the program to be processed, and what methods and algorithms can be applied to perform data process. Taking a reference of the generic FEA modeling in Chapter 1, the corresponding data types and

Table 7.1 Examples of Commercial Finite Element Analysis Software Tools[a]

Software	Feature	Developer	Recent Version	Platform
Abaqus	A unified product suite offers powerful and complete solutions for both routine and sophisticated engineering problems covering a vast spectrum of industrial applications.	Abaqus Inc.	6.14-AP in 2014-11	Linux, Windows
ANSYS	A general-purpose software used to simulate interactions of all disciplines of physics, structural, vibration, fluid dynamics, heat transfer, and electromagnetic for engineers	Ansys Inc.	17.0 in 2016-01	Windows, Linux
COMSOL Multiphysics	A suite of add-on products expands this multiphysics simulation platform for modeling specific application areas as well as interfacing with third-party software and their capabilities	COMSOL Inc.	5.2 on 2015-11-16	Linux, Mac OS X, Windows
SolidWork Simulation	Computer-aided design and computer-aided engineering (CAD and CAE) integrated package with SolidWorks	Dassault Systèmes SolidWorks Corp.	2017 version	Windows
LS-DYNA	The code's origins lie in highly nonlinear, transient dynamic finite element analysis using explicit time integration	Livermore Software Technology Corporation	R9.0.1 in 2016	Linux, Windows
Nastran	Originally developed for National Aeronautics and Space Administration, now available commercially from several software companies	MSC Software	2015	Linux, Mac OS X, Windows
RFEM	A sophisticated software for quick and easy modeling, structural analysis and design of two-dimensional and three-dimensional models consisting of member, plate, wall, folded plate, shell, solid, and contact elements	Dlubal Software	5.0.6 in 2016-02	Windows

https://en.wikipedia.org/wiki/List_of_finite_element_software_packages.
http://www.globalstressengineers.info/2013/02/list-of-finite-element-software-packages/.
[a]Please see the references for the links of these software tools.

Table 7.2 Examples of Open Access Finite Element Analysis (FEA) Tools[a]

Software	Feature	Developer	Recent Version	Platform
Agros2D	In C++, two-dimensional coupled problems, sophisticated tools for building geometrical models and input of data, generators of meshes, tables of weak forms for the partial differential equations, and tools for evaluating results and drawing graphs and maps	University of West Bohemia	3.2 on 2014-03-03	Linux, Windows
Hermes Project	Modular C/C++ library for rapid development of space- and space—time adaptive hp- finite element method (FEM) solvers	hp-FEM group	3.0 on 2014-03-01	Linux, Unix, Mac OS X, Windows
CalculiX	The solver uses a partially compatible ABAQUS file format. The pre-/postprocessor generates input data for many FEA and computational fluid dynamics (CFD) applications.	Guido Dhondt, Klaus Wittig	2.1 on 2016-07-31	Linux, Windows
CALFEM	An interactive computer program for teaching the FEM. The name CALFEM is an abbreviation of "computer aided learning of the finite element method." The program can be used for different types of structural problems and field problems.	Lund University	3.4 on 2015-12-15	Linux, Unix, Mac OS X, Windows
Code Aster	Civil and structural engineering finite element analysis and numeric simulation in structural mechanics in Python and Fortran.	EDF	12.5 on 2015-12-14	Linux, FreeBSD
Deal.II	Written in C++, comprehensive set of tools for finite element codes, scaling from laptops to clusters with 10,000 + cores.	Wolfgang Bangerth, Timo Heister, Guido Kanschat, Matthias Maier et al.	8.3.0 on 2015-08-01	Linux, Unix, Mac OS X, Windows
DUNE	Written in C++, distributed and unified numeric environment	Christoph Grüninger	2.4.1 on 2016-02-29	Linux, Unix, Mac OS X
Elmer	Multiphysical simulation software developed written primarily in Fortran (written in Fortran 90, C and C++)	Finnish Ministry of Education's CSC	8.0 on 2015-05-05	Linux, Mac OS X, Windows
FEATool	Easy to use Matlab and Octave Multiphysics FEM Solver and GUI Toolbox	Precise Simulation	1.3 on 2015-09-16	Linux, Unix, Mac OS X, Windows

Table 7.2 Examples of Open Access Finite Element Analysis (FEA) Tools[a] *continued*

Software	Feature	Developer	Recent Version	Platform
FEAP	Designed for research and educational use that was derived from the FEAP [2] program	Robert L. Taylor	8.4 on 2015-09-19	Linux, Unix, Mac OS X, Windows
FEBio	Finite elements for biomechanics	University of Utah, MRL	2.5 on 2015-10-09	Linux, Mac OS X, Windows
FreeFem++	C++ for rapid testing and finite element simulations. The problem is defined in terms of its vibrational formulation	Université Pierre et Marie Curie and Laboratoire Jacques-Louis Lions	3.45 on 2016-03-11	Linux, Mac OS X, Windows, Solaris
GetFEM++	A generic finite element library written in C++ with interfaces for Python, Matlab, and Scilab. It focuses on modeling of contact mechanics and discontinuities (e.g., cracks)	Yves Renard, Julien Pommier	5.0 in 2015-07	Unix, Mac OS X, Windows
OOFEM	Object Oriented Finite EleMent solver, written in C++	Bořek Patzák	2.4 on 2016-02-15	Unix, Windows
OpenFOAM	Originally for CFD only, but now includes finite element analysis through tetrahedral decomposition of arbitrary grids	OpenCFD Ltd	3.0.1 on 2015-12-15	Unix, Linux
SimScale	100% web-based CAE platform in German	SimScale GmbH	14 in 2013-07	Web browser
VisualFEA	Finite element software for structural, geotechnical, heat transfer, and seepage analysis	Intuition Software	5.11 in 2016-01	Mac OS X, Windows

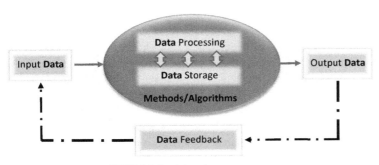

■ **FIGURE 7.1** Description of a computer program.

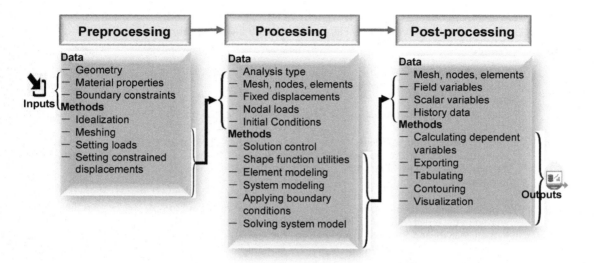

■ FIGURE 7.2 Typical data and methods in the procedure of finite element analysis modeling.

methods can be identified as shown in Fig. 7.2. An FEA consists of three basic phases, i.e., *the preprocessing phase* where data is prepared in the format of an FEA model, *the processing phase* where a system model is developed and solved, and *the postprocessing phase* where the solution data is visualized and exported.

The primary data at the preprocessing phase includes (1) the geometrics and material properties of solid objects and (2) the boundary constraints occurring to the interfaces of solid object in its residential environment. Taking an example of a static analysis problem, boundary constraints can be further classified into external loads and imposed displacements. At the preprocessing phase, typical tasks consist of cleaning the geometry of a solid model, generating a mesh, and setting boundary conditions and loads. Idealization is to convert a physical system to a mathematical model. This is a critical step in defining design problem, and it relies heavily on designers' expertise. The *idealization* of a solid model is to eliminate unnecessary small features such as holes and chamfers. These features will have negligible impact to the solution, but would affect the quality of FEA model and convergence of solving processes. A continuous domain has infinite number of degrees of freedom due to the continuity of computational domain. To make numerical simulation practical, a continuous domain must be discretized into a mesh, so that a finite number of degrees of freedom can be applied to approximate the behaviors of the original solid model. The process for the discretization is called *meshing*. In addition, the constraints of a design problem are represented by external loads and imposed displacements at the boundaries, these loads and displacement have to be defined in the FEA model.

The outcomes of preprocessing are converted as the input data of processing. Therefore, the basic data types at the processing phase are *analysis type*, *mesh*, *nodes*, *elements*, specified *loads* and *displacements*. If the solution has to be spanned along the time frame, data at the processing phase also includes initial conditions. The processing phase involves in a number of tasks as below. (1) *Solution control* is to determine solution types and implement solving process. Fig. 7.3 shows an example of available analysis types in the Ansys workbench. Applicable analysis types vary from static structural analysis, dynamics, thermal analysis, fluid flow, magneto-static analysis, and multiphysics simulation. (2) *Shape function utilities* are for the representation of behavior of a continuous domain based on the variables for nodal degrees of freedom. The value of a state variable at any position in an element can be evaluated from nodal values of variables. (3) *Element modeling* is to define a relational model of external loads and nodal degrees of freedom which are expressed in the matrix form. (4) *System modeling* is to assembly all element models together as a whole with respect to a global coordinate system (GCS). A system model usually consists of a vector of design variables, a vector of nodal loads, and a matrix to represent the corresponding of these two vectors. (5) *Applying boundary conditions*. A system model without constraints cannot be solved due to insufficient constraints to eliminate the rigid motion of nodes. Applying boundary control is to ensure that a unique solution can be obtained. Boundary constraints of a mechanical system will include the

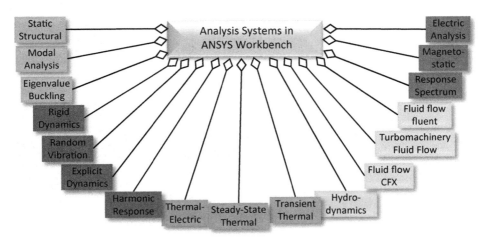

■ **FIGURE 7.3** Exemplifying analysis types in ANSYSs workbench 17.1.

external loads and imposed displacements. (6) *Solving a system model.* After boundary conditions are applied in, the system model can be solved to calculate the distributions of field variables.

The outcomes from the processing phase include the distributions of state variables over nodes and elements. This data set becomes the input data for the postprocessing phase. Tasks in postprocessing include the evaluation of dependent variables based on obtained independent variables, exporting and visualizing FEA results.

7.2.3 **Programming techniques**

Fig. 7.1 has shown that a computer program is about *data* and *data processing*. In the history of information technology (IT), different techniques were developed to represent, store, and process data in computers. As shown in Fig. 7.4, programming techniques can be classified into four types according to the level of abstract. (1) In *unstructured programming*, execution commands are run directly on specified values. Data is given in the form of values and data was processed by the commends in a commend line interactively. The commends are unstructured, and the sequence of commends is given by the order of inputs. This technique is applicable only for basic calculation. (2) In *procedural programming*, variables or dimensions are used as containers to keep data, and scripts are used to gather a sequence of commends as a batch to process data. A computer program is divided

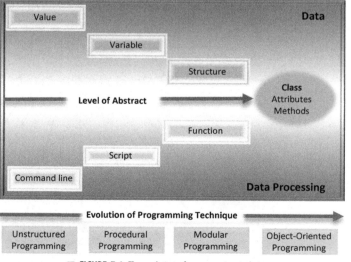

■ **FIGURE 7.4** The evolution of programming technique.

into small blocks of code as routines, which can be called repetitively. The methods and algorithms in procedural programming are structured based on the data flow where data is processed and passed from one routine to another. (3) *Modular programming* divides the functionality of a program into a set of independent and interchangeable modules; each module contains and maintains its own data, which is needed to perform its own functionality. In modular programming, diversified and relevant data can be defined as *structures*. (4) *Object-oriented programming* (OOP) differs radically from other programming techniques in sense that data and the methods for data processing are unified in new entities called as *classes*. In OOP, *objects* are the instances from classes where data is assigned to *attributes*, and the implementation of procedures is assigned to *methods*. An important feature of the object definition is that an object's procedures can access and often modify the data fields of the object with which they are associated.

OOP paradigm is used in this chapter. Because most of FEA codes are developed using structural or modular programming paradigms, a brief comparison of procedural program and OOP in Table 7.3 helps readers to understand the rationales why OOP is used here to illustrate computer implementation of FEA modeling.

7.2.4 **Matlab as programming language**

Different programming techniques and languages have been adopted to solve complex engineering models, and it is critical to design and implement computer solutions in formal and systematic ways. FEA software tools are not only versatile tools to deal with a wide scope of engineering problems, but also pedagogical methods in engineering education (Austrell et al., 2004). The FEA methodology divides a complex design problem into a set of simple subproblems, and the system-level model can be defined by an assemblage of element models together with applied boundary conditions. Because FEA solutions are developed for differential or integral equations, which are applicable to different physical applications; same solvers can be used to solve a variety of design problems.

In procedural programming, a software tool is developed by a stepwise refinement of the system's abstract function. Such type of abstract is well supported by some programming languages such as C, Fortran, and Matlab. In OOP, the implementation of data abstraction relies on objects instantiated from classes and the methods associated with classes. This type of abstraction is well supported by OO languages including C++ and Java. Matlab began to fully support OOP since its R2008a version; new functionalities were developed to be compatible with an OOP standard (Scardapane, 2015).

Table 7.3 A Comparison of Procedural Programming and Object-Oriented Programming

	Procedural Programming	**Object-Oriented Programming**
Data	Represented by variables, dimensions, or structures	Represented by attributes in classes
Data processing	Program-level function of data processing is divided into pieces, each piece corresponds to a function, which can be implemented by a routine or subroutine	Data processing is performed by the methods defined in class
Security	It is less secure since the programming paradigm does not have a proper way to hiding data.	It is more secure since the class definition provides the mechanism to support data hiding.
Structure	Top-down design	Use objects to keep data and define interactions of objects to pass over data
Reuse	Limited code reuse	Excellent code reuse
Complexity	Complex in coding	Complex in defining classes
Transparency of methods	Poor since most of data is global data, and it is hard to identify the input and output of a method in a program.	Good observe only relevant data in a class is involved in the definition of a method.

However, Matlab has been developed to perform scientific computation. Because vectors and matrices are widely used in defining and solving various engineering problems, Matlab has a vast number of functions, which can be used directly for the manipulation of matrices. Matlab program code could be much concise and shorter for equivalent implementations in other languages such as C, Fortran, or Java. It allows one to code numerical methods by accessing a vast library of functional for manipulation of matrices. In addition, Matlab provides an integrated programming platform with a large number of specialized toolboxes and graphical interfaces (Ferreira, 2009). The extensive mathematics and graphics functions can relieve a programmer's burden in developing these functions. Therefore, Matlab is chosen here to illustrate the computer implementation of FEA modeling; readers are able to focus on the understanding of FEA procedure and principles rather than the programming of operations in linear matrix algebra.

7.2.5 **System architecture**

System architecture is a concept model that define components, behaviors, structure, and views of a system. Architecture is often defined as a formal representation of a system. Fig. 7.5 shows the system architecture of the computer implementation of FEA modeling in this chapter. The inputs of a computer program consist of two catalogues, i.e., *inputs of design problems* and *inputs for controls* for modeling and solving process. The outputs are the solutions of state variables, the evaluated dependent quantities, the visualization, and the exported data. The modules in *system architecture* include the methods and algorithms to process data, store data, and control the data flow and modeling process. Correspondingly, the hardware architecture includes computing resources such as central processing units (CPUs) for data processing, internal and external memories for data storage, motherboards, and other computer modules for process control. Both of hardware and software architectures must include the interfaces to receive inputs and export outputs of FEA programs.

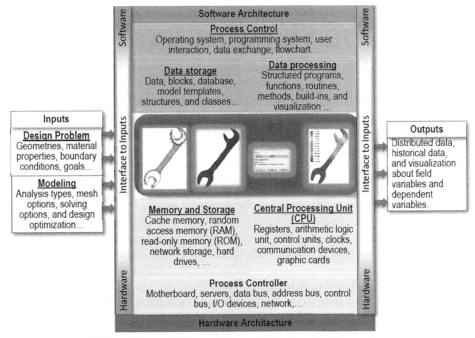

■ **FIGURE 7.5** Hardware and software architecture of finite element analysis computer program.

7.3 **UML FOR OOP**

A model is a representation of a computer program to capture important aspects of the program to be implemented. The unified modeling language (UML) is a generic modeling language to specify, visualize, construct, and document the artifacts of a software system. An artifact is a physical unit of information or behavior description in a computer system. UML itself is not a programming language, but it can be used to understand and support programming. UML is developed to simplify and consolidate the OOP.

An UML describes the decisions and understanding of a system to be constructed. The purpose of a UML-model is to understand, design, browse, configure, maintain, and control the information of a software system. UMI includes semantic concepts, notation, and guidelines. The UML specification does not define a standard process but is intended to be useful with an iterative development process. It is intended to support most existing OO development processes (Rumbaugh et al., 2004).

In UML, a system is modeled as a collection of discrete objects that interact with each other to perform the work that eventually leads to the solutions of design problems. The static structure defines the types of objects and their relations, which are important to a software system and its application. The dynamic behavior defines the history of objects over time and the communications among objects to accomplish goals (Rumbaugh et al., 2004).

The UML provides an organizational view to divide a complex system into workable pieces. It helps to understand and control the dependencies of workable pieces. In a complex system, the model information must be divided into coherent pieces, so that programmers can deal with manageable tasks at a time. Even for a small system, human designer requires to organize the contents of model as a set of packages with a reasonable size. Packages are general-purpose hierarchical units of UML model structures.

In any computer model, one has to begin with the definition of key concepts from the applications, internal properties of key concepts as well as their mutual relations. This group of constructs is called the static view of an UML model. Key concepts are defined as *classes*. Each class describes certain type of discrete objects that hold certain information and communicate to implement the behaviors that objects can perform. The information an object holds is called as *attributes*; behaviors an object can perform are modeled as *methods*. If one class includes common attributes and methods of another class, this class can be defined as a *child* class of the other class (i.e., a *parent* class). The mechanism of defining a child class from a parent class is called *generalization*. A child class can add new attributes and

Table 7.4 List of Some Common Types of Relations (Wikipedia, 2017)

Relation	Symbol
Association	⟶
Inheritance	⟶▷
Realize/implementation	---------▷
Dependency	---------⟶
Aggregation	⟶◇
Composition	⟶◆

methods to existing ones inheriting from its parent class. Objects have run-time connections to other individual objects, such object–object relations are modeled as *associations* of classes. Table 7.4 lists some common types of relations (Wikipedia, 2017).

7.4 STATIC STRUCTURE OF FEA COMPUTER IMPLEMENTATION

The data involved in an FEA program is sorted and treated as attributes in different classes, and methods in these classes are defined for data processing. Fig. 7.6 shows a static structure of FEA computer implementation for one analysis type, for example, the stress analysis for a truss structure. A root class *CModel* is defined to include all of the relevant data and the methods for data processing. A *CModel* class consists of a *CDomain* class for the description of a computational domain, a *CMaterial* class for material properties and constitutive models, a *CBoundaryLoad* class for external loads on object, a *CBoundaryDoF* class for imposed displacements, a *CNode* class and a *CElement* class for attributes and methods of objects of nodes and elements, respectively. A *CEProperties* class is defined to include extra properties for elements; for example, the cross-section area of an axial member. The *CModel* class must also include a *CSolver* class to generate and solve system models, and a *CPostPro* class to calculate dependent variables and scalars and visualize analyzed data for the design model. The author has developed a prototype of a computer program, and as follows, the details of each class are explained.

7.4.1 CModel class

Depending on the functionalities of a computer program to perform different analysis type, a CModel class aggregates a number of subclasses for different FEA models. Note that the purpose of this computer program

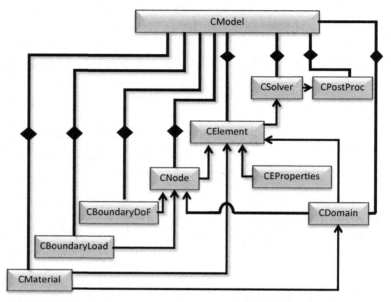

■ **FIGURE 7.6** Static structure of general finite element analysis computer implementation.

is to illustrate the application of OOP in the computer implementation. It is not our goal to develop a comprehensive FEA package, which is competent to other commercial FEA packages. Therefore, some basic analysis types are discussed, the developed computer program is capable of analyzing spring systems, truss structures in 2D and 3D, frame structure, 2D heat transfer problems, planar stress problems, and planar strain problems. As shown in Fig. 7.7, the CModel class consists of *C1DSpringModel*, *C2DTrussModel*, *C3DTrussModel*, *C2DBeamModel*, *C3DFrameModel*,

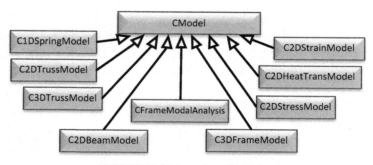

■ **FIGURE 7.7** Subclasses of root CModel class.

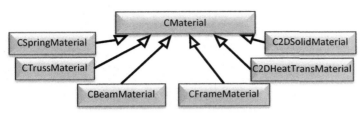

■ **FIGURE 7.8** Subclasses of root CMaterial class.

CFrameModelAnalysis, *C2DHeatTrans*, *C2DStressModel*, and *C2DStrain-Model*. These analysis types are sufficient to readily solve the majority of homework problems in conventional FEA textbooks by computers.

7.4.2 **CMaterial class**

The number of attributes and the constitutive model vary from one application to another. Therefore, a set of classes for various materials have been developed for different analysis types in Fig. 7.8. Readers can easily insert new types of materials such as nonlinear materials or composite materials if this is required by applications.

7.4.3 **CDomain class**

In OOP, objects are the instantiations from classes. An object instanced from a CDomain defines a computational domain, and the methods included in a CDomain are to discretize the computational domain into nodes and elements. The critical attributes of a CDomain object are a set of nodes and elements from meshing processes. As shown in Fig. 7.9, a variety of subclasses in a CDomain class include an assembly of parts, a geometry of a part in one- (1D), two- (2D), and three-dimensional (3D). Because meshing is a task of preprocessing, which needs special software tools, it has not been covered here. For simple parts, readers are encouraged to define a CDomain objects directly by specifying its attributes (e.g., nodes and elements) manually.

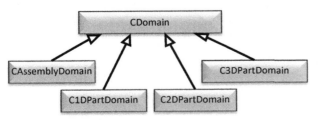

■ **FIGURE 7.9** Subclasses of root CDomain class.

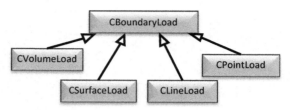

■ **FIGURE 7.10** Subclasses of root CBoundaryLoad class.

7.4.4 **CBoundaryLoad class**

In establishing an FEA model, all external loads are applied on degrees of freedom of nodes. Depending on the types of a continuous domain, external loads can be *volume*, *surface*, *line*, or *point loads*; the methods in the CBoundaryLoad class are to transfer external loads into nodal loads on the specified degrees of freedom. Fig. 7.10 gives a list of subclasses of CBoundaryLoad class where additional attributes and methods are needed to find the corresponding nodal loads for external volume, surface, line, and point loads.

7.4.5 **CBoundaryDoF class**

A CBoundaryDoF class is to define the constraints on degrees of freedom on nodes. Note that a degree of freedom can be a displacement along certain direction or another physical quantity such as temperature in heat transfer or pressure in fluid flow. In Fig. 7.11, the subclasses of CBoundaryDoF are created based on (1) the number of affected nodes, the constraints can involve in single point (*CSinglePointDoF*) or multiple points (*CMultiPointDOF*), (2) if the number of DoF at a node is fully or partially fixed, i.e., *CFFixedDoF* and *CPFixedDoF*, respectively.

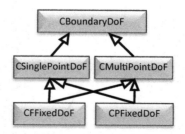

■ **FIGURE 7.11** Subclasses of root CBoundaryDoF class.

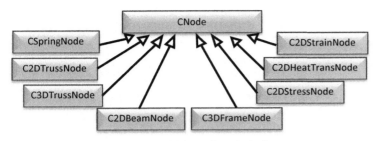

■ **FIGURE 7.12** Subclasses of root CNode class.

7.4.6 **CNode class**

The computational domain of a design problem is represented by a set of discretized nodes and elements. Fig. 7.12 shows the defined CNode class with the attributes associated to nodes. Besides the coordinates for nodal positions, and the attributes also include the quantities of state variables on the node, for example, the displacement in static analysis, temperature in heat transfer, and pressure in fluid flow. In addition, equivalent nodal loads are also treated as the attributes of CNode class. After the FEA is solved, some methods in a CModel class will update the solutions of state variables to the attributes of state variables.

7.4.7 **CElement class**

Similar to CNode classes, the definitions of CElement classes are directly associated with the available types of analysis types in the CModel class. It is desirable that each analysis type of CModel class corresponds to a subclass of CElement. Objects of both of CNode and CElement classes are instantiated when the meshing method of a CDomain is performed. As shown in Fig. 7.13, the attributes in a CElement class include its nodal information, some specific properties of element defined as the attributes of

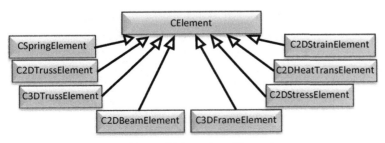

■ **FIGURE 7.13** Subclasses of root CElement class.

a CEProperties, for example, the cross-section of a truss element. Some dependent variables are also treated as the attributes whose values can be obtained from postprocessing. It is critical that a CElement class includes some methods to define element models.

7.4.8 **CEProperties class**

In some simple types where the solid geometry of element is not given, additional system parameters have to be specified for elements. For examples, cross-section area must be given to a truss member, one or three moments of areas must be given to a beam or frame members, respectively; so is the thickness for a plate member. An exemplifying list of subclasses under the CEProperties class is illustrated in Fig. 7.14.

7.4.9 **CSovler class**

A CSolver class corresponds to the activities involved at the phase of processing. The method to solve different types of system model must be defined. For numerical computation of an FEA model, three basic types of analysis problems are equivalence problems, eigenvalue problems, and time-variant transient problems. Therefore, the root CSolver class includes *CEquivalenceSolver*, *CEigenSolver*, and *CTransSolver* for three types of analysis problems, respectively. As shown in Fig. 7.15, a CSolver class is associated with objects instantiated from the CElement class to obtain the information about the system model and in turn, pass over the obtained solutions to the objects of elements and nodes.

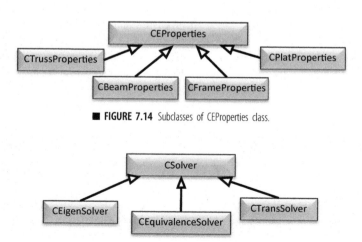

■ **FIGURE 7.14** Subclasses of CEProperties class.

■ **FIGURE 7.15** Subclasses of root CSolver class.

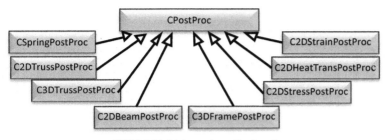

■ **FIGURE 7.16** Subclasses of root CPostProc class.

7.4.10 **CPostProc class**

A CPostProc class includes the attributes of the solved FEA model, and it also includes the methods to determine dependent variables and other scalar quantities based on obtained state variables. Because the attributes of dependent variables vary with the analysis types of element types, the subclasses of a CPostProc class can be defined based on the element types in a computer program. Corresponding to element subclasses in Fig. 7.16, the sub-CPostProc classes in our computer program include *CSpringPostProc*, *C2DTrussPostProc*, *C3DTrussPostProc*, *C2DBeam-PostProc*, *C3DFramePostProc*, *C2DStressPostProc*, *C2DHeatTransPost-Proc*, and *C2DStrainPostProc*.

7.5 **COMPUTER IMPLEMENTATION OF 1D SPRING SYSTEMS**

To illustrate OOP for various applications, the computer implementations for a few FEA applications are introduced. The classes are defined locally so that the code for each analysis type is self-contained. The static structure for 1D spring systems is shown in Fig. 7.17, where the subclasses from their generic classes are directly defined. Because Matlab is a very powerful language to deal with matrix algorithms, the implementations of methods involved in the *CEuivalenceSolver* and *CSpringPostProc* are very straight-forward. Therefore, these two classes are not defined separately; their methods will be moved into the definition of *CSpringModel* class to avoid the complexity in passing data of attributes from *CSpringModel* to *CSpring-PostProc* or *CEquivalenceSolver*.

A CSpringMaterial class is simple because it requires only one attribute for the representation of equivalent stiffness coefficient for element. Besides the method to construct an object from the class, other two methods are defined for setting and getting stiffness coefficient,

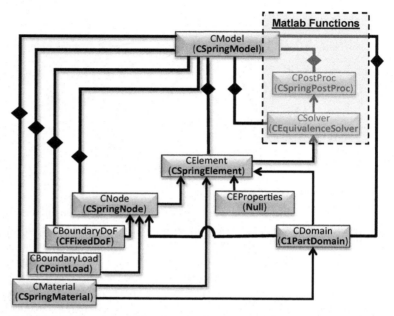

■ **FIGURE 7.17** Static structure of finite element analysis computer implementation for spring systems.

respectively. Table 7.5 shows the representation of CSpringMaterial Class and the corresponding Matlab code.

Two important methods in *C1DSpringModel* are (1) *create_sys_model* () to define a system model and (2) *apply_BCs* () to apply boundary conditions and modify the system model. The implementations of these methods for a 1D spring system are illustrated in Tables 7.6A and B.

7.6 APPLICATIONS OF FEA PROGRAM FOR 1D SPRING SYSTEMS

Due to the similarity of governing mathematical models, 1D linear systems in many different systems can be readily solved by the program in Section 7.4. To apply the program, Table 7.7 shows the equivalent stiffness coefficient (k) for different physical systems.

7.6.1 Spring system example

Fig. 7.18 shows a spring assemblage, which consists of five springs. The left side is fixed and the external load is applied on the right side. The lateral supports (*supports* 2, 3, and 4) ensure that all springs are deformed along

Table 7.5 Definition of Classes for One-Dimensional Spring Systems

Class Definition	Description
Class Name attributes/properties; methods/functions;	In OOP, a class can be visualized by a block with three elements: ■ The upper field gives a class name as an identifier of the class, ■ The middle field lists the attributes associated with the data of class. *Attributes* are called as *Properties* of a class in Matlab, and ■ The last field gives a list of *Methods* for data processing. Methods are defined as *Functions* in Matlab. A class definition always includes a method (i.e., *constructor*) for an instantiation of an object from the class. The function of a constructor has the same name with the class.

CSpringMaterial
 	
k;
 	
CSpringMaerial (val);
set_k(val);
get_k();
cal_stress();
	CSpringMaterial class is to define a type of spring material. The only attribute property is the stiffness coefficient *k*.
	The methods for CSpringMaterial include (1) a constructor to create an object by specifying *k*, (2) *set_k(val)* to set the stiffness coefficient *k* for a class object, (3) *get_k()* to get the stiffness k from an class object, and (4) *cal_stress ()* to define the relation of stress and strain relation for materials under the investigation.

C1DPartDomain
 	
n_points;
coords=[];
 	
C1DPartDomain (n_points, coords);
get_n_points();
get_coord_i(i);
get_coords();
	C1DPartDomain class is to specify the domain to be analyzed. Here, we assume that the discretization is completed manually. Therefore, a one-dimensional part domain can be defined as a set of discrete points along *x*- axis.
	This class includes two attributes (1) the number of discrete points *n-points* and (2) the coordinates of these discrete points *coords* []. The methods for CSpringMaterial include (1) a constructor to create an object by specifying number of discrete points and their coordinates, (2) *get_n_points ()* to read the number of points, (3) *get_coord_i(i)* to get the coordinates for the *i*-th point, and (4) *get_coords ()* to get a dimension of coordinates for all discrete points.

CFFixedDoF
 	
DoF_i;
dis;
 	
CFFixedDoF(DoF_i, dis);
set_DoF_i(val);
get_DoF_i();
set_dis(val);
get_dis();
	CFFixedDoF class is to define a type of boundary condition for fixed displacement. Two attributes of a CFFixedDoF object are the index of DoF (*DoF_i*) and the corresponding value of fixed displacement (*dis*). The methods for CFFixedDoF include (1) a constructor to create an object by specifying the index of fixed DoF and the given displacement, (2) *set_DoF_i (val)* to set the index of the fixed DoF, (3) *get_DoF_i (i)* to read the index of fixed DoF, (4) *set_dis(val)* to set the fixed displacement, and (5) *get_dis()* to read the fixed displacement.

Continued

Table 7.5 Definition of Classes for One-Dimensional Spring Systems *continued*

Class Definition	Description

CPointLoad
 	
DoF_i;
load;
 	
CPointLoad(DoF_i, load);
set_DoF_i(val);
get_DoF_i();
set_load(val);
get_load();
	CPointLoad class is to define a type of concentrated load. Its attributes include the index of DoF (*DoF_i*) where the load is applied and the load (*load*).
	The methods for CPointLoad include (1) a constructor to create an object by specifying the index of DoF where the load is applied, (2) *set_DoF_i* (val) to set the index *DoF_i* for an applied load (3) *get_DoF_i* (*i*) to read the index *DoF_i*, (4) *set_load* (val) to set the load value, and (5) *get_load* () to read the applied load.

CSpringElement
 	
k;
matrix_k=zero2(2,2);
i_node;
j_node;
force=0;
 	
CSpringElement(val);
get_k_item(i, j);
set_nodes(i, j);
get_i_node();
get_j_node();
get_force(val)
	$i \overset{\wedge}{\underset{k}{\text{—WW—}}} \bullet j \quad \begin{bmatrix} k & -k \\ -k & k \end{bmatrix}$
	CSpringElement class is to define a type of spring element. As shown above, the attributes of a spring element are (1) the stiffness coefficient *k*, (2) the indices of two nodes (*i_node* and *j_node*), (3) the stiffness matrix for element model (*matrix_k*), (4) the internal force under given deflection (*force*).
	The methods for *CSpringElement* are (1) a constructor to create an object with the given stiffness coefficient *k*, (2) *get_k_item* (*i*, *j*) to read an element at (*i*, *j*) in *matrix_k*, (3) *set_nodes* (*i*, *j*) to specify the indices of two nodes, (4) *get_i_node* () and *get_j_node* () to read the indices of two nodes, respectively, and (5) *get_force* () to calculate the force based on the given deflection.

CSpringNode
 	
x=0;
i_force=0;
i_dis=0;
external_f=0;
fixed_x=0;
 	
CSpringNode(val);
setd_node_1D(dis);
setf_node_1D(force);
getd_node_1D();
getf_node_1D();
get_i_dis();
	CSpringNode class is to define a node type for a spring element. Besides one attribute (*x*) for its location, this class has indicators (*i_force* and *i_dis*) to identify if there is an external load or fixed displacement on the node. Correspondingly, the attributes of *external_f* and *fixed_x* are the values of external load or fixed displacement if the node has.
	An object can be constructed from *CSpringNode* by specifying the nodal position *x*. The class also includes (1) *setd_node_1D* (*dis*) to set the fixed displacement, (2) *setf_node_1D* (*force*) to set an applied load, (3) *getd_node_1D* () to get the displacement, (4) *getf_node_1D* () to get the applied force, and (5) *get_i_dis* () to read the indicator for a fixed displacement.

Table 7.5 Definition of Classes for One-Dimensional Spring Systems *continued*

Class Definition	Description

C1DSpringModel
 	
n_materials;
material_ks=[];
n_points;
n_nodes;
coords=[] ;
node_dof=1;
n_elements;
ele_materialtype[];
n_dis;
node_dis=[];
set_dis=[];
n_load;
node_load=[];
set_load=[];
connect_m=[];
OSpringMaterial=CSpringMaterial.empty;
O1DPartDomain=C1DPartDomain.empty;
ONdes=CSpringNode.empty;
OElements=CSpringElement.empty;
OFixedDoF=CFFixedDoF.empty;
OLoad=CPointload.empty;
del_dof=[];
sys_k=[];
sys_f=[];
m_sys_k=[];
m_sys_f=[];
sys_u=[];
m_sys_u=[];
 	
C1DSpringModel(n_materials, n_points,
    n_nodes, n_elements, node_dof,
    coords, n_dis, node_dis, n_load,
    node_load, set_dis, set_load,
    material_ks, connect_m);
create_sys_model();
apply_BCs();
sys_solver();
post_proc();
	C1DSpringModel class is to define a FEA model of a one-dimensional spring system. It includes the attributes of all input data and intermediate data, and it has the most important methods for data processing and data postprocessing. As shown on the left, the attributes in this class can be classified into three groups, which relate to the data at the phases of *preprocessing*, *processing*, and *post-processing*, respectively.
	Because the program assumes that the proprocessing is completed manually, all information of the discretized model must be available. Input data of the program includes the number of material types (*n_materials*), a dimension for stiffness coefficients of these materials (*material_ks[]*), the number of discrete points (*n_points*) and corresponding coordinates (*cords[]*), the number of nodes (*n_nodes*), the degrees of freedom on each node (*node_dof=1* for 1-D spring), the number of elements (*n_elements*), types of materials of elements (*ele_materialtype[]*), the number of fixed displacements (*n_dis*), nodal indices (*node_dis[]*) and fixed displacements (*set_dis[]*), the number of applied loads (*n_load*), nodal indices (*node_load[]*) and applied loads (*set_load[]*), and the relations of nodes and elements (*connect_m[]*). The input data to the processing is stored in the objects instantiated from the classes of *CSpringMaterial*, *C1DPartDomain*, *CSpringNode*, *CSpringElement*, *CFFixedDoF*, and *CPointLoad*. The output data from data processing includes the system model (*sys_k[]*, *sys_f[]*, *sys_u[]*, *m_sys_k[]*, *m_sys_f[]*, *m_sys_u[]*), and the postprocessing is performed on the same amount of data to obtain some dependent attributes in *CSpringNode* and *CSpringElement* such as internal force in an object from *CSpringElement*. In addition, the order of DoF in the model is determined, and *del_dof[]* is introduced to record the number of fixed DoF in front of a specified index of DoF; it is used to modify system model when the boundary conditions are applied.
	The constructor of *C1DSpringModel* is to create an object of the spring system model with all input data. The input data is then processed by other methods. Namely, *create_sys_model* () is to generate a system model including the stiffness matrix *sys_k[]*, the vector of loads *sys_f[]*. The method of *apply_BCs* () is to modify the system model by applying boundary conditions; the outputs are *m_sys_k[]* and *m_sys_f []*. The modified system model is solved to obtain *m_sys_u[]* by *sys_solver* (), which is implemented by calling Matlab functions. Finally, *post-proc* () is performed to evaluate dependent variables and scalars on nodes or elements.

Table 7.6A *Create_sys_model* () in C1DSpringModel

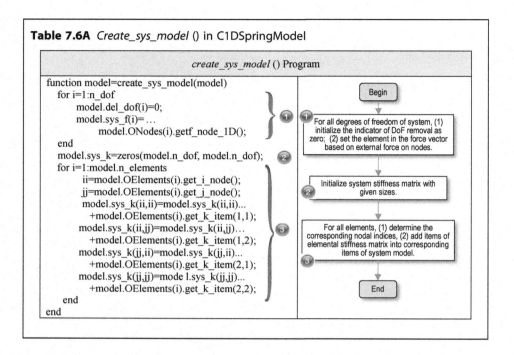

create_sys_model () Program

```
function model=create_sys_model(model)
    for i=1:n_dof
        model.del_dof(i)=0;
        model.sys_f(i)= ...
            model.ONodes(i).getf_node_1D();
    end
    model.sys_k=zeros(model.n_dof, model.n_dof);
    for i=1:model.n_elements
        ii=model.OElements(i).get_i_node();
        jj=model.OElements(i).get_j_node();
        model.sys_k(ii,ii)=model.sys_k(ii,ii)...
            +model.OElements(i).get_k_item(1,1);
        model.sys_k(ii,jj)=model.sys_k(ii,jj)...
            +model.OElements(i).get_k_item(1,2);
        model.sys_k(jj,ii)=model.sys_k(jj,ii)...
            +model.OElements(i).get_k_item(2,1);
        model.sys_k(jj,jj)=mode l.sys_k(jj,jj)...
            +model.OElements(i).get_k_item(2,2);
    end
end
```

Begin

① For all degrees of freedom of system, (1) initialize the indicator of DoF removal as zero; (2) set the element in the force vector based on external force on nodes.

② Initialize system stiffness matrix with given sizes.

③ For all elements, (1) determine the corresponding nodal indices, (2) add items of elemental stiffness matrix into corresponding items of system model.

End

the *x*-axis. Use the computer program to determine the displacements of three intersections.

Solution. The spring assemblage has a domain determined by four points ($n1$, $n2$, $n3$, and $n4$); each point is treated as a node. Each spring is modeled as a spring element. Each element has its own material type with given stiffness coefficient. In addition, there is one fixed displacement on *support* 1 and one external load on *support* 4. The connection matrix of elements and nodes can be determined based on the illustration in Fig. 7.18. Therefore, the main program to solve the problem can be given as Table 7.8.

Running the computer program gives,

$$
K = \begin{bmatrix} 55 & -30 & -25 & 0 \\ -30 & 65 & -15 & -20 \\ -25 & -15 & 55 & -15 \\ 0 & -20 & -15 & 35 \end{bmatrix} (\text{lbf/in}), \quad F = \begin{Bmatrix} -20 \\ 0 \\ 0 \\ 0 \end{Bmatrix} (\text{lbf}),
$$

$$
U = \begin{Bmatrix} 0 \\ 0.3700 \\ 0.3560 \\ 0.9354 \end{Bmatrix} (\text{in}), f^{(e)} = \begin{Bmatrix} 11.0995 \\ 8.9005 \\ -0.2094 \\ 11.3089 \\ 8.6911 \end{Bmatrix} (\text{lbf})
$$

Table 7.6B *Apply_BCs* () in C1DSpringModel

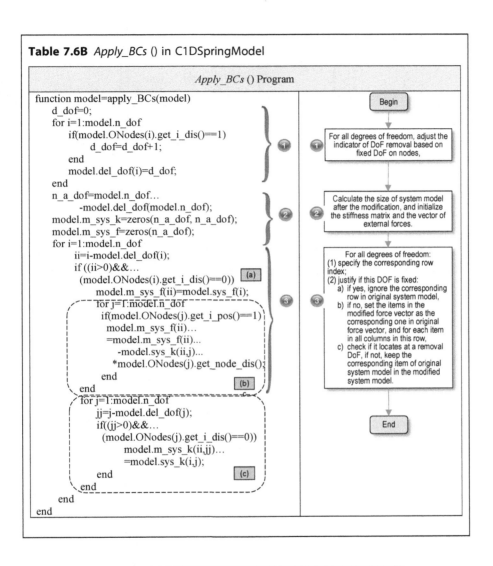

Apply_BCs () Program

```
function model=apply_BCs(model)
    d_dof=0;
    for i=1:model.n_dof
        if(model.ONodes(i).get_i_dis()==1)
            d_dof=d_dof+1;
        end
        model.del_dof(i)=d_dof;
    end
    n_a_dof=model.n_dof...
            -model.del_dof(model.n_dof);
    model.m_sys_k=zeros(n_a_dof, n_a_dof);
    model.m_sys_f=zeros(n_a_dof);
    for i=1:model.n_dof
        ii=i-model.del_dof(i);
        if ((ii>0)&&...
          (model.ONodes(i).get_i_dis()==0))    (a)
            model.m_sys_f(ii)=model.sys_f(i);
            for j=1:model.n_dof
                if(model.ONodes(j).get_i_pos()==1)
                    model.m_sys_f(ii)...
                    =model.m_sys_f(ii)...
                        -model.sys_k(ii,j)...
                        *model.ONodes(j).get_node_dis();
                end
            end                              (b)
            for j=1:model.n_dof
                jj=j-model.del_dof(j);
                if((jj>0)&&...
                  (model.ONodes(j).get_i_dis()==0))
                    model.m_sys_k(ii,jj)...
                    =model.sys_k(i,j);
                end
            end                              (c)
        end
    end
end
```

Flowchart:

① For all degrees of freedom, adjust the indicator of DoF removal based on fixed DoF on nodes,

② Calculate the size of system model after the modification, and initialize the stiffness matrix and the vector of external forces.

③ For all degrees of freedom:
(1) specify the corresponding row index;
(2) justify if this DOF is fixed:
  a) if yes, ignore the corresponding row in original system model,
  b) if no, set the items in the modified force vector as the corresponding one in original force vector, and for each item in all columns in this row,
  c) check if it locates at a removal DoF, if not, keep the corresponding item of original system model in the modified system model.

Begin → End

Table 7.7 Equivalent Stiffness Coefficient (*k*) of Different Physical Systems

Element Type	Equivalent Stiffness Coefficient (*k*)	Field Variable	Load
Axial member	$\frac{EA}{L}$	Displacement u	Force F
Heat transfer	$\frac{\lambda A}{L}$	Temperature T	Heat H
Electrical circuit	$\frac{1}{R}$	Voltage U	Current I
Pipe flow	$\frac{\pi D^4}{128\mu L}$	Pressure p	Flow rate Q

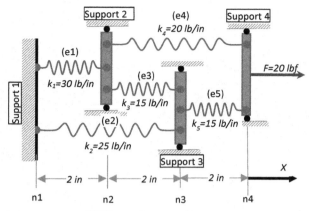

■ FIGURE 7.18 Spring assemblage example.

Table 7.8 Main Program to Calculate Displacement of Spring Assemblage

Data Inputs

n_materials=4;	% type of materials
material_ks=[30 25 15 20];	% types of materials of elements
n_points=4;	% number of points
n_nodes=4;	% number of nodes
n_elements=5;	% number of elements
node_dof=1;	% degrees of freedom (DoF) on each node
coords=[0.0 2.0 4.0 6.0];	% coordinates of nodes
n_dis=1;	% number of fixed degrees of freedom
node_dis=[1];	% nodes with fixed degrees of freedom
set_dis=[0.0];	% values of fixed degrees of freedom
n_load=1;	% number of specified loads
node_load=4;	% nodal indices with applied loads
set_load=20.0;	% values of applied loads
ele_materialtype=[1 2 3 4 3];	% vector of equivalent ks of spring elements
connect_m=[1 2; 1 3; 2 3; 2 4; 3 4];	% connection matrix of elements and nodes

model=C1DSpringModel(n_materials,...
   material_ks, n_points, n_nodes,...
     coords,node_dof, n_elements,...
     ele_materialtype, n_dis,...
     node_dis,set_dis,n_load,...
     node_load, set_load, connect_m);
	% create an object of C1DSpringModel

Data Processing

model=model.create_sys_model();
   model=model.apply_BCs();
   model=model.sys_solver();
	% create system model for finite element analysis problem
	 % apply boundary conditions
	 % solve system model

Postprocessing

model=model.post_proc(); system_stiffness=model.sys_k system_f=model.sys_f system_u=(model.sys_u)' sys_e_dltx=(model.sys_e_dltx)' sys_e_force=(model.sys_e_force)'	% postprocessing to recover full system model and find dependent variables % print out the solution % system stiffness matrix % vector of external load % vector of nodal displacements % vector of deflections of elements % vector of internal forces of elements

The free body diagrams (FBDs) of supports 2, 3, and 4 have shown that the forces on these supports are fully balanced. This validates the results of the computer program for 1D spring systems.

7.6.2 **Electric circuit example**

Fig. 7.19 shows an electric circuit example, which consists of eight resistors and seven nodes. One node ($n7$) is grounded, and the other one ($n1$) is powered with 15 (V). Use the computer program to determine the voltages of all intersecting as well as the currents passing through resistors.

Solution. For the inputs of the computer program, this electric circuit has a domain determined by 7 points/nodes ($n1$, $n2$, ..., and $n7$); each point is treated as a node. Each resistor is treated as a spring element, whose equivalent stiffness coefficient is determined by its resistance R as $k_e = 1/R$. Because the circuit has three types of resistors (50, 100, and 200 Ω), the FEA model has three types of materials. In addition, there is two fixed displacements (15 V and 0 V) on $n1$ and $n7$, respectively. The connection matrix of elements and nodes can be determined based on the illustration in Fig. 7.19. Therefore, the main program to solve the problem can be given as Table 7.9.

Running the computer program gives,

$$[K] = \begin{bmatrix} 0.01 & -0.01 & 0 & 0 & 0 & 0 & 0 \\ -0.01 & 0.035 & -0.02 & -0.005 & 0 & 0 & 0 \\ 0 & -0.02 & 0.03 & 0 & -0.01 & 0 & 0 \\ 0 & -0.005 & 0 & 0.035 & -0.02 & -0.01 & 0 \\ 0 & 0 & -0.01 & -0.02 & 0.05 & -0.02 & 0 \\ 0 & 0 & 0 & -0.01 & -0.02 & 0.04 & -0.01 \\ 0 & 0 & 0 & 0 & 0 & -0.01 & -0.01 \end{bmatrix} (l1/\Omega)$$

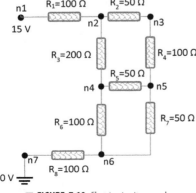

■ **FIGURE 7.19** Electric circuit example.

Table 7.9 Main Program to Calculate Voltage of Electric Circuit
Data Inputs

n_materials=3;	% type of materials
material_ks=[1/50 1/100 1/200];	% types of materials of elements
n_points=7;	% number of points
n_nodes=7;	% number of nodes
n_elements=8;	% number of elements
node_dof=1;	% degrees of freedom (DoF) on each node
coords=zeros(n_points);	% coordinates of nodes
n_dis=2;	% number of fixed degrees of freedom
node_dis=[1 7];	% nodes with fixed degrees of freedom
set_dis=[15 0];	% values of fixed degrees of freedom
n_load=0;	% number of specified loads
node_load=0;	% nodal indices with applied loads
set_load=0;	% values of applied loads
ele_materialtype=[2 1 3 2 1 2 1 2];	% vector of equivalent ks of spring elements
connect_m=[1 2; 2 3; 2 4; ...	% connection matrix of elements and nodes
3 5; 4 5; 4 6; 5 6; 6 7];	% create an object of C1DSpringModel
model=C1DSpringModel(n_materials,...	
material_ks, n_points, n_nodes,...	
coords,node_dof, n_elements,...	
ele_materialtype, n_dis,...	
node_dis,set_dis,n_load,...	
node_load, set_load, connect_m);	

Data Processing

model=model.create_sys_model();	% create system model for FEA problem
model=model.apply_BCs();	% apply BCs
model=model.sys_solver();	% solve system model

PostProcessing

model=model.post_proc();	% postprocessing to
system_stiffness=model.sys_k	recover full system model and find dependent
system_f=model.sys_f	variables
system_u=(model.sys_u)'	% print out the solution
sys_e_dltx=(model.sys_e_dltx)'	% system stiffness matrix
sys_e_force=(model.sys_e_force)'	% vector of external load
	% vector of nodal displacements
	% vector of deflections of elements
	% vector of internal forces of elements

$$\{F\} = \begin{Bmatrix} 15 \\ 10.303 \\ 8.9394 \\ 6.3636 \\ 6.2121 \\ 4.6970 \\ 0 \end{Bmatrix} (V) \quad \{I\} = \begin{Bmatrix} 0.047 \\ 0 \\ 0 \\ 0 \\ 0 \\ 0 \\ -0.047 \end{Bmatrix} (A)$$

The results can be validated by checking the summed currents at internal nodes $n2$, $n3$..., and $n6$ and it shows that the in- and out-currents are balanced.

7.6.3 **Pipe flow network example**

Fig. 7.20 shows a pipe flow example with the laminate flow. The oil has its dynamic viscosity of $\mu = 0.3$ N s/m^2 and the density of $\rho = 900$ kg/m^3. The pipe network consists of 10 branches with given diameters on the right. Assume the flow rates at the inlet of $Q_1 = 3.0$ (m^3/s) and three outlets of $Q_4 = 1.0$ (m^3/s). Outlets 6 and 7 have the air pressure of $p_6 = p_8 = 10,350$ Pa. Determine the pressure distributions over all intersecting nodes.

Branch Node	1	2	3	4	5	6	7	8	9	10
Length (L in m)	300	250	350	125	350	125	300	125	350	125
Diameter (D in m)	0.3	0.25	0.2	0.2	0.2	0.2	0.2	0.15	0.2	0.15

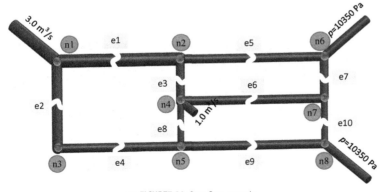

■ **FIGURE7.20** Pipe flow example.

The main program for this example is given in Table 7.10. Running the computer program gives,

$$[K] = 1.0e-05 \begin{bmatrix} 0.3487 & -0.2209 & -0.1278 & 0 & 0 & 0 & 0 & 0 \\ -0.2209 & 0.2957 & 0 & -0.0374 & 0 & -0.0374 & 0 & 0 \\ -0.1278 & 0 & 0.2326 & 0 & -0.1047 & 0 & 0 & 0 \\ 0 & -0.0374 & 0 & 0.1753 & -0.0331 & 0 & -0.1047 & 0 \\ 0 & 0 & -0.1047 & -0.0331 & 0.1753 & 0 & 0 & -0.0374 \\ 0 & -0.0374 & 0 & 0 & 0 & 0.081 & -0.0436 & 0 \\ 0 & 0 & 0 & -0.1047 & 0 & -0.0436 & 0.1815 & -0.0331 \\ 0 & 0 & 0 & 0 & -0.0374 & 0 & -0.0331 & 0.0705 \end{bmatrix}$$

$$[Q] = \begin{bmatrix} 3.0 & 0 & 0 & -1.0 & 0 & -1.21 & 0 & -0.79 \end{bmatrix}^T$$

$$[p] = 1.0e6 * \begin{bmatrix} 3.7195 & 2.8546 & 2.8674 & 0.5901 & 1.8271 & 0.0104 & 0.3449 & 0.0104 \end{bmatrix}^T$$

The flow rates at inlet and three outlets are balanced, which validate the simulation result.

7.7 COMPUTER IMPLEMENTATION OF 3D TRUSS SYSTEMS

The static structure of C3DTrussModel for 3D truss systems is shown in Fig. 7.21, where the subclasses from their generic classes in Fig. 7.7 are directly defined. Main functions for solving and processing of a system model are performed by Matlab functions. Therefore, *C3DTrussPostProc* and *CEquivalenceSolver* are not defined separately. A few of functions are defined in *C3DTrussModel* as the interfaces to access Matlab functions.

7.7.1 Definition of classes

In contrast to the static structure of *CSpringModel* in Fig. 7.17, subclasses *CPointLoad* and *CFFixedDoF* are the same, which can be reused directly. Although other classes have been expanded to include more attributes for truss members. The definitions of these classes are provided in Table 7.11.

7.7.2 Example of 3D truss system

Fig. 7.22 shows an example of truss structure, which consists of four truss members. Four nodes are used to join the trusses and the coordinates of nodes n1, n2, n3, and n4 are given in inch. The materials of trusses have the Young's modulus of $E = 3.0 \times 10^7$ lb/in^2, Poisson ratio of $\nu = 0.3$,

Table 7.10 Main Program to Calculate Displacement of Spring Assemblage

Data Inputs

n_materials=10;	% type of materials
ele_materialtype=[1 2 3 4 5... 6 7 8 9 10];	% types of materials of elements % number of points
n_points=8;	% number of nodes
n_nodes=8;	% number of elements
n_elements=10;	% degrees of freedom (DoF) on each
node_dof=1;	node
coords=zeros (n_elements);	% coordinates of nodes
n_dis=2;	% number of fixed degrees of freedom
node_dis=[6 8];	% nodes with fixed degrees of freedom
set_dis=10350*[1 1];	% values of fixed degrees of freedom
n_load=2;	% number of specified loads
node_load=[1 4];	% nodal indices with applied loads
set_load=[3.0−1.0];	% values of applied loads
L=[300 250 350 125 350 125 300 125 350 125];	% Lengths of pipe branches % connection matrix of elements and
D=[0.3 0.25 0.2 0.2 0.2 0.2 0.2 0.15 0.2 0.15]; material_ks=zeros(n_elements);	nodes % diameters of pipe branches
mu=0.3;	% create a dimension for materials ks
for i=1:n_elements material_ks(i)=... 3.1415926*D(i)^4/(128.0*mu*L(i));	% coefficient mu of fluid %calculate equivalent ks for pipe flow
end	% define the connection matrix of nodes
connect_m=[1 2; 1 3; 2 4; 3 5; 2 6; 4 7; 6	and elements
7; 4 5; 5 8; 7 8];	% create an object of C1DSpringModel

model=C1DSpringModel(n_materials,...
   material_ks, n_points, n_nodes,...
      coords,node_dof, n_elements,...
      ele_materialtype, n_dis,...
      node_dis,set_dis,n_load,...
      node_load, set_load, connect_m);	

Data Processing

model=model.create_sys_model();	% create system model for FEA problem
model=model.apply_BCs();	% apply BCs
model=model.sys_solver();	% solve system model

Postprocessing

model=model.post_proc();	% postprocessing to recover full system
system_stiffness=model.sys_k	model and find dependent variables
system_f=model.sys_f	% print out the solution
system_u=(model.sys_u)'	% system stiffness matrix
sys_e_dltx=(model.sys_e_dltx)'	% vector of external load
sys_e_force=(model.sys_e_force)'	% vector of nodal displacements
	% vector of deflections of elements
	% vector of internal forces of elements

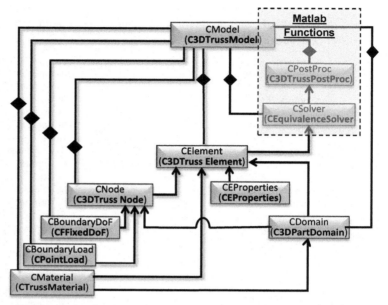

■ FIGURE 7.21 Static structure of finite element analysis computer implementation for truss structure.

and the cross-section area A for all truss members is 2.0 in^2. A load of 500 lbf is applied on $n1$ along y-axis. The restraints at $n2$, $n3$, and $n4$ are given in Fig. 7.22, respectively. Determine the displacement at $n1$.

The main program in given in Table 7.12. Running the computer program gives.

```
system_stiffness =

  1.0e+06 *

   0.6000        0        0  -0.3000  -0.5196        0        0        0        0  -0.3000   0.5196        0
        0   2.1966  -0.5633  -0.5196  -0.9000        0        0  -0.3966   0.5633   0.5196  -0.9000        0
        0  -0.5633   0.7999        0        0        0        0   0.5633  -0.7999        0        0        0
  -0.3000  -0.5196        0   1.7975   0.6913   0.4878  -0.2975  -0.1717  -0.4878  -1.2000        0        0
  -0.5196  -0.9000        0   0.6913   0.9991   0.2816  -0.1717  -0.0991  -0.2816        0        0        0
        0        0        0   0.4878   0.2816   0.8000  -0.4878  -0.2816  -0.8000        0        0        0
        0        0        0  -0.2975  -0.1717  -0.4878   0.5949        0        0  -0.2975   0.1717   0.4878
        0  -0.3966   0.5633  -0.1717  -0.0991  -0.2816        0   0.5948  -0.0001   0.1717  -0.0991  -0.2816
        0   0.5633  -0.7999  -0.4878  -0.2816  -0.8000        0  -0.0001   2.4000   0.4878  -0.2816  -0.8000
  -0.3000   0.5196        0  -1.2000        0        0  -0.2975   0.1717   0.4878   1.7975  -0.6913  -0.4878
   0.5196  -0.9000        0        0        0        0   0.1717  -0.0991  -0.2816  -0.6913   0.9991   0.2816
        0        0        0        0        0        0   0.4878  -0.2816  -0.8000  -0.4878   0.2816   0.8000


system_f =

   0.0000   0.0000 -500.0000  -0.0000 264.0244 250.0000        0 -528.0488  -0.0000   0.0000 264.0244 250.0000


system_u =

   0.0001  -0.0003  -0.0012        0        0        0   0.0001        0  -0.0004   0.0002        0        0
```

Table 7.11 Class Definitions for 3D Truss Systems

Class Definition	Description

CTrussMaterial
 	
E;
Mu;
 	
CTrussMaterial (val);
set_E(val);
get_E();
set_mu(val);
get_mu();
cal_ss();
	CTrussMaterial class is to define a type of truss material. Two main attributes are Young's modulus E and Poisson ratio v.
	The methods for CTrussMaterial include (1) a constructor to create an object by specifying E; (2) *set_E(val)* to set Young's modulus E for a class object; (3) *get_E ()* to get the Young's modulus E from a class object; (4) *set_mu (val)* to set the Poisson ratio v for a class object; (5) *get_mu ()* to get the Poisson Ratio v from an class object, and (6) *cal_ss ()* to define the constitutive model of materials

C3DPartDomain
 	
n_points;
coords=[];
 	
C1DPartDomain (n_points,
    coords);
get_n_points();
get_coord_i(i);
get_coords();
	C3DPartDomain class is to specify the domain to be analyzed. Here, we assume that the discretization is completed manually. A 3-D part domain can be defined as a set of discrete points in a 3D space.
	This class includes two attributes (1) the number of discrete points *n-points* and (2) the coordinates of these discrete points *coords []*.
	The methods for *C3DParDomain* include (1) a constructor to create an object by specifying number of discrete points and their coordinates, (2) *get_n_points ()* to read the number of points, (3) *get_coord_i(i)* to get the coordinate for the *i*-th point, and (4) *get_coords ()* to get a dimension of coordinates for all discrete points.

CEProperties
 	
A;
E;
L=0;
 	
CEProperties (val);
set_E(val);
get_E();
set_A(val);
get_A();
set_L(val);
get_L();
	CEProperties class is to define a type of additional properties for the chose truss materials. The key attribute is cross-section area A since the truss stress. Other two attributes are the Young's modulus E from *CTrussMaterial* and the length L from *C3DTrussElement* for an easy access of these attributes.
	The methods for *CEProperties* include (1) a constructor to create an object by specifying the cross-section A of element; (2) *set_A (val)* and *get_A ()* to set and read the cross-section area of element; (3) *set_E (val)* and *get_E ()* to set and read Young's modulus of element; (4) *set_L (val)* and *get_L ()* to set and read the length of truss element.

C3DTrussElement
 	
E;mu;
A;L;
matrix_k=zeros(6,6);
stiffness_k;
l;m;n;
i_node;j_node;
 	
C3DTrussElement(E, A, p1, p2);
get_k_item(i, j);
set_nodes(i, j);
get_i_node();
get_j_node();
get_k ();
get_A();
get_ length ();
	C3DTrussElement class is to define a type of 3D truss element. Its attributes include (1) Young's modulus E and Poisson ratio v; (2) the cross-section area A and length L of element; (3) the stiffness coefficient k and corresponding stiffness matrix *matrix_k*; (4) the direction of truss member represented by the cosine vector (l, m, n), and (5) the indices of two nodes (*i_node* and *j_node*).
	The methods for *C3DTrussElement* are (1) a constructor to create an object with the given Young's modulus E, cross-section area A, the coordinates p_1 and p_2 of two nodes; (2) *get_k_item (i, j)* to read an element at (i, j) in *matrix_k*; (3) *set_nodes (i, j)* to specify the indices of two nodes; (4) *get_i_node ()* and *get_j_node ()* to read the indices of two nodes, respectively; (5) *get_length ()* to get the length of element; (6) *get_k ()* to get the value of equivalent *stiffness_k*; and (7) *get_A ()* to get the cross-section area of element.

Continued

Table 7.11 Class Definitions for 3D Truss Systems *continued*

Class Definition	Description

C3DTrussNode
 	
p=[0 0 0];
i_force=[0 0 0];
i_p=[0 0 0];
external_f=[0 0 0];
fixed_p=[0 0 0];
 	
CSpringNode(val);
setd_node_1D(dis);
setf_node_1D(force);
getd_node_1D();
getf_node_1D();
get_i_dis();
	C3DTrussNode class is to define a node type for a truss element. Besides one attribute (*p*) for its 3D location, this class has two vectors of indicators (*i_force* and *i_p*) to identify if there is an external load or fixed displacement on a specified DoF. Correspondingly, the attributes of *external_f* and *fixed_p* are the vectors to contain the values of external loads or fixed displacements if the node has.
	An object can be constructed from *C3DTrussNode* by specifying its coordinate *p*. The class also includes (1) *setd_node_1D* (*dis*) to set the fixed displacement; (2) *setf_node_1D* (*force*) to set an applied load; (3) *getd_node_1D* () to get the displacement; (4) *getf_node_1D* () to get the applied force, and (5) *get_i_dis* () to read the indicator for a fixed displacement.

C3DTrussModel
 	
n_materials;
n_EProperties;
val_E=[];
val_A=[];
n_nodes;
n_elements;
iE=[];
iA=[];
node_dof;
n_dof;
coords=[];
n_fdof;
dof_dis=[];
n_load;
dof_load=[];
set_dis=[];
set_load=[];
connect_m=[];
OTrussMaterial=CTrussMaterial.empty;
O3DPartDomain=C3DPartDomain.empty;
OEProperties=CEProperties.empty;
OFixedDoF=CFFixedDoF.empty;
OLoad=CPointLoad.empty;
ONodes=C3DTrussNode.empty;
OElements=C3DTrussElement.empty;
del_dof=[];
sys_k=[];
sys_f=[];
m_sys_k=[];
m_sys_f=[];
sys_u=[];
m_sys_u=[];
 	
C3DTrussModel(n_materials,
n_EProperties, val_E, val_A, n_nodes,
n_elements, iE, iA, node_dof, coords,
n_fdof, dof_dis, n_load, dof_load,
set_dis, set_load, connect_m);
create_sys_model();
apply_BCs();
sys_solver();
post_proc();
	C3DTrussModel class is to define an FEA model of a 3D truss structure. It includes the attributes of all input data and intermediate data; it includes the most important methods for data processing and data postprocessing. As shown on the left, the attributes in this class can be classified into three groups, which relate to the data at the phases of *preprocessing*, *processing*, and *post-processing*, respectively.
	Since the program assumes that the proprocessing is completed manually, all information of the discretized model must be available. Input data of the program includes the number of material types (*n_materials*), the number of elemental properties (*n_EProperties*), the values of the Young's modulus *val_E* [] and cross-section areas *val_A()* for all section types, the numbers of nodes (*n_nodes*) and elements (*n_elements*), the indices of section properties for elements (*iE[]* and *iA[]*), the DoF on each node (*node_dof=3* for 3D truss), the number of elements (*n_elements*), the number of fixed displacements (*n_dis*), nodal indices (*node_dis[]*) and fixed displacements (*set_dis[]*), the number of applied loads (*n_load*), nodal indices (*node_load[]*) and applied loads (*set_load[]*), and the relations of nodes and elements (*connect_m[]*).
	The input data to the processing is stored in the objects instantiated from the classes of *CTrussMaterial, CEProperties, C3DPartDomain, C3DTrussNode, C3DTruss Element, CFFixedDoF,* and *CPointLoad*. The output data from data processing includes the system model (*sys_k[], sys_f[], sys_u[], m_sys_k[], m_sys_f[], m_sys_u[]*), and the postprocessing is performed on the same amount of data to obtain some dependent attributes in *C3DTrussNode* and *C3DTrussElement* such as internal force in an object from *CTrussElement*. In addition, the order of DoF in the model is determined, and *del_dof[]* is introduced to record the number of fixed DoF in front of a specified index of DoF; it is used to modify system model when the boundary conditions are applied.
	The constructor of *C3DTrussModel* is to create an object of the 3D truss system model with all input data. The input data is then processed by other methods. Namely, *create_sys_model* () is to generate a system model including the stiffness matrix *sys_k[]*, the vector of loads *sys_f[]*. The method of *apply_BCs* () is to modify the system model by applying boundary conditions; the outputs are *m_sys_k[]* and *m_sys_f[]*. The modified system model is solved to obtain *m_sys_u[]* by *sys_solver* (), which is implemented by calling Matlab functions. Finally, *post-proc* () is performed to evaluate dependent variables and scalars on nodes or elements.

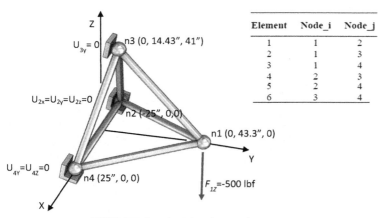

Element	Node_i	Node_j
1	1	2
2	1	3
3	1	4
4	2	3
5	2	4
6	3	4

■ **FIGURE 7.22** Example of three-dimensional truss structure.

Table 7.12 Main Program to Calculate Displacement of Spring Assemblage

Data Inputs

```
n_materials=1;                        % type of materials
n_EProperties=1; val_E=3.0e7;         % types of materials of elements
val_A=2.0;                            % value of Young's modulus
n_nodes=4;                            % value of cross-section area
n_elements=6;                         % number of nodes
iE=[1 1 1 1 1 1];                     % number of elements
iA=[1 1 1 1 1 1];                     % index of Young's modulus for elements
node_dof=3;                           % index of cross-section areas for elements
coords=[ 0.0 43.3 0.0;...             % degrees of freedom (DoF) on each node
    −25.0 0.0 0.0;...                 % coordinates of nodes
    0.0 14.43 41.0;...                % number of fixed degrees of freedom
    25.0 0.0 0.0; ];                  % nodes with fixed degrees of freedom
n_fdof=6;                             % values of fixed degrees of freedom
dof_dis=[4 5 6 8 12 12];             % number of specified loads
set_dis=[0 0 0 0 0 0];               % nodal indices with applied loads
n_load=1;                             % values of applied loads
dof_load=2;                           % define the connection matrix of nodes and
set_load=−500.0;                      elements
connect_m=[1 2; 1 3; 1 4;...          % create an object of C3DTrussModel
    2 3; 2 4; 3 4; ];
C3DTrussModel(n_materials, n_EProperties,
val_E, val_A, n_nodes, n_elements, iE, iA,
node_dof, coords, n_fdof, dof_dis, n_load,
dof_load, set_dis, set_load, connect_m);
```

Data Processing

```
model=model.create_sys_model();      % create system model for FEA problem
  model=model.apply_BCs();           % apply BCs
  model=model.sys_solver();          % solve system model
```

Postprocessing

```
model=model.post_proc();             % postprocessing to recover full system model
system_stiffness=model.sys_k         and find dependent variables
system_f=model.sys_f                 % print out the solution
system_u=(model.sys_u)'              % system stiffness matrix
                                     % vector of external load
                                     % vector of nodal displacements
```

It is seen from the result that due to the symmetric setup, the external force along y-axis is balanced. In additions, verifying the moments along $n2-n4$ finds that the truss structure is fully balanced as well.

7.8 COMMERCIAL CAD/CAE SOFTWARE TOOL—SOLIDWORKS

The introduction of programming implementation in above sections aims to understand the foundation of FEA theory, it does not intend to replace the role of any commercial FEA tools. In this section, SolidWorks is used as the vehicle to illustrate how a commercial software tool can be adopted to fulfill various tasks involved in the proprocessing, processing, and postprocessing of FEA. Fig. 7.23 gives an overview of the mappings between available functional modules in SolidWorks and major tasks involved in FEA modeling.

No matter what type of software architecture (i.e., structural, procedural, or OO architecture) is adopted in programming, graphic user interfaces (GUIs) of a commercial FEA tool are usually user friendly, which allow users to access functional modules for FEA modeling interactively. In the following, some typically modules to support the preprocessing, processing, and post-processing are introduced.

7.8.1 CAD/CAE interface

FEA is a general tool to evaluate the distributions of state variables over a continuous domain, a continuous domain can be in one-dimensional (1D), two-dimensional (2D), or three-dimensional (3D). To model a design problem, the domain must be represented by a virtual model in computer. For some simple objects, such as springs and truss members, the domain to be analyzed can be directly described by specifying coordinates of their intersections as *nodes* and a connection matrix of nodes for *elements*. However, objects in the real-world applications are usually very complex, sophisticated computer aided design (CAD) tool is required to create the computer models of objects. A computer aided engineering (CAE) tool such as FEA must have an interface which allows to import and export virtual models of objects directly from and to CAD tools.

The vendor-neutral file formats in Table 7.13 provide basic representations of geometries such as vertices, edges, and boundary surfaces of solid bodies; the models of solid objects in these formats do not include all information about how solid objects are created, such as the information about sketches and the parameters for different features of objects. This causes the difficulty

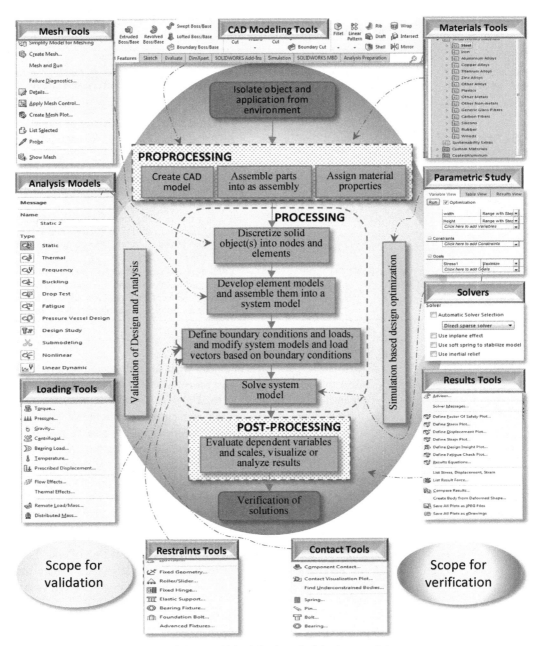

■ **FIGURE 7.23** SolidWorks/simulation for finite element analysis.

Table 7.13 The Common Formats of Solid Models

Format	Developer	Description
IGES—Initial Graphics Exchange Specification (.igs)	U.S. National Bureau of Standards in 1080	IGES is a vendor-neutral file format that allows the digital exchange of information among computer-aided design (CAD). The information of geometries is not complete; therefore, tolerances of feature vary from one system to another.
STEP—The Exchange of Product model data (.step, .stp)	International Organization for Standardization, 1994	STEP files improve the IGES format in sense that the tolerance data is included along with significant amount of meta-data such as product structure and the definition of solid features. STEP files can be geometry based or product structure based.
STL—STereoLithography (.stl)	3D Systems, 1988	STL is a polygon file format which has been widely used for rapid prototyping. It is a low fidelity in terms of geometric representation of solid object.
VRML—The Virtual Reality Modelling Language (.wrl, .X3D)	Web3D Consortium, 1994	VRML files are text based and designed in a structured human readable manner for web browsers. VRML files are widely used to transport 3D models between graphics applications.

to revise the geometries of objects when needed. Therefore, it is desirable to use a solid model in the native format where all of the information about the parameters and relations of solid objects is sustained. Some leading CAD software suppliers have developed their own formats to support CAD data exchanges and parametric designs. For example, *.x_t* and *.x_b* from Parsolid, *.sat* and *.sldlfp* from Dassault Systems, *.dxf* and *.dwg* from Autodesk, *.jt* from Siemens PLM Software. Due to an increasing need of data exchange across the products from different vendors, direct converters of a CAD file from one native format to another are available. For example, Fig. 7.24 gives a list of 31 formats, which are compatible to SolidWorks 2017 for importing and exporting a CAD model.

In an integrated CAD/CAE software tool such as SolidWorks, FEA tools are packed as an embedded functional module in the software platform. Therefore, parametrized solid models in a native format can be accessed by FEA tools seamlessly.

7.8.2 **Materials library**

The data about material properties is essential to numerical simulation. Before running an FEA simulation, one must define all the necessary

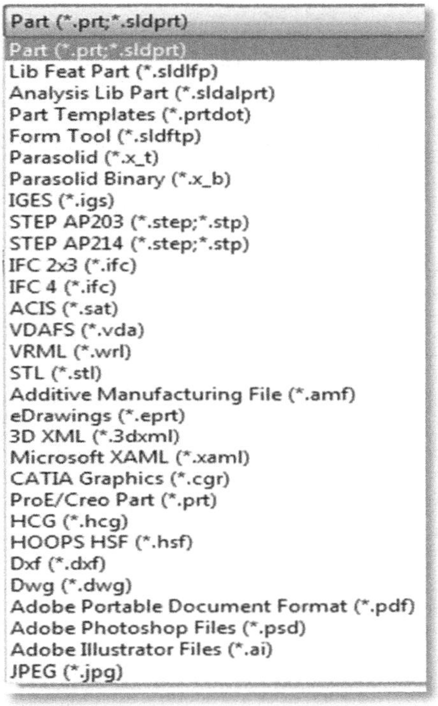

Part (*.prt;*.sldprt)
Part (*.prt;*.sldprt)
Lib Feat Part (*.sldlfp)
Analysis Lib Part (*.sldalprt)
Part Templates (*.prtdot)
Form Tool (*.sldftp)
Parasolid (*.x_t)
Parasolid Binary (*.x_b)
IGES (*.igs)
STEP AP203 (*.step;*.stp)
STEP AP214 (*.step;*.stp)
IFC 2x3 (*.ifc)
IFC 4 (*.ifc)
ACIS (*.sat)
VDAFS (*.vda)
VRML (*.wrl)
STL (*.stl)
Additive Manufacturing File (*.amf)
eDrawings (*.eprt)
3D XML (*.3dxml)
Microsoft XAML (*.xaml)
CATIA Graphics (*.cgr)
ProE/Creo Part (*.prt)
HCG (*.hcg)
HOOPS HSF (*.hsf)
Dxf (*.dxf)
Dwg (*.dwg)
Adobe Portable Document Format (*.pdf)
Adobe Photoshop Files (*.psd)
Adobe Illustrator Files (*.ai)
JPEG (*.jpg)

■ **FIGURE 7.24** A list of file formats compatible to SolidWorks 2017.

material properties specified by the given analysis type. For example, the modulus of elasticity is required for static and modal analysis; while thermal conductivity is needed for a heat transfer problem.

Most products use common industrial materials such as iron, steel, concrete, plastics, and aluminium; the properties of those materials are well documented. A commercial FEA software is usually equipped with a materials library for commonly used industrial materials. It also provides users with the interface and template to customize material properties for solid objects. Common properties of solid objects can be classified into (1) **physical properties** such as density and melting temperature, (2) **mechanical properties** such as the Young's modulus, the Poisson's ratio, yield strength, and hardness, (3) **thermal properties** such as thermal conductivity, specific hear, and coefficient of thermal expansion, (4) **electric properties** such as resistivity, and (5) **acoustic properties** such as compression wave velocity, shear ware velocity, and bar velocity. Depending on the types of applications, corresponding material properties are essential. For example, mechanical

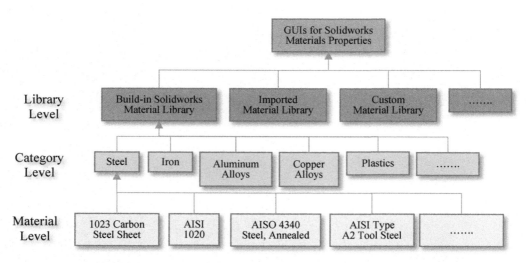

■ **FIGURE 7.25** Typical structure of materials library in finite element analysis package.

properties of a solid object must be given if static analysis or modal analysis is performed on object.

As shown in Fig. 7.25, the materials library in the SolidWorks simulation is organized in the levels of *Library*, *Category*, and *Material*. In the structure of SolidWorks material library, *custom material* has to be placed in a category of a custom material library. Therefore, one has to start by creating a custom material library, then a new category under custom library, and finally, a new material under the custom new category.

A commercial FEA tool usually provides users with material template when new material model is needed. As shown Fig. 7.26, basic physical, mechanical, and thermal properties such as density, elastic modules, and thermal conductivity can be input directly. If composite materials or nonlinear material models have to be defined, the software tool allows one to use custom curves or even measurement data as inputs. Fig. 7.27 shows an example interface for a user to input Strength to Number of Cycles (S–N) curves, which is essential to fatigue analysis of solid objects.

For fatigue analysis, dynamic analysis, or large displacement under plastic deflection, raw data of material properties is unlikely from single source. The SolidWorks provide the tool called Materials Web Portal to collect material properties from the third party (Matereality LLC). This portal offers the data on nonlinear materials and fatigue curves, which can be difficult to find from other places.

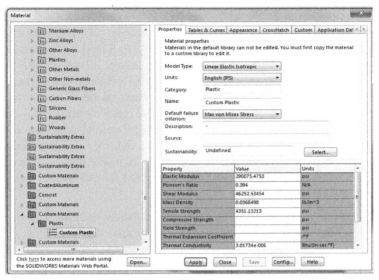

■ **FIGURE 7.26** Interface to create custom material model.

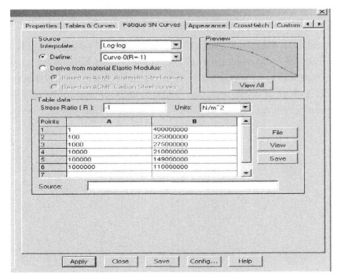

■ **FIGURE 7.27** Generate S—N curve for fatigue analysis.

7.8.3 **Meshing tool**

A continuous domain with infinite degrees of freedom (DoF) is represented by discretized nodes and elements with a finite DoF. A meshing tool aims to convert a continual domain into nodes and elements. It is critical step in FEA modelling. An FEA tool is equipped with an *automatic meshing* module. It is used to estimate a global element size based on the volume, surface areas and geometric details of a solid object, and create elements and nodes based on global element size, tolerance, and local mesh control. Note that the scale of system model directly relates to the element's size; the smaller elements are, the more number of DoF a system model has. *Local mesh control* allows one to specify divisions of selected features such as edges, faces, and components.

An FEA tool supports many element types for different analyses. As shown in Fig. 7.28, in meshing, element types must be specified based on object shapes to avoid distorted elements. For *bulky* objects, solid elements are suitable. For *thin* objects, shell elements should be used. For extruded or revolved trusses and beams with a constant cross-section, truss or beam elements are appropriate.

When the object to be analyzed is an assembled solid, the meshing tool might need manual interventions to (1) ensure no interference occurs at the interfaces of two objects, and (2) set up mesh control to achieve a compatible mesh if it is feasible.

It is desirable to run interface check and eliminate all possible interferences at interfaces of solid components. For example, the SolidWorks software has an *Interference detection* tool to evaluate if there are interferences in an object group. If there are, the CAD models or assembly relations of corresponding objects have to be revised to eliminate physical intrusions

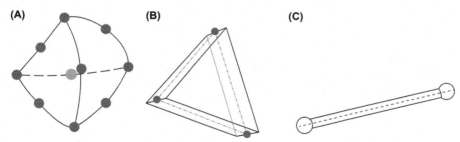

■ **FIGURE 7.28** Exemplified element types for different shapes: (A) bulky element; (B) shell element; (C) truss or beam element.

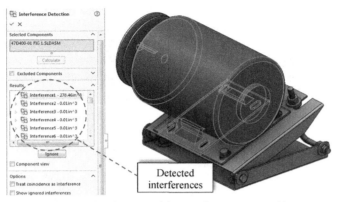

■ **FIGURE 7.29** Detecting and fixing interferences in an assembly.

of two solid bodies. Fig. 7.29 shows an example where an interface detection is performed to identify interferences in an assembly.

As an assembled object involved in a number of interfaces where two or multiple solids are jointed together as bonded contact. During the meshing process, nodes from different solids can be joined differently to generate either of a *compatible mesh* or *incompatible mesh*. As shown in Fig. 7.30, the nodes from two of more solids have one-to-one correspondences in a compatible mesh and such correspondences are not satisfied in an incompatible mesh. In a "bonded" contact, nodes on two contact surfaces can be merged or superposed. In a "no penetration" contact, two contact surfaces with the node-to-node correspondence become source and target faces.

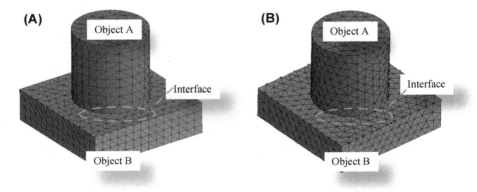

■ **FIGURE 7.30** Comparison of compatible and incompatible meshes: (A) compatible mesh (node-node at interface); (B) incompatible mesh (nonnode—node at interface).

Because nodes in an incompatible mesh are restrained only by constraint equations, incompatible mesh causes potential issues of stress concentration at the bonded contact. Computation on a compatible mesh leads to better accuracy than that of incompatible mesh. In using a commercial FEA tool for a complex assembled model, a user should refine meshing parameters to obtain a compatible mesh as much as possible.

For an actual part or assembly, it is not rare that the first run of meshing processes does not succeed. If a failure occurs to the meshing process, the failure diagnostics module in SolidWorks can be applied to diagnose the causes, a trial-and-error process is deployed to adjust element sizes, define mesh controls on critical features, and activate automatic remeshing until the mesh is generated for the entire domain.

To increase the accuracy of solution, *mesh refinement* is an effective means. A mesh can be refined in two alternative ways, i.e., *h-adaptive meshing* and *p-adaptive meshing*. H-adaptive meshing is to refine the mesh by reducing element sizes at critical areas; while the p-adaptive meshing is to insert more nodes in existing elements without the changes of element sizes. Increasing nodes in an element leads to a high order of polynomial interpolation in an element. Both the ways of mesh refinement can be performed automatically to meet the expected meshing accuracy given by users; the refinements are performed iteratively on an FEA model without a manual intervention by users.

7.8.4 **Analysis types**

An FEA software tool is a generic tool to solve differential or integral equations with given boundary conditions. As long as the governing equations for a design problem are covered by the software, this problem can be solved readily by the FEA tool. Therefore, most of the FEA tools are applicable to a variety of *Analysis Types*. Table 7.14 lists some common analysis types in the SolidWorks Simulation.

7.8.5 **Tools for boundary conditions**

To solve a set of differential equations, boundary conditions must be given in the forms of restraints and loads. The *fixtures* module in the PropertyManager of the SolidWorks simulation provides the interfaces to specify the restraints of displacements on vertices, edges, or faces. Restraints can be zero or nonzero displacements. Boundary conditions of displacements are essential to analysis types for the deformation of a solid object such as static, frequency, dynamic, or nonlinear studies. Table 7.15 lists some common options of restraints in the SolidWorks Simulation.

Table 7.14 Common Analysis Types of Simulation in SolidWorks

Analysis Type	Descriptions
Static analysis	Static analysis is suitable to the cases of small deformations of elastic materials subject to static loads. It is assumed that the materials of an object behaves in its elastic range, i.e., the strain has a linear relation with stress. It is used to evaluate stress distribution thus predict the static failure where the maximum stress exceeds the yield strength of materials. Static analysis will not model the behavior of plastic deformation of the materials appropriately.
Nonlinear Analysis	Nonlinear analysis aims to model the scenarios where (1) both of elastic and plastic deformation occur to objects, or (2) the properties of materials are nonlinear. For the first type of scenario, an object deforms in a way that the shape or stiffness of object is changed significantly depending on the stress state. For second type of scenario, the applied materials such as plastics, rubbers, elastomers can be represented with linear stress—strain curve due to possible large deformation.
Frequency Analysis	Frequency analysis is for a modal analysis to evaluate natural frequencies of structure; an excitation at one of these critical frequencies likely causes problematic vibrational response to the structure. The outcomes of a frequency analysis include (1) a list of natural frequencies and (2) mode shapes corresponding to the frequencies.
Dynamic Analysis	Dynamic analysis takes into account of the dynamics of loading conditions, such as a shock or a vibration. Time-dependent loads can be defined in stress analysis. The method call *modal superposition* is used to analyze the responses of structure to individual inputs. The system behavior is defined by adding individual responses together. Three types of dynamic loads are (1) time-dependent acceleration or load, (2) a load or acceleration with frequency change, and (3) nondeterministic excitation including random vibration expressed by a power spectrum density (PSD) curve.
Thermal Analysis	Thermal analysis is to analyze a solid model with three types of heat transfer behaviors: conduction, convection and radiation. A heat transfer model is developed based on the energy conservation where heat is transferred by conduction in a solid body, and the convection and radiation on boundary surfaces. Heat convection is measured by a convective coefficient. Note that convective coefficient is not a material property, it is affected by many environmental parameters. An input convective coefficient commonly corresponds to a laminate air or fluid flow under the natural convection. If the surrounding atmosphere is complicated, the heat transfer coefficient should be obtained separately from experiment or flow simulation.
Flow Simulation	Flow Simulation is a computational fluid dynamics (CFD) tool that allows one to model air and fluid flow around and through solid objects. Flow simulation can perform detailed thermal analysis on a variety of heating and cooling scenarios; it can perform conductive, convective and radiative calculation simultaneously without an input of convective coefficient.
Fatigue Analysis	Repeated loading and unloading causes the damage of objects over time even when the stresses are considerably less than yield strengths. This damage is called as fatigue. Fatigue is the prime cause of the failure of most of metal objects. Fatigue analysis investigate the accumulated damage on solid object caused by repeated or random load cycles; the more cycles a load applies on object, the more significant the fatigue damage is. A failure caused by repetitive loads is called fatigue failure. The capability of materials to resist fatigue failure is characterized by the plot of strength and number of cycles (S—N curve). Once dynamic loads are given, a fatigue analysis uses S—N curve to predict fatigue life or safety factor of design of solid objects.
Drop Test	A drop test analysis aims to calculate time-dependent stresses and deformations caused by an initial impact of an object with a rigid or flexible planar surface. In a drop test, it is desirable to define the materials as *elasto-plastic* one; this enables the software to account for energy lost in the dynamic simulation.

Table 7.15 Displacement Boundary Conditions for Structural Analysis

Restraints	Description
Fixed Geometry	In a fixed geometry, DoF of the restraints vary with element types. For solid or truss elements, three translational DoF are fixed. For shell and beam elements, both translational and rotational DoF are fixed. No reference geometry is needed to define a fixed geometry.
Immovable	Immovable restrains all translational motions for whatever types of elements. It is applicable to vertices, edges, faces or nodes of beam elements. For solids, immovable and fixed geometry have the same functions.
Roller/Sliding	Roller/sliding defines a planar face where nodes on a contact surface can move freely into its plane; it allows the contact face be shrunk or expanded under loading. However, the motions in the normal direction of plane are constrained.
Fixed Hinge	Fixed hinge defines a round face where nodes on this face are free to rotate about its rotational axis. During the deformation, the radius and the length of the round are set.
Symmetry	When both of geometries and loads are symmetric about a reference, numerical simulation can be performed only on a portion of the whole solid model to reduce computation. Symmetry applies is to define the constraints of symmetric reference to replace a full model by a partial model. For a solid mesh, it constrains one translation, for a shell mesh, it set the displacement of a translation and two rotations. Symmetry is applicable only on a flat face.

Table 7.16 Loads Boundary Conditions for Structural Analysis

Loads	Description
Pressure	Pressure is a type of surface load. It applies uniform or varying pressure on edges or surfaces of a physical structure. If the pressure varies, an analytic function must be defined to calculate pressure value for corresponding nodes.
Force	Force can be used to define forces, moments, or torques. Force is a type of concentrated load; however, it will be modeled as a uniformly distributed load on the nodes of selected faces, edges, vertices.
Gravity	Gravity is a type of body load. It applies a linear acceleration to a solid object. It is a common load type in structural analysis and nonlinear analysis.
Centrifugal	Centrifugal is a type of body load. It applies an initial force caused by angular velocity and acceleration of solid object. The software calculates the loads based on the specified angular velocity, acceleration and the mass density of materials.
Remote loads and restraints	As far as an assembled model is analysed, it is unnecessary to include all parts or components in the finite element analysis model. When some parts or components are excluded, the loads and constraints on those components can be converted as equivalent ones on the simplified model by using remote loads and restraints; the converted loads or constraints can be *remote load*, *remote displacement*, and *remote mass*.

Table 7.16 lists common types of loads for structural analysis in the Solid-Works Simulation.

Note that types of restraints relate to analysis types. If another analysis type is defined, the types of restraints and loads can be very different.

For example, in an FEA model of heat transfer, the restraints and loads on solid objects are related to temperature, convection, radiation, heat flux, and heat power, which are defined on vertices, edges, faces, or components.

7.8.6 **Solvers to FEA models**

An FEA model is eventually formulated as a mathematical model. A mathematic model describes design variables, their relations, and constraints. Design variables are system parameters of interest, which are solved. As shown in Fig. 7.31, mathematical models can be classified (Bender, 2000; Aris, 1994; Bokil, 2009) based on the criteria of probability, linearity, time-dependence, and continuity. As a generic tool, FEA can be applied to solve all of them except a probabilistic mode.

- **Deterministic and probabilistic model**: In a deterministic model, every set of variable states is uniquely determined by parameters in the model as well as the previous states of these variables. The model generates same results if the initial conditions are given. A statistical model involves a number of uncertainties where the states of variables are probability distributions of mean values.
- **Linear and nonlinear models**: If all of the relations among design variables are linear, the corresponding model is linear; otherwise, it is a nonlinear model. Nonlinear systems are generally difficult to be solved. A common approach to solve a nonlinear problem is the linearization.
- **Static and dynamic models**: A dynamic model treats the time as another dimension of design problems; it takes into account of

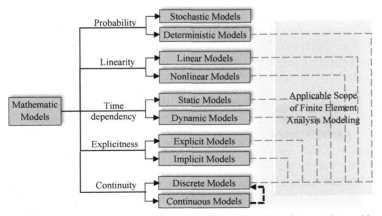

■ **FIGURE 7.31** Classification of mathematic models and application of finite element analysis modeling.

time-dependent changes of system states. A static model is also called as steady-state model; design variables in such a model is time invariants, and the system is in equilibrium.

■ **Explicit and implicit models**: When all of model inputs are known; if the system outputs can be calculated in a sequence of steps explicitly; the corresponding model is called as an explicit model. Otherwise, if system outputs have to be solved iteratively, such a model is called as an implicit model.

■ **Discrete and continual models**: A discrete model treats objects as discrete and continuous model treats objects as continuous domains.

SolidWorks Simulation provides four solvers to a formulated FEA model: *Auto*, *FFEPlus*, *Direct Sparse*, and *Large Problem Direct Sparse* (Reuss, 2014). *Fast Finite Elements* (FFEPlus) is an iterative solver that uses an implicit integration method; the solution is evolved iteratively under computational errors are small enough to meet terminating conditions. FFEPLus is efficient when the number of DOF of a system model is large, in particular, a model more than 100,000 DOF; it becomes more efficient when the model size is larger. FFEPLus can be suspicious if (1) incompatible mesh is applied and any local bonded contact is not covered by global bonded contact, (2) external forces or gravity is applied in frequency analysis, (3) base excitation is considered in linear dynamic study, (4) elasticity moduli vary greatly from one solid to another, (5) the boundary conditions of pressure or temperature are imported and circular/cyclic symmetry boundary conditions are applied, and (6) nonlinear analysis.

Direct Sparse finds a solution directly using exact numerical techniques. "Sparse" refers to the sparsity (zeroes) of the matrices that represents the relations of design variables and loads. Direct Sparse achieves a good accuracy in solving small or medium sized problems. It is faster if a computer has large memory. The Direct Sparse solver may be applied in a small FEA model, nonlinear analysis, or more accurate result. The size of an FEA model is confined according to the analysis type, 100,000 DOF for general analysis, 50,000 DOF for nonlinear analysis, and 500,000 DOF in heat transfer problems.

Large Problem Direct Sparse (LPDS) is an enhanced Direct Sparse solver for a large FEA model. High computation is addressed by using multiple cores. An LPDS solver can be applied when the Direct Sparse solver is required but the computer does not possess random access memory (RAM). LPDS can be used as a last resort to solving an FEA model. In addition, SolidWorks Simulation has an "*auto*" option that an appropriate solver can be automatically be selected to solve practical problems.

7.8.7 **Postprocessing**

The results of an FEA model usually include a considerable large amount of data. It is difficult and tedious to review and understand the meanings of calculated results. Postprocessing tools are used to sort, visualize, and output the data such as the distribution of stress, strain, safety factor, and temperature. Postprocessing helps in visualizing simulation results and identifying the weakest locations in computational domains.

Even though the result of an FEA model is available at the phase of postprocessing, critical thinking is needed to review and understand results. SolidWorks Simulation provides many postprocessing tools for users to understand simulation results: (1) visualize the distributions and contours of state variables, (2) animate the responses of objects such as deformations, vibrational models, and contact behaviors, (3) create flow trajectories in flow simulation, (4) make slides and create sectional views to visualize the distribution of state variables internally, and (5) use the probe tools to retrieve data at specified vertices, edges, faces, or components.

7.8.8 **Design optimization**

Optimization is a key ingredient in many engineering disciplines; the innovations in optimization algorithms and techniques have been experiencing a fast growth in recent years due to the rapid development in computer hardware and software capabilities. The implementation of any optimization algorithm requires a design analysis module where the performances of design candidates are evaluated (Bi and Zhang, 2001). For the designs of complex systems, FEA has been widely used as an analysis tool in simulation-based optimization.

Many CAE software tools provide the functionalities of *Design Study* to implement simulation-based optimization. *Design Study* in SolidWorks Simulation is developed to evaluate and optimize models. In a design study, a number of design variables are selected, each design variable corresponds to a set of discrete values, the simulations are conducted on the models with different combinations of values. The software is able to record the simulation results of all models and determine the model with the best result according to the specified design goals or criteria. *Sensor* tools are applied to specify constraints and goals.

7.9 **SUMMARY**

The computer implementation of FEA methods has been discussed. To make programs flexible and reusable, a software architecture based on the OO programming paradigm is proposed. The programs for 1D spring systems and 3D truss systems have been introduced to illustrate the computer implementation of the FEA methods. For existing FEA codes, both of commercial and open-source FEA packages have been reviewed and compared. The simulation of SolidWorks is discussed in details to explain the basic interfaces for users to access the functionalities of a commercial FEA code. It is emphasized that the modularized architecture of computer implementation makes it possible for an FEA code to deal with the versatility and complexity of a broad scope of FEA applications.

REFERENCES

Aris, R., 1994. Mathematical Modelling Techniques. Dover, New York, ISBN 0-486-68131-9.

Austrell, P.-E., Dahlblom, O., Lindemann, J., Olsson, A., Olosson, K.-G., Persson, K., Petersson, H., Ristinmaa, M., Sandberg, G., Wernberrg, P.-A., 2004. CALFEM a Finite Element Toolbox. The Division of Structural Mechanics, LTH, Sweden, ISBN 91-8855823-1. http://www.solid.lth.se/fileadmin/hallfasthetslara/utbildning/kurser/FHL064_FEM/calfem34.pdf.

Bender, E.A., 2000. An Introduction to Mathematical Modeling. Dover, New York, ISBN 0-486-41180-X [1978].

Bokil, V.A., 2009. Introduction of Mathematic Modelling. http://math.oregonstate.edu/~gibsonn/Teaching/MTH323-001S09/Supplements/IntroToModel.pdf.

Ferreira, A.J.M., 2009. Matlab Codes for Finite Element Analysis. Springer, ISBN 978-1-4020-9199-5.

Reuss, B., 2014. SolidWorks Simulation 2015 — FEA Solvers. https://www.3dvision.com/blog/entry/2014/11/26/solidworks-simulation-2015-fea-solvers.html.

Rumbaugh, J., Jacobson, I., Booch, G., 2004. The Unified Modelling Language Reference Manual, second ed. Addison-Wesley, Boston, U.S.A., ISBN 0-321-24562-8

Scardapane, S., 2015. Adaptive Algorithms and Parallel Computing: Object-Oriented Programming in Matlab (Primer). http://ispac.diet.uniroma1.it/scardapane/wp-content/uploads/2015/04/Object-Oriented-Programming-in-MATLAB.pdf.Wilson.

Wikipedia, 2017. Class Diagram. https://en.wikipedia.org/wiki/Class_diagram.

FURTHER READING

Archer, G.C., 1996. Object-Oriented Finite Element Analysis (Doctoral thesis). University of California At Berkeley.

Arudchelvam, T., Ratnajeevan, S., Hoole, H., Wijayakulasooriya, J., 2014. Component-based design from finite element Software written in the FORTRAN language. International Journal of Computer Theory and Engineering 6 (2), 124—128.

Austin, M.A., 2000. Engineering Programming in Matlab: A Primer. University of Maryland.

Balmes, E., Leclere, J.M., Chapelle, D., Delforge, C., Hassim, A., Vidrascu, M., 2005. Open FEM — an Open Source Finite Element Toolbox. http://www.cmap.polytechnique.fr/~allaire/gamni/OpenFEM_05.pdf.

Bi, Z.M., Zhang, W.J., 2001. Modularity technology in manufacturing: taxonomy and issues. International Journal of Advanced Manufacturing Technology 18 (5), 381—390.

Briggs, J.C., 2013. Developing an Architecture Framework for CloudBased, Multi-user, Finite Element Pre-processing (All theses and dissertations). Brigham Young University — Provo. Paper 3813.

Chessa, J., 2002. Programming the Finite Element Method with Matlab. McMaster University, Canada. http://ms.mcmaster.ca/ ~ bprotas/MATH745b/matlab_fem.pdf.

College Park, Maryland 20742, U.S.A. http://ac.aua.am/skhachat/Web/MATLAB/MATLAB%20ebooks/matlab%20engineering%20book.pdf.

Craig, D., 2004. An Introduction to Object Oriented Analysis and Design Using UML. Matincor, Inc.

Crandall, S.H., 1956. Engineering analysis: a survey of numerical procedures. McGraw-Hill Book Co., Inc.

Dari, J.S., 2010. An Object-Oriented Approach to the Finite Element Method.

Hededal, O., 1994. Object-oriented structuring of finite elements (Doctoral thesis). Aalborg University, Denmark.

Krysl, P., 2015. Thermal and Stress Analysis with the Finite Element Method. Pressure Cooker Press.

Krysl, P., Trivedi, A., 2005. Instructional use of Matlab software components for computational structural engineering applications. International Journal of Engineering Education 21 (5), 778—783.

Kumar, S., 2010a. Object-oriented finite element analysis of metal working processes. Journal of Software Engineering & Applications 3 (2010), 572—579.

Kumar, S., 2010b. Object-oriented finite element programming for engineering analysis in C++. Journal of Software 5 (7), 689—696.

Ma, Y.-Q., Fen, W., 2002. Object-oriented finite element analysis and programming in VC++. Applied Mathematics and Mechanics 23 (12), 0253—4827.

Martin, J., 2008. Introduction to object-oriented programming in Matlab. MathWorks Aerospace and Defence Conference '08.

McKenna, F., Scott, M.H., Fenves, G.L., January/February 2010. Nonlinear finite-element analysis software architecture using object composition. Journal of Computing in Civil Engineering 95—107.

Mueller, D.W., 2003. Introducing the finite element method to mechanical engineering students using Matlab. Proceedings of the 2003 American Society for Engineering Education Annual Conference & Exposition, Session 3566.

Nikishkov, G.P., NIkishkov, Y.G., 1992. FEATR-finite element educational code. Computers Education 19 (3), 237—245.

Pantale, O., Caperaa, S., Rakotomalala, R., 2004. Development of an object-oriented finite element program: application to metal-forming and impact simulations. Journal of Computational and Applied Mathematics 168 (2004), 341—351.

Patzák, B., 2012. OOFEM-an object-oriented simulation tool for advanced modeling of materials and structures. Acta Ploytechnica 52 (6), 59—66.

Phongthanapanich, S., Dechaumphai, P., 2006. EasyFEM—an object-oriented graphics interface finite element/finite volume software. Advances in Engineering Software 37 (2006), 797—804.

Rumbaugh, J., Jacobson, I., Booch, 2005. The Unified Modeling Language Reference Manual, second ed. Addison-Wesley, Boston, MA 02116, ISBN 0-321-24562-8.

Wu, K.-W., Hsieh, S.-H., Tsai, I.-C., Yang, Y.-S., 2002. An object-oriented design for modeling multi-point constraints in finite element analysis. Bulletin of the College of Engineering, NTU 85 (2002), 69—86.

Yaghoobi, A., 2012. Implementation of nonlinear finite element using object-oriented design patterns. Journal of Mechanical Engineering and Technology 4 (1), 61–79.

Zahr, M.J., 2014. Lecture 5 Advanced Matlab Object-Oriented Programming, CME 292: Advance Matlab for SC. Standford University.

■ PROBLEMS

7.1. Following the guides in Section 7.4, develop a computer program to model and analyze a 1D spring system. Use the developed program to calculate the deflections of springs in Fig. 7.32.

7.2. Use the computer program for 1D spring system from problem 7.1 to model and analyze the deflection of a tapered plate dimensioned in Fig. 7.33. Assume the plate is divided into five elements and six nodes, determine the displacements under an external load F.

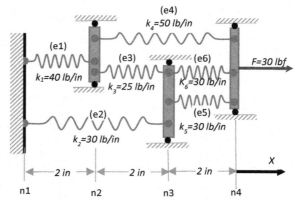

■ **FIGURE 7.32** Spring assemblage for problem 7.1.

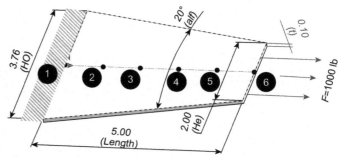

Materials: 1060 Alloy, $E = 1.0 \times 10^7$ psi, Poisson's Ratio 0.33, Density 0.0975 lb/in^3, Yield Strength 4000 psi

Thickness: $t = 0.1$ in, $L=5$-in, $H_e=2.0$ in, $H_0=3.76$ in, tapered angle=20°
Force: $F = 1000$ lbf

■ **FIGURE 7.33** Tapered plate subjected to $F = 1000$ lbf.

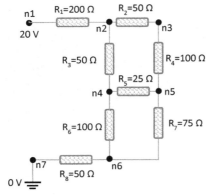

■ **FIGURE 7.34** Electric circuit for problem 7.3.

7.3. Using the computer program in problem 7.1 for 1D spring systems to model and analyze the electrical circuit in Fig. 7.34, determine the voltages on all nodes.

7.4. Develop a computer program to model and analyze 2D beam systems. Use the developed program to calculate the deflections of node *B* in Fig. 7.35; *A* and *C* in the beam structure is fixed.

7.5. Develop a computer program to model and analyze a 2D heat transfer system. Use the developed program to calculate the temperature distribution on the cross-section of chimney shown in Fig. 7.36. Note that the length unit is meter, only one-eighth of continuous domain is modeled, and the discretized model includes nine nodes, three rectangle elements, and two triangular elements.

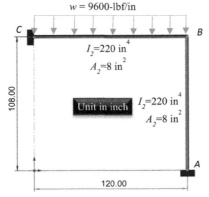

■ **FIGURE 7.35** 2D beam structure for problem 7.4.

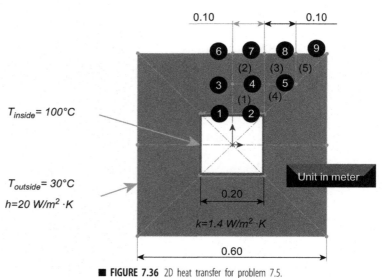

■ **FIGURE 7.36** 2D heat transfer for problem 7.5.

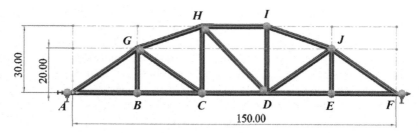

Materials: Gray Cast Iron, E=9.6×10^6 psi, Poisson's Ratio 0.27,
Density 0.26 lb/in^3, Yield Strength 21900 psi
Geometry: cross-section is a circle with the diameter of 1.0-in,
A = 0.7854 in^2, I_z=0.0491/ in^4 , mass/per inch=0.2042 lb/in; L=30×5-in
Boundaries: Fixed as A and F

■ **FIGURE 7.37** Natural frequencies for a frame structure in problem 7.6.

7.6. Following the guides in Section 7.6 to develop a computer program
for modal analysis of 2D frame structure. Use the developed program
to calculate the first three natural frequencies of the 2D frame illustrated in Fig. 7.37.

7.7. Develop a program which can assist participators in a bridge contest
to analyze their bridge designs based on the given boundary and
loading conditions.

Applications—Solid Mechanics Problems

8.1   INTRODUCTION

Mechanical design is to design parts, components, products, or systems of mechanical nature. For example, designs of various machine elements such as *shafts*, *bearings*, *clutches*, *gears*, and *fasteners* fall into the scope of mechanical design. Numerous criteria have been proposed in mechanical design processes, some primary design criteria include functions, safety, reliability, manufacturability, weight, size, wear, maintenance, and liability. In general, a mechanical design problem should be formulated with clear and complete statements of *functions*, *specifications*, and evaluation *criteria* (Mott, 2014):

- **Functions** are specified for what a product can fulfill. Functions are usually described by nonquantitative statements. Exemplifying product functions are to charge power on electronics (charger), clean floors (vacuum), transport objects (mobile platform), or support loads (structure).
- **Specifications** are detailed requirements described by quantitative statements. For example, product specifications can be defined in terms of size, weight, precision, working volume, speed, or load capacity. Specifications turn into design constraints in problem-solving processes.
- **Evaluation criteria** are the statements of desirable qualitative characteristics. Evaluation criteria are treated as design objectives to optimize the solutions in problem-solving processes. Evaluation criteria are set to maximize benefits and minimize disadvantages of mechanical designs.

Although the numbers and priorities of specifications and criteria vary from one product design to another; some common design considerations are applicable to any mechanical systems. These considerations include loading

Finite Element Analysis Applications. http://dx.doi.org/10.1016/B978-0-12-809952-0.00008-X

capability, deformation, stability, and durability. The dependence of these evaluation criteria on design variables must be modeled and analyzed to optimize products. In this chapter, the governing mathematical models of common engineering design problems are discussed, and finite element analysis (FEA) is used to analyze mechanical systems from the perspective of loading capability, stability, and fatigue life.

8.2 STRUCTURAL ANALYSIS

A mechanical system is usually subjected to different types of loads such as force, pressure, heat, temperature, or constraints at supports. *Structural analysis* is to determine the effect of loads on physical structures and their components. Structural analysis is required for all of the physical systems, such as machinery, furniture, transportation, and constructions. These systems are designed to withstand specified loads. Structural design must ensure the sufficient rigidity and strength to carry given loads under specified boundary conditions. A number of critical tasks in a structural analysis include to (1) model and analyze the response of system to external loads, (2) identify critical areas of loading conditions, (3) evaluate corresponding stresses, and (4) determine whether or not the obtained stresses at critical areas exceed the strength of selected materials.

As shown in Fig. 8.1, a structural system consists of structural elements with given materials. The materials decide structural strengths, and the structural elements decide stress distribution of system subjected to given loads. The stresses and strengths over the domain are then compared to see if the material at any position has the sufficient strength to withstand external loads. A stress causes strain, and the strain corresponds to the deformation at the system level. The evaluated deformation is compared to the required accuracy to see if the structure has the sufficient rigidity subjected to the given loads. The loading capability of a mechanical

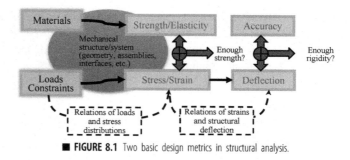

■ **FIGURE 8.1** Two basic design metrics in structural analysis.

structure/system is determined by the stress level at its weakest position. Although the weakest position is affected by many factors such as types of materials, types and magnitudes of loads, constraints at supports, geometric shapes, and assemblies and interfaces of physical bodies. Besides the constitutive stress–strain models determined by materials, other two relations in structural analysis are (1) the relations of stress distribution and loads and (2) the relations of strain distribution and structural deflection. Despite their dependence, it is convenient to treat nodal displacements as state variables in a system model. The stress distribution and deflections over the structure are treated as dependent variables, which can be derived from nodal displacements after the system model is solved.

Fig. 8.2 describes an FEA model for structural analysis. For a continuous domain, the applied external forces can be classified as *volume force* such as a weight caused by gravity, a *surface force* such as a drag force by pressure, and a *concentrated load* such as a point load on beam. Any one of external loads will affect the stress distribution over the domain. For any position with an infinitesimal volume, its stress state can be described as,

$$\boldsymbol{\sigma} = \begin{bmatrix} \sigma_x, & \sigma_y, & \sigma_z, & \tau_{xy}, & \tau_{yz}, & \tau_{xz} \end{bmatrix}^T \tag{8.1}$$

where

σ_x, σ_y, and σ_z are the components of normal stresses and
τ_{xy}, τ_{xz}, τ_{yz} are the components of shear stresses over y–z, x–z, and x–y planes respectively.

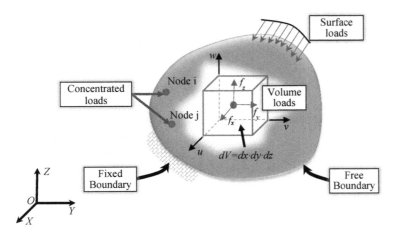

■ **FIGURE 8.2** Description of a finite element analysis model for structural analysis.

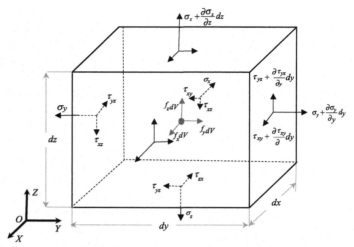

■ **FIGURE 8.3** Stress equilibrium at an infinitesimal volume.

Fig. 8.3 shows the force equilibrium at an infinitesimal volume $(dx \times dy \times dz)$. The force with six components in the stress state of Eq. (8.1) is balanced by body force. There are three equations of force equilibrium along x-, y-, and z-axis.

Take an example of the stress equilibrium over the Y–Z plane,

$$\sum F_y = \left(\sigma_y + \frac{\partial \sigma_y}{\partial y} dy \right) dxdz - \sigma_y dxdz + \left(\tau_{xy} + \frac{\partial \tau_{xy}}{\partial x} dx \right) dydz - \tau_{xy} dydz$$
$$+ \left(\tau_{zy} + \frac{\partial \tau_{zy}}{\partial z} dz \right) dxdy - \tau_{zy} dxdy - f_y dxdydz = 0$$

(8.2)

$$\sum F_y = \frac{\partial \tau_{xy}}{\partial x} + \frac{\partial \sigma_y}{\partial y} + \frac{\partial \tau_{yz}}{\partial z} + f_y = 0 \qquad (8.3)$$

Because the selection of X, Y, and Z is arbitrary, the same force equilibrium condition applies to X–Z plane and X–Y plane. As a result, the conditions of the force equilibrium on three planes are

$$\left. \begin{array}{l} \sum F_x = \dfrac{\partial \sigma_x}{\partial x} + \dfrac{\partial \tau_{xy}}{\partial y} + \dfrac{\partial \tau_{xz}}{\partial z} + f_x = 0 \\[2mm] \sum F_y = \dfrac{\partial \tau_{xy}}{\partial x} + \dfrac{\partial \sigma_y}{\partial y} + \dfrac{\partial \tau_{yz}}{\partial z} + f_y = 0 \\[2mm] \sum F_z = \dfrac{\partial \tau_{xz}}{\partial x} + \dfrac{\partial \tau_{yz}}{\partial y} + \dfrac{\partial \sigma_z}{\partial z} + f_z = 0 \end{array} \right\} \qquad (8.4)$$

Note that external forces turn into the distributed stress over object, and the response of the materials is quantified by the strain state $\boldsymbol{\varepsilon} = \begin{bmatrix} \varepsilon_x, & \varepsilon_y, & \varepsilon_z, & \gamma_{xy}, & \gamma_{yz}, & \gamma_{xz} \end{bmatrix}^T$ as,

$$\left.\begin{aligned}
\varepsilon_x &= \frac{1}{E}(\sigma_x - v(\sigma_y + \sigma_z)) \\
\varepsilon_y &= \frac{1}{E}(\sigma_y - v(\sigma_x + \sigma_z)) \\
\varepsilon_z &= \frac{1}{E}(\sigma_z - v(\sigma_x + \sigma_y)) \\
\gamma_{xy} &= \frac{\tau_{xy}}{G}, \quad \gamma_{yz} = \frac{\tau_{yz}}{G}, \quad \gamma_{xz} = \frac{\tau_{xz}}{G}
\end{aligned}\right\} \tag{8.5}$$

where

E is elastic or Young's modulus,
v is Poisson's ratio, and
G is shear modulus or modules of rigidity.

Shear modulus G depends on elastic modulus E by the relation of,

$$G = \frac{E}{2(1 + v)} \tag{8.6}$$

Alternatively, Eq. (8.5) can also be reformatted to determine the state of stresses based on the given strains as,

$$\{\sigma\} = [D]\{\varepsilon\} \tag{8.7}$$

where $[D]$ is the matrix form of the Hooke's law in a three-dimensional space, i.e.,

$$[D] = \frac{E}{(1 + v)(1 - 2v)} \begin{bmatrix}
1 - v & v & v & 0 & 0 & \\
v & 1 - v & v & 0 & 0 & 0 \\
v & v & 1 - v & 0 & 0 & 0 \\
0 & 0 & 0 & \frac{1}{2} - v & 0 & 0 \\
0 & 0 & 0 & 0 & \frac{1}{2} - v & 0 \\
0 & 0 & 0 & 0 & 0 & \frac{1}{2} - v
\end{bmatrix} \tag{8.8}$$

Depending on the type of design problems, the stress and strain relations can be simplified if not all of the directions have nonzero stresses. In the following, the physical behaviors for a few of classic mechanical design problems are discussed, the procedures in developing corresponding FEA models are illustrated.

8.2.1 **Truss structures**

A mechanical structure is often needed to support loads in a large space. To fully utilize materials, a set of spatially structured members are assembled to replace bulk bodies. Fig. 8.4 shows some examples of applications of truss structures in construction, factory, and transportation. The majority of components in these applications are *trusses*, which can be idealized as binary elements in a system model. In design a truss structure, the number, types, materials, configurations, and cross-sections of trusses are optimized to maximize the material utilization and the capability of structure.

8.2.1.1 Description of truss elements

A truss structure consists of a number of *joints*, connected by two-force binary members. Each truss member in the structure carries either *compressive* or *tensile* force. A joint confines the relative displacement of the connection

■ FIGURE 8.4 Truss-structure examples with distributed loads in large space.

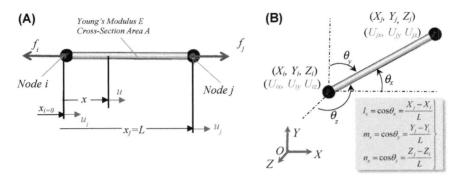

■ **FIGURE 8.5** One-dimensional truss element in global coordinate system (LCS) and GCS: (A) local coordinate system; (B) global coordinate system.

between two members, but it allows a free rotation of two members. To simplify the analyze of a truss structure, it is assumed that external loads are all concentrated forces and these forces apply only on joints. Therefore, a truss member can be treated as an axial member whose behavior can be modeled as a spring element.

As shown in Fig. 8.5, a linear truss member consists of two nodes (node i and node j). Its material property is given by Young's modulus E, and its structural parameters are cross-section area A and length L. The element behavior is represented by the displacements of nodes u_i and u_j. In the local coordinate system $x \in (0, L)$ of element, the interpolation in the element can be performed using the shape functions of element as,

$$\{u\} = \begin{bmatrix} S_i(x) & S_j(x) \end{bmatrix} \begin{Bmatrix} u_i \\ u_j \end{Bmatrix} \text{ and } \{\boldsymbol{\varepsilon}\} = \begin{bmatrix} \frac{-1}{l} & \frac{-1}{l} \end{bmatrix} \begin{Bmatrix} u_i \\ u_j \end{Bmatrix} \quad (8.9)$$

where

$$S_i(x) = \frac{l-x}{l}, S_j(x) = \frac{x}{l}$$

The potential energy of the truss member can be evaluated as,

$$\Pi = \int_V \frac{1}{2} E \frac{du}{dx} dV - (f_i \cdot u_i + f_j \cdot u_j) = \frac{EA}{2} \int_V \frac{du}{dx} dx - (f_i \cdot u_i + f_j \cdot u_j) \quad (8.10)$$

Using the condition for the minimal potential energy $\frac{\partial \Pi}{\partial u_i} = \frac{\partial \Pi}{\partial u_j} = 0$ yields,

$$\begin{bmatrix} k_e & -k_e \\ -k_e & k_e \end{bmatrix} \begin{Bmatrix} u_i \\ u_j \end{Bmatrix} = \begin{Bmatrix} f_i \\ f_j \end{Bmatrix} \quad (8.11)$$

where $k_e = \frac{EA}{L}$ is the equivalent stiffness coefficient.

As shown in Fig. 8.5, even though a truss member only has its displacements along the axial direction in its LCS; a truss member can be an arbitrary position in a two-dimensional or three-dimensional space, the coordination transformation must be performed to transform an element model from LCS to a GCS. Therefore, Eq. (8.11) has to be expended to represent the relation of the forces and the displacements in GCS as,

$$[K_{L,e}]\{u\} = \begin{bmatrix} k_e & 0 & 0 & -k_e & 0 & 0 \\ 0 & 0 & 0 & 0 & 0 & 0 \\ 0 & 0 & 0 & 0 & 0 & 0 \\ -k_e & 0 & 0 & k_e & 0 & 0 \\ 0 & 0 & 0 & 0 & 0 & 0 \\ 0 & 0 & 0 & 0 & 0 & 0 \end{bmatrix} \begin{Bmatrix} u_i \\ 0 \\ 0 \\ u_j \\ 0 \\ 0 \end{Bmatrix} = \begin{Bmatrix} f_i \\ 0 \\ 0 \\ f_j \\ 0 \\ 0 \end{Bmatrix} \qquad (8.12)$$

where $[K_{L,e}]$ is the three-dimensional stiffness matrix of a truss element with respect to GCS.

Similarly, the nodal displacements of element have to be transformed from LCS to GCS as well. Taking the origin of GCS as a reference and using Eq. 3.19 for the coordinate transformation for two nodes yield,

$$\begin{Bmatrix} U_{ix} \\ U_{iy} \\ U_{iz} \end{Bmatrix} = [T] \cdot \begin{Bmatrix} u_i \\ 0 \\ 0 \end{Bmatrix}, \quad \begin{Bmatrix} U_{jx} \\ U_{jy} \\ U_{jz} \end{Bmatrix} = [T] \cdot \begin{Bmatrix} u_j \\ 0 \\ 0 \end{Bmatrix} \qquad (8.13)$$

where $[T] = \begin{bmatrix} l_x & l_y & l_z \\ m_x & m_y & m_z \\ n_x & n_y & n_z \end{bmatrix}$ and $\begin{Bmatrix} l_x \\ m_x \\ n_x \end{Bmatrix}, \begin{Bmatrix} l_y \\ m_y \\ n_y \end{Bmatrix}, \begin{Bmatrix} l_z \\ m_z \\ n_z \end{Bmatrix}$ are the vectors of directional cosines of x, y, z in GCS.

The x-axis must be selected to be aligned with the axial direction of a truss member; although the directions of y- and z-axes can be arbitrary as long as the perpendicular relations of x-with y- and z- are satisfied.

The coordinate transformation from GCS to LCS can be obtained by reformatting, Eq. (8.13) as,

$$\{u_i \ 0 \ 0 \ u_j \ 0 \ 0\}^T = [T_{\text{global\_local}}]\{U_{ix} \ U_{iy} \ U_{iz} \ U_{jx} \ U_{jy} \ U_{jz}\}^T \qquad (8.14)$$

where

$$
[T_{\text{global\_local}}] = \begin{bmatrix} l_x & m_x & n_x & 0 & 0 & 0 \\ l_y & m_y & m_y & 0 & 0 & 0 \\ l_z & m_z & n_z & 0 & 0 & 0 \\ 0 & 0 & 0 & l_x & m_x & n_x \\ 0 & 0 & 0 & l_y & m_y & n_y \\ 0 & 0 & 0 & l_z & m_y & n_z \end{bmatrix}
\tag{8.15}
$$

Substituting Eq. (8.15) into Eq. (8.12) gets,

$$
[K_{G,e}] \cdot \begin{Bmatrix} U_{ix} \\ U_{iy} \\ U_{iz} \\ U_{jx} \\ U_{jy} \\ U_{jz} \end{Bmatrix} = \begin{Bmatrix} F_{ix} \\ F_{iy} \\ F_{iz} \\ F_{jx} \\ F_{jy} \\ F_{jz} \end{Bmatrix}
\tag{8.16}
$$

where

$$
[K_{G,e}] = [T]' \cdot [K_{L,e}] \cdot [T]
$$

$$
= \begin{bmatrix} l_x & l_y & l_z & 0 & 0 & 0 \\ m_x & m_y & m_z & 0 & 0 & 0 \\ n_x & n_y & n_z & 0 & 0 & 0 \\ 0 & 0 & 0 & l_x & l_y & l_z \\ 0 & 0 & 0 & m_x & m_y & m_z \\ 0 & 0 & 0 & n_x & n_y & n_z \end{bmatrix} \cdot \begin{bmatrix} k_e & 0 & 0 & -k_e & 0 & 0 \\ 0 & 0 & 0 & 0 & 0 & 0 \\ 0 & 0 & 0 & 0 & 0 & 0 \\ -k_e & 0 & 0 & k_e & 0 & 0 \\ 0 & 0 & 0 & 0 & 0 & 0 \\ 0 & 0 & 0 & 0 & 0 & 0 \end{bmatrix} \cdot \begin{bmatrix} l_x & m_x & n_x & 0 & 0 & 0 \\ l_y & m_y & m_y & 0 & 0 & 0 \\ l_z & m_z & n_z & 0 & 0 & 0 \\ 0 & 0 & 0 & l_x & m_x & n_x \\ 0 & 0 & 0 & l_y & m_y & n_y \\ 0 & 0 & 0 & l_z & m_y & n_z \end{bmatrix}
$$

$$
= k_e \begin{bmatrix} l_x^2 & l_x m_x & l_x n_x & -l_x^2 & -l_x m_x & -l_x n_x \\ l_x m_x & m_x^2 & m_x n_x & -l_x m_x & -m_x^2 & -m_x n_x \\ l_x n_x & m_x n_x & n_x^2 & -l_x n_x & -l_x m_x & -n_x^2 \\ -l_x^2 & -l_x m_x & -l_x n_x & l_x^2 & l_x m_x & l_x n_x \\ -l_x m_x & -m_x^2 & -m_x n_x & l_x m_x & m_x^2 & m_x n_x \\ -l_x n_x & -m_x n_x & -n_x^2 & l_x n_x & l_x m_x & n_x^2 \end{bmatrix}
$$

Note that the deviated stiffness matrix relates only to the directional cosines of local x-axis where the deformation occurs.

If all of the truss members are in the same plane, element models in a two-dimensional (2D) space are sufficient. The model can be simplified by removing the columns and rows in Eq. (8.16) related to z-axis as,

$$[K_{G,e}] \cdot \begin{Bmatrix} U_{ix} \\ U_{iy} \\ U_{jx} \\ U_{jy} \end{Bmatrix} = \begin{Bmatrix} F_{ix} \\ F_{iy} \\ F_{jx} \\ F_{jy} \end{Bmatrix} \tag{8.17}$$

where the stiffness matrix for a 2D truss member is

$$[K_{G,e}] = [T]' \cdot [K_{L,e}] \cdot [T]$$

$$= \begin{bmatrix} l_x & l_y & 0 & 0 \\ m_x & m_y & 0 & 0 \\ 0 & 0 & l_x & l_y \\ 0 & 0 & m_x & m_y \end{bmatrix} \begin{bmatrix} k_e & 0 & -k_e & 0 \\ 0 & 0 & 0 & 0 \\ -k_e & 0 & k_e & 0 \\ 0 & 0 & 0 & 0 \end{bmatrix} \begin{bmatrix} l_x & m_x & 0 & 0 \\ l_y & m_y & 0 & 0 \\ 0 & 0 & l_x & m_x \\ 0 & 0 & l_y & m_y \end{bmatrix}$$

$$= k_e \begin{bmatrix} l_x^2 & l_x m_x & -l_x^2 & -l_x m_x \\ l_x m_x & m_x^2 & -l_x m_x & -m_x^2 \\ -l_x^2 & -l_x m_x & l_x^2 & l_x m_x \\ -l_x m_x & -m_x^2 & l_x m_x & m_x^2 \end{bmatrix}$$

8.2.1.2 Boundary conditions and loads

Truss members are two-force members; a connection of two members does not restrain any rotation. Each node in a truss element has three degrees of freedom (DOF) for translations; the rotations are free and not treated as design variables. Fig. 8.6 shows the types of boundary conditions for displacements. The imposed constraints at a node can be one, two, or three DOF. In the case of Fig. 8.6D, the imposed DOF is not aligned with any axis of a coordinate system. A new reference plane has to be created, so that the boundary condition of a roller can be defined to restrain the motion perpendicular to that reference plane.

As show in Fig. 8.7, truss members may be assembled by various of techniques such as *riveting*, *fastening*, *welding*, or *mechanical joints* in the

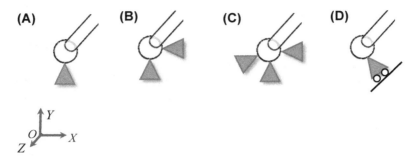

■ **FIGURE 8.6** Types of boundary conditions for displacements. (A) DOF restrained displacement (U_x, U_y, or U_z); (B) 2-DOF restrained displacement (1) U_x and U_y, (2) U_x and U_z, or (3) U_y and U_z; (C) 3-DOF restrained displacement (U_x, U_y, and U_z); (D) 1-DOF restrained displacement in any arbitrary direction.

■ **FIGURE 8.7** Various joints in truss structures: (A) restraints on rotations and translations; (B) restraints on translations.

real-world applications. Because the mathematic models of truss members are applied, any connections in the truss structure are assumed to be free of rotations. However, one must ensure that the impact of rotational restraints is ignorable in contrast to that of translational restraints. Otherwise, the element types such as bending or frame members have to be used.

Only nodal forces are applicable to a truss structure. Therefore, external loads on trusses must be converted to equivariant nodal forces. Taking an example of the gravity force, the equivalent nodal loads must be determined based on the force or moment equilibrium or energy conservation of truss members.

8.2.1.3 Truss structure examples

Example 8.1

Fig. 8.8 shows a truss structure in a two-dimensional space with the specified loads $F = 1000$ lbf at two sides (A and B). All of the truss members use the same materials AISI 1020. It has the Young's modulus $E = 2.9 \times 10^7$ psi and the Poisson' ratio of 0.29. The yield strength is $S_y = 5.10 \times 10^4$ psi. The cross-section areas of all truss members is given as $A = 2$ in^2. Use the SolidWorks simulation to predict the maximal stress and deflection of the structure.

Solution

(1) Create a Computer Aided Design (CAD) model. It is convenient to use an integrated simulation tool such as the SolidWorks, so that the CAD model of a design problem can be readily created for engineering analysis in the same design platform. The SolidWorks has many modeling tools for users to create complex geometric shapes of structures or systems. The software supports different unit systems; users can select and switch unit systems to provide inputs with different units. Besides standard unit systems such as meter, kilogram, second and inch, pound, second, Fig. 8.9 shows that a custom unit system can be used by mixing the types of units in inputs. The software takes care of the conversions

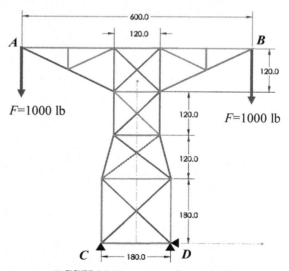

■ **FIGURE 8.8** A truss structure for example 8.1.

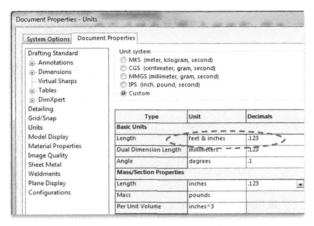

■ **FIGURE 8.9** Definition of a custom unit system.

between different units. The truss structure is represented by a sketch on a reference plane (i.e., default front plane). As shown in Fig. 8.10, each truss member is modeled as a line segment and an intersection of two lines is treated as a node later. In the developed FEA model, the orientations, locations, and lengths of all truss members are defined according to the sketch of structure in Fig. 8.8. (2) Applying material properties: the Solid-Works includes a comprehensive library of materials for most of the commonly used materials. If the applied material of a solid object is included in the library, one can simply specify and apply it on the selected object(s). Otherwise, a new model of material properties can be defined by setting major attributes of materials such as Young's modulus, Poisson

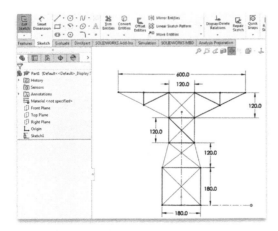

■ **FIGURE 8.10** Creation of sketch for truss structure.

■ **FIGURE 8.11** Definition of structural members from a sketch.

ratio, and so on. (3) Creating truss elements. Truss elements are the simplest element types. The system parameters of a truss member are Young's modulus E, cross-section area A, and length L. As shown in Fig. 8.11, the SolidWorks has the tool so called *Weldments* to define structure members including trusses with a given cross-section area. In addition, the nodes of an FEA model can be detected automatically for all of the intersections. Fig. 8.12 shows the truss structure with defined structure members. (4) Defining a static analysis. In a SolidWorks simulation, one can choose *an analysis type* for an FEA model. As shown in Fig. 8.13, a static analysis is defined for the developed truss structure. Because default types of structural members are beam elements, these structural members have to be selected and redefined as truss members in Fig. 8.14. The *Simulation Manager* provides a list of components in an FEA model including *geometry*, *joints*, *connections*, *fixtures*, *meshes*, and *results*, which are shown in Fig. 8.15. The geometric attributes with given material

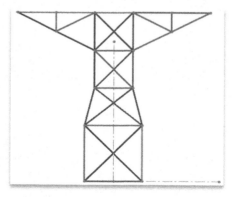

■ **FIGURE 8.12** Truss members in the structure.

■ **FIGURE 8.13** Definition of a static analysis.

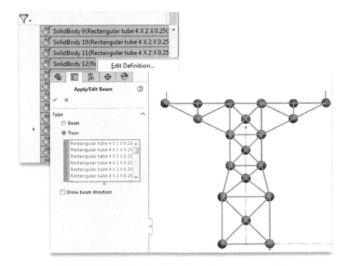

■ **FIGURE 8.14** Assigning structural members as truss members.

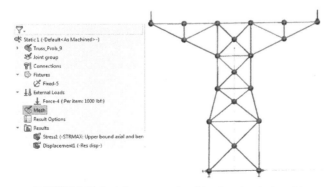

■ **FIGURE 8.15** Simulation manager of an finite element analsysis model.

properties, joints, and connections are imported directly from the CAD model of the truss structure. (5) Setting boundary conditions and loads. For the lowest truss, the *x*-and *y*-displacements of node **D** are fixed, so is the *y*-displacement of node **C** (See Fig. 8.16). In addition, the whole truss structure lies in the same plane; the *z*-displacements must be fixed for all of the nodes (See Fig. 8.17). Finally, the loads on two nodes are defined at Nodes **A** and **B**. Running the simulation leads to the solution of nodal displacements, which are displayed in Fig. 8.18.

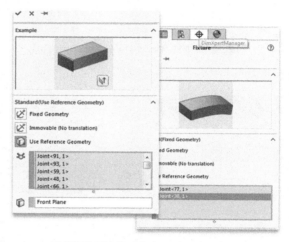

■ **FIGURE 8.16** Fixed displacement on boundaries.

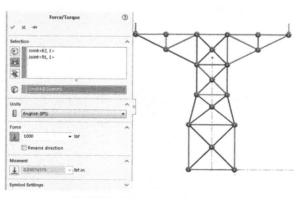

■ **FIGURE 8.17** Applying loads on nodes.

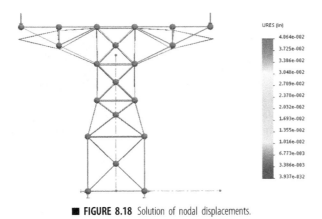

■ **FIGURE 8.18** Solution of nodal displacements.

Example 8.2

Fig. 8.19 shows a truss structure in 3D space with the specified loads $F = (100\ \text{lb},\ 300\ \text{lb},\ 100\ \text{lb})$ at node B. Truss members use AISI 304 with $E = 2.76 \times 10^7$ psi and Poisson' ratio $v = 0.29$, the yield strength is $S_y = 2.9 \times 10^4$ psi, and the cross-section area $A = 1.5\ \text{in}^2$. Use the Solid-Works simulation to predict the maximal deflection of truss structure.

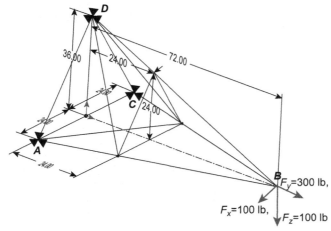

■ **FIGURE 8.19** A truss structure for example 8.2.

Solution

The procedure of FEA modeling consists of the following steps. (1) Creating a CAD model: seven reference key points are defined as the nodes of the structure in Fig. 8.20; line segments are defined by connecting nodes as truss elements. Because the truss structure is in a 3D space, it is convenient to use a 3D sketch to include these truss members in different spatial planes. The orientations, locations, and lengths of all truss members are defined based on the given condition in Fig. 8.19. (2) Applying material properties: a custom material model can be created for a solid object in the SolidWorks. One can simply specify the attributes of materials such as Young's modulus and Poisson ratio for truss members. (3) Applying displacement boundary conditions and loads. As shown in Fig. 8.21, the displacements ($U_x = U_y = U_z = 0$) on A, C, and D are fixed,

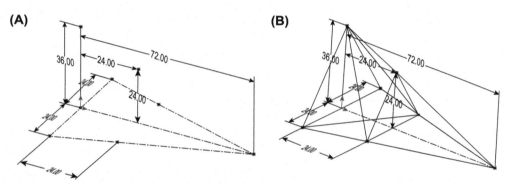

■ **FIGURE 8.20** Creating a truss structure from reference nodes and planes: (A) reference points and line segments; (B) structural members from reference points and lines.

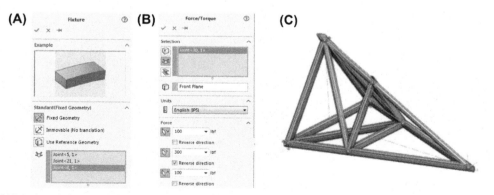

■ **FIGURE 8.21** Applying displacement boundary conditions and load on truss structure: (A) fixed displacements on A, C, and D; (B) load on node B; (C) the three-dimensional truss structure with boundary conditions.

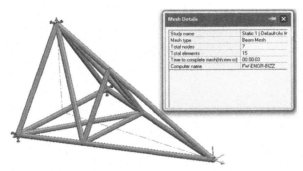

■ **FIGURE 8.22** Meshing of three-dimensional truss structure.

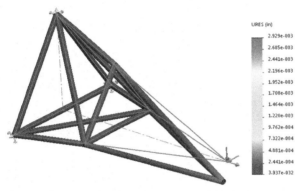

■ **FIGURE 8.23** Deflection of three-dimensional truss structure.

and the given load is applied on node *B*. (4) The meshing process in Fig. 8.22 shows that the SolidWorks is available to map the intersections into nodes, and connect nodes as elements based on the topology of structure. For this design problem, 7 nodes and 15 truss members have been defined in the mesh. (5) The FEA model is solved, and Fig. 8.23 shows that the maximized deflection of 2.292×10^{-3} (in) occurs to the tip of structure.

8.2.2 **Plane stress problems**

Many problems in a structural analysis can be solved satisfactorily by a 2D simulation model. Two general types in the plane theory of elasticity are *plane stress* and *plane strain*. Both of them are defined by specifying certain restraints on stress or strain fields. An object is said to be in a plane stress

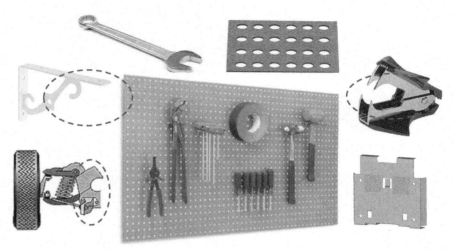

■ **FIGURE 8.24** Examples of plane stress applications.

state if the stress vector is zero across a plane. If such a case occurs to the entire domain of a structure, for example of a thin plate, the structural analysis of the object can be simplified considerably by representing the stress state as a 2D tensor. Fig. 8.24 gives some examples of the applications where the products or parts can be analyzed by a plane stress model.

Under a plane stress state, all of stresses occur on the same plane. Assume that all of stresses occur on $X-Y$ plane, i.e., $\sigma_z = 0$, and $\tau_{xz} = \tau_{yz} = 0$, Eqs. (8.7) and (8.8) can be simplified as,

$$\{\boldsymbol{\sigma}\} = \begin{Bmatrix} \sigma_x \\ \sigma_y \\ \tau_{xy} \end{Bmatrix} = [\boldsymbol{D}]\boldsymbol{\varepsilon} = \frac{E}{1-v^2} \begin{bmatrix} 1 & v & 0 \\ v & 1 & 0 \\ 0 & 0 & \frac{1-v}{2} \end{bmatrix} \begin{Bmatrix} \varepsilon_x \\ \varepsilon_y \\ \gamma_{xy} \end{Bmatrix} \qquad (8.18)$$

where $[\boldsymbol{D}]$ is a symmetric matrix, i.e., $[\boldsymbol{D}] = [\boldsymbol{D}]^T$.

The principle of the minimized potential energy is applied to formulate element models about nodal displacements. The potential energy of an element consists of the strain energy of materials and the work made by external loads. Assume that external loads are applied on nodes, the potential energy of an element can be evaluated as,

$$\Pi = \Lambda - \sum_{i=1}^{n} F_i \cdot U_i \qquad (8.19)$$

where

$\prod$ is the total potential energy,

Λ is the strain energy,

n is the DOF of element,

F_i is an external load over the ith degree of freedom, and

U_i is the displacement on the ith degree of freedom.

The total strain energy of element can be evaluated as,

$$\Lambda = \int_V \frac{1}{2}[\sigma]^T[\varepsilon]dV = \int_V \frac{1}{2}[\varepsilon]^T[D]^T[\varepsilon]dV = \int_V \frac{1}{2}[\varepsilon]^T[D][\varepsilon]dV \qquad (8.20)$$

In an FEA model, the behavior of a continuous domain is collectively represented by state variables of nodes. An element model describes the relations of external forces and state values on nodes. An element model depends on the number of nodes, DOF on each node, geometric shape, and the governing equations of physical behaviors. In this section, we discuss two basic types of 2D plane elements, i.e., linear triangle element and rectangle element.

Fig. 8.25 shows a linear triangle element in a GCS. It consists of three nodes $(i, j, \text{ and } k)$, whose coordinates in GCS are given as (X_i, Y_i), (X_j, Y_j), and (X_k, Y_k), respectively. The element behavior is represented by the displacements on three nodes, namely (U_{ix}, U_{iy}), (U_{jx}, U_{jy}), and (U_{kx}, U_{ky}). To derive the model of a triangle element in Fig. 8.25, the interpolation is performed, so that the displacement (u, v) in an arbitrary position (X, Y) can be derived from the displacements at nodes using shape functions.

$$\left\{ \begin{array}{c} u \\ v \end{array} \right\} = \begin{bmatrix} S_i & 0 & S_j & 0 & S_k & 0 \\ 0 & S_i & 0 & S_j & 0 & S_k \end{bmatrix} \left\{ \begin{array}{c} U_{ix} \\ U_{iy} \\ U_{jx} \\ U_{jy} \\ U_{kx} \\ U_{ky} \end{array} \right\} \qquad (8.21)$$

where the shape functions of a triangle element are given in Eq. (3.73).

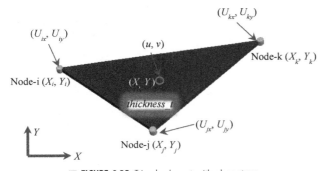

■ **FIGURE 8.25** Triangle element with plane stress.

The plane strain can be derived from Eq. (8.21) as,

$$
\left\{ \begin{array}{c} \varepsilon_x \\ \varepsilon_y \\ \gamma_{xy} \end{array} \right\} = \left\{ \begin{array}{c} \dfrac{\partial u}{\partial x} \\[6pt] \dfrac{\partial v}{\partial y} \\[6pt] \dfrac{\partial u}{\partial y} + \dfrac{\partial v}{\partial x} \end{array} \right\} = \left[\begin{array}{cccccc} \dfrac{\partial S_i}{\partial x} & 0 & \dfrac{\partial S_j}{\partial x} & 0 & \dfrac{\partial S_k}{\partial x} & 0 \\[6pt] 0 & \dfrac{\partial S_i}{\partial y} & 0 & \dfrac{\partial S_j}{\partial y} & 0 & \dfrac{\partial S_k}{\partial y} \\[6pt] \dfrac{\partial S_i}{\partial y} & \dfrac{\partial S_i}{\partial x} & \dfrac{\partial S_j}{\partial y} & \dfrac{\partial S_j}{\partial x} & \dfrac{\partial S_k}{\partial y} & \dfrac{\partial S_k}{\partial x} \end{array} \right] \left\{ \begin{array}{c} U_{ix} \\ U_{iy} \\ U_{jx} \\ U_{jy} \\ U_{kx} \\ U_{ky} \end{array} \right\}
$$

$$(8.22)$$

Substituting Eq. (3.73) into Eq. (8.22) yields,

$$\{\varepsilon\} = [B] \cdot \{U\} \qquad (8.23)$$

Expanding Eq. (8.20) by substituting Eq. (8.23) yields its matrix form as,

$$\Lambda = \frac{1}{2} \int_V \{U\}^T [B]^T [D][B]\{U\} dV \qquad (8.24)$$

where

$$\{\varepsilon\} = \left\{ \begin{array}{c} \varepsilon_x \\ \varepsilon_y \\ \gamma_{xy} \end{array} \right\} \text{ is the vector of 2D strain,}$$

$$[B] = \frac{1}{2A} \left[\begin{array}{cccccc} \beta_i & 0 & \beta_j & 0 & \beta_k & 0 \\ 0 & \delta_i & 0 & \delta_j & 0 & \delta_k \\ \delta_i & \beta_i & \delta_j & \beta_j & \delta_k & \beta_k \end{array} \right] \text{ is the matrix for the strain-}$$

displacement relation,
$(\beta_i, \beta_j, \beta_k, \delta_i, \delta_j, \delta_k)$ are the coefficients defined in Eq. (7.71), and
A is the area of the triangle element.

For the triangle element, applying the principle of minimum potential energy $\frac{\partial \Pi}{\partial U} = 0$ in Eq. (8.19) gets,

$$(A \cdot t)[B]^T [D][B] \cdot \{U\} = \{F\} \qquad (8.25)$$

where

A and t are the area and thickness of the triangle element,
$[D]$ and $[B]$ are the matrices of constants defined in Eqs. (8.18) and (8.24), and
$\{U\}$ and $\{F\}$ are the vectors of state variables and loads of element, respectively.

Note that if the distributed load is applied on an edge of a triangle element, it should also be included as a part of the external load in Eq. (8.25). The external load by the distributed load has been discussed in Chapter 5.

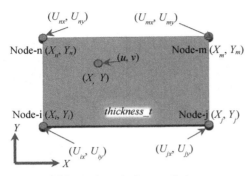

■ **FIGURE 8.26** Rectangle element with plane stress.

Fig. 8.26 shows a rectangle element in GCS. It consists of four nodes (i, j, m, and n), whose coordinates are given as (X_i, Y_i), (X_j, Y_j), (X_m, Y_m), and (X_n, Y_n), respectively. The element behavior is represented by displacements on four nodes, namely, (U_{ix}, U_{iy}), (U_{jx}, U_{jy}), (U_{mx}, U_{my}), and (U_{nx}, U_{ny}). To derive the model of a rectangle element in Fig. 8.26, the interpolation is performed, so that the displacement (u, v) in an arbitrary position (X, Y) can be derived from the displacements at nodes using shape functions.

$$\left\{ \begin{array}{c} u \\ v \end{array} \right\} = \begin{bmatrix} S_i & 0 & S_j & 0 & S_m & 0 & S_n & 0 \\ 0 & S_i & 0 & S_j & 0 & S_m & 0 & S_n \end{bmatrix} \left\{ \begin{array}{c} U_{ix} \\ U_{iy} \\ U_{jx} \\ U_{jy} \\ U_{mx} \\ U_{my} \\ U_{nx} \\ U_{ny} \end{array} \right\} \tag{8.26}$$

where the shape functions of a rectangle element are given in Eq. (3.54).

The model for the plane strain state can be derived from Eq. (8.26) as,

$$\left\{ \begin{array}{c} \varepsilon_x \\ \varepsilon_y \\ \gamma_{xy} \end{array} \right\} = \left\{ \begin{array}{c} \dfrac{\partial u}{\partial x} \\[2mm] \dfrac{\partial v}{\partial y} \\[2mm] \dfrac{\partial u}{\partial y} + \dfrac{\partial v}{\partial x} \end{array} \right\} = \begin{bmatrix} \dfrac{\partial S_i}{\partial x} & 0 & \dfrac{\partial S_j}{\partial x} & 0 & \dfrac{\partial S_m}{\partial x} & 0 & \dfrac{\partial S_n}{\partial x} & 0 \\[2mm] 0 & \dfrac{\partial S_i}{\partial y} & 0 & \dfrac{\partial S_j}{\partial y} & 0 & \dfrac{\partial S_m}{\partial y} & 0 & \dfrac{\partial S_n}{\partial y} \\[2mm] \dfrac{\partial S_i}{\partial y} & \dfrac{\partial S_i}{\partial x} & \dfrac{\partial S_j}{\partial y} & \dfrac{\partial S_j}{\partial x} & \dfrac{\partial S_m}{\partial y} & \dfrac{\partial S_m}{\partial x} & \dfrac{\partial S_n}{\partial y} & \dfrac{\partial S_n}{\partial x} \end{bmatrix} \left\{ \begin{array}{c} U_{ix} \\ U_{iy} \\ U_{jx} \\ U_{jy} \\ U_{mx} \\ U_{my} \\ U_{nx} \\ U_{ny} \end{array} \right\}$$

$$\tag{8.27}$$

Substituting Eq. (3.54) into Eq. (8.27) gets,

$$\{\varepsilon\} = [B] \cdot \{U\} \tag{8.28}$$

Using Eq. (8.20) for the strain in Eq. (8.28) yields,

$$\Lambda = \frac{1}{2} \int_V \{U\}^T [B]^T [D] [B] \{U\} dV \tag{8.29}$$

where

$$[B] = \begin{bmatrix} \dfrac{-(w-y)}{lw} & 0 & \dfrac{(w-y)}{lw} & 0 & \dfrac{y}{lw} & 0 & -\dfrac{y}{lw} & 0 \\[2mm] 0 & \dfrac{-(l-x)}{lw} & 0 & \dfrac{-x}{lw} & 0 & \dfrac{x}{lw} & 0 & \dfrac{(l-x)}{lw} \\[2mm] \dfrac{-(l-x)}{lw} & \dfrac{-(w-y)}{lw} & \dfrac{-x}{lw} & \dfrac{(w-y)}{lw} & \dfrac{x}{lw} & \dfrac{y}{lw} & \dfrac{(l-x)}{lw} & -\dfrac{y}{lw} \end{bmatrix}$$

For a rectangle element, applying the principle of minimum potential energy $\frac{\partial \Pi}{\partial U} = 0$ in Eq. (8.19) gets,

$$\left[\int_A [B]^T [D] [B] t dA \right] \cdot \{U\} = \{F\} \tag{8.30}$$

where

[D] and [B] are the matrices defined in Eqs. (8.18) and (8.29), and {U} and {F} are the vectors of field variables and loads of element, respectively.

Note that if the distributed load is applied on an edge of a rectangle element, it should also be included as a part of the external load in Eq. (8.30). The external load by the distributed load has been discussed in Chapter 5.

Example 8.3

Fig. 8.27 shows a thin plate with a thickness of 0.2-in. The plate includes three discontinuities of geometries, a Φ 1.5-in hole, a Φ 2.0-in hole, and a 0.5-in fillet at the shoulder. The plate uses aluminum 1060 alloy with the elastic modulus of $E = 1.0 \times 10^7$ psi, the Poisson's ratio is 0.33, yield strength $S_y = 3.999 \times 10^3$ psi. Determine stress concentration factors at three discontinuity sections under a tensile or compression load.

Solution

The solving process consists of the following steps. (1) Creating a Computer Aided Design (CAD) model. The computer model of a thin plate is created in the CAD environment. The model is imported into the simulation environment. (2) Defining a static analysis. Only the stress concentration is concerned, so static analysis under a constant load is sufficient. However, the option for *2D Simplification* is activated to model the object. As shown in Fig. 8.28, there are three types under 2D simplification, that is, *plane stress*, *plane strain*, and *axis-symmetric*; the plane stress type is selected for this example. The SolidWorks identifies a neutral plane from the flat model in the meshing process. (3) Applying boundary conditions and loads. According to the definition of the stress concentration, a stress concentration factor depends on three main factors: (a) the type of discontinuity, (b) the size of discontinuity, and (c) the load type. Stress concentration has no relation to material properties or the magnitude of load. Therefore, the magnitude of an applied load can be arbitrary as long as the resulted stress in the plane is within the elastic range of deformation. As shown in Fig. 8.29, the edge of the stress plane on the left is fixed and 500-psi pressure is applied on the right edge; after the FEA model is solved, Fig. 8.30 illustrates the stress distribution over the stress plane.

As shown in Fig. 8.30, the *probe* tool is applied to retrieve the stress at three critical positions (A, B, and C). The stresses at three positions are found as,

$$\sigma_{max,A} = 1201 \text{ psi}, \quad \sigma_{max,B} = 1313 \text{ psi}, \quad \sigma_{max,C} = 1147 \text{ psi}$$

The average stresses at three sections passing A, B, and C are,

$$\overline{\sigma}_A = \frac{500 \times 3.75}{(5-2)} = 625 \text{ psi}, \quad \overline{\sigma}_B = \frac{500 \times 3.75}{(5-1.5)} = 535.71 \text{ psi},$$

$$\overline{\sigma}_C = \frac{500 \times 3.75}{3.75} = 500 \text{ psi}$$

The stress concentration factors on three critical sections passing A, B, and C, can be found as,

$$k_{c,A} = \frac{\sigma_{max,A}}{\overline{\sigma}_A} = \frac{1201}{625} = 1.92, \quad k_{c,B} = \frac{\sigma_{max,B}}{\overline{\sigma}_B} = \frac{1313}{535.71} = 2.45,$$

$$k_{c,C} = \frac{\sigma_{max,C}}{\overline{\sigma}_C} = \frac{1147}{500} = 2.29$$

It indicates that the geometric discontinuities may increase stress concentrations significantly.

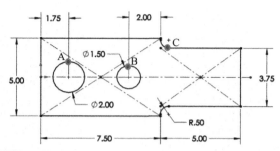

■ **FIGURE 8.27** A plane stress plate with a thickness of 0.2-in (aluminum 1060 alloy).

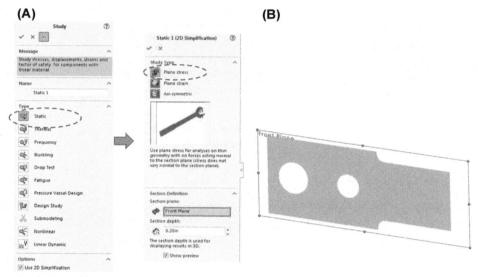

■ **FIGURE 8.28** Creating a plane stress model for finite element analysis modeling: (A) define a plane stress analysis; (B) definition of neutral plane for a plane stress object.

8.2.3 **Plane strain problems**

If the dimensions in one direction (z-axis) are extremely large compared to those in other two directions (x- and y-axes), the deformation in z-axis is restrained. Therefore, the corresponding principal strain (ε_z) is zero. Even though all of three principal stresses are nonzero components, the principal stress in z-axis depends on the principal stresses in x- and y-axes, which will not be included in the plane strain model. Fig. 8.31 gives some examples of applications where a structure or product can be analyzed by a plane strain model.

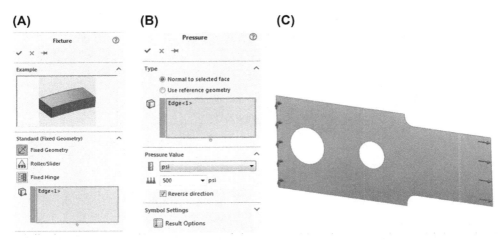

■ **FIGURE 8.29** Applying boundary conditions and loads on thin plate: (A) fixed displacements on left edge; (B) pressure on right edge; (C) the stress plane with boundary conditions.

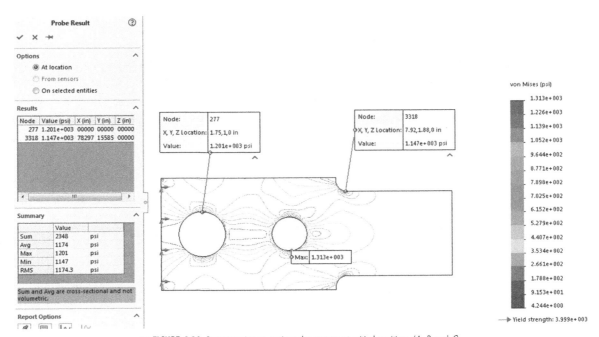

■ **FIGURE 8.30** Postprocessing to retrieve the stresses at critical positions (*A*, *B*, and *C*).

■ **FIGURE 8.31** Examples of plane strain parts.

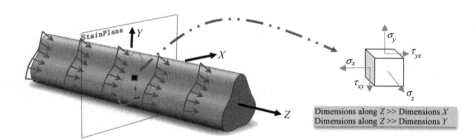

■ **FIGURE 8.32** Stress state in a plane strain model.

As shown in Fig. 8.32, the plane strain state corresponds to the case where $\varepsilon_z = 0$, and $\gamma_{xz} = \gamma_{yz} = 0$. Therefore, Eq. (8.7) can be simplified as,

$$\{\boldsymbol{\sigma}\} = \begin{Bmatrix} \sigma_x \\ \sigma_y \\ \tau_{xy} \end{Bmatrix} = [\boldsymbol{D}]\{\boldsymbol{\varepsilon}\} = \frac{E}{(1+v)(1-2v)} \begin{bmatrix} 1-v & v & 0 \\ v & 1-v & 0 \\ 0 & 0 & \dfrac{1-v}{2} \end{bmatrix} \begin{Bmatrix} \varepsilon_x \\ \varepsilon_y \\ \gamma_{xy} \end{Bmatrix}$$

(8.31)

The plane strain model is similar to a plane stress model except for the constitutive model [**D**]. The models for triangular and rectangle elements are derived in Eqs. (8.25) and (8.30) and are applicable to plane strain elements as well; however, the constitutive model [**D**] is defined in Eq. (8.31).

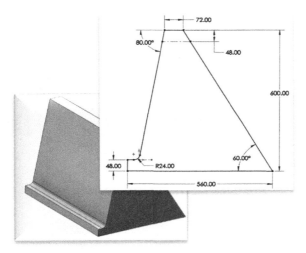

■ **FIGURE 8.33** The dam as a plane strain model.

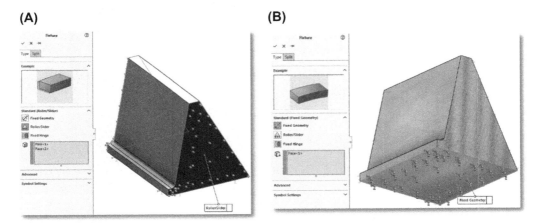

■ **FIGURE 8.34** Displacement boundary conditions on dam: (A) roller constraints on side surfaces; (B) fixed constraints on bottom surface.

(A) **(B)** **(C)**

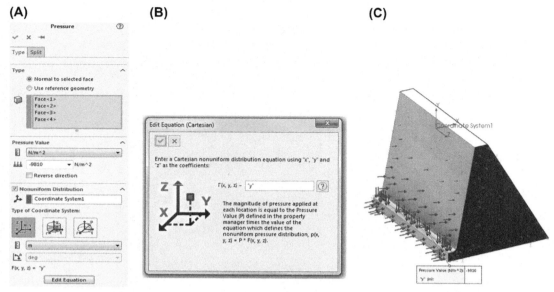

■ **FIGURE 8.35** Definition of load boundary conditions: (A) pressure load; (B) function of *y*-altitude; (C) hydrostatic load on dam.

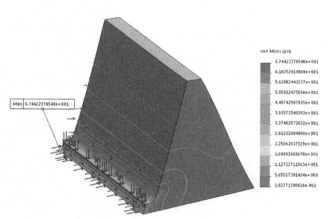

■ **FIGURE 8.36** Stress distribution over dam.

Example 8.4

Fig. 8.33 shows a long dam subjected to the hydrostatic pressure. The dimensions on the cross-section are given. The water level is assumed to be 48-in from the top. Establish an FEA model to determine the maximized stress in the dam.

Solution

The solving procedure of FEA modeling consists of the following steps. (1) Creating Computer Aided Design (CAD) model. The dam model is defined based on the dimensions on the cross-section; it is extruded with a unit length. The model can be imported into the simulation environment. (2) Defining a static analysis. The design problem can be formulated as a plane strain model. However, the dam is under a hydrostatic load; it will be easy to be defined in 3D model. Therefore, we define a static analysis model in a 3D space instead. (3) Applying boundary conditions and loads. Because it is a plane strain problem, the length direction does not allow any deflection. The roller constraints are applied on lateral surfaces. The bottom surface is fixed as a boundary surface in Fig. 8.34. The hydrostatic pressure is a function of the *y*-altitude; the reference for zero pressure is 48-in from the top plane of dam.

The definition of the hydrostatic pressure is shown in Fig. 8.35; a custom function can be defined as varying hydrostatic pressure with respect to the selected coordinate system. After the FEA model is solved, Fig. 8.36 illustrates the stress distribution over the cross-section of the strain plane. It is found the maximized stress is 67.44 psi, which is close to the maximized pressure applied by water.

8.3 MODAL ANALYSIS

Modal analysis concerns the dynamic response of a structure subjected to vibrational excitations. The goal of modal analysis is to determine the natural frequencies and corresponding mode shapes of an object or structure subjected to boundary conditions. The mathematical model for a modal analysis is called as an eigenvalue system, and the solution to such a model is represented by eigenvalues and eigenvectors, which correspond to the natural frequencies and mode shapes, respectively. Fig. 8.37 shows a number of engineering design problems where modal analyses are essential to ensure the safe applications of structures or systems. All of these applications involve dynamic loads or excitations whose frequencies must be away from any of natural frequencies of products.

■ **FIGURE 8.37** Examples of products where modal analyses are needed.

In this section, 2D frame elements are used as an example to develop element models for modal analysis. A node in a 2D frame element may have loads and displacements along x- and y-axes, and a rotational displacement around z-axis. We assume that small deformations are applied in the element; in other words, the model of a 2D frame can be treated as a combined model from 1D axial member and 2D beam member.

8.3.1 **2D truss member in LCS**

2D truss member is its LCS can be described in Fig. 8.38. If the linear approximation is used, it includes two nodes (i and j).

The interpolations in 2D truss member for *displacements* and *velocities* are performed separately as,

$$u_x = [S]\{u\} = \begin{bmatrix} \dfrac{L-x}{L} & \dfrac{x}{L} \end{bmatrix} \begin{Bmatrix} u_{ix} \\ u_{jx} \end{Bmatrix} \tag{8.32}$$

$$\dot{u}_x = [S]\{\dot{u}\} = \begin{bmatrix} \dfrac{L-x}{L} & \dfrac{x}{L} \end{bmatrix} \begin{Bmatrix} \dot{u}_{ix} \\ \dot{u}_{jx} \end{Bmatrix} \tag{8.33}$$

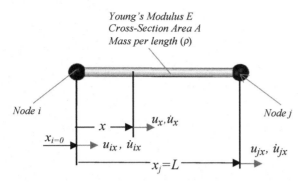

■ FIGURE 8.38 Two-dimensional truss member in LCS for modal analysis.

Assume that a system includes potential energy Λ, kinematic energy T, and the work done by external force, the application of the potential minimum energy principle yields the Lagrange's equation as,

$$\frac{d}{dt}\left(\frac{\partial T}{\partial \dot{q}_i}\right) - \frac{\partial T}{\partial q_i} + \frac{\partial \Lambda}{\partial q_i} = Q_i \quad (n = 1, 2 \cdots n) \tag{8.34}$$

where

t, is time variable,
T, is kinetic energy of system,
q_i, is an independent DOF of system $(i = 1, 2, \ldots n)$,
$\dot{q}_i$, is the velocity along a DOF of system $(i = 1, 2, \ldots n)$,
Λ, is the potential energy of system, and
Q_i, is the external load along one DOF $(i = 1, 2, \ldots n)$.

A truss member has two displacements (u_{ix}, u_{jx}) and corresponding velocities $(\dot{u}_{ix}, \dot{u}_{jx})$. The potential and kinetic energies are evaluated, respectively, as,

$$\Lambda = \int_V \frac{1}{2} E \left(\frac{du}{dx}\right)^2 dV = \frac{EA}{2} \int_L \{u\}^T \left[\frac{\partial \{S\}}{\partial x}\right]^T \left[\frac{\partial \{S\}}{\partial x}\right] \{u\} dl \tag{8.35}$$

$$T = \int_V \frac{1}{2} \rho (\dot{u})^2 dV = \frac{\rho A}{2} \int_L \{\dot{u}\}^T [S]^T [S] \{\dot{u}\} dl \tag{8.36}$$

Substituting Eqs. (8.35) and (8.36) into Eq. (8.34) gets,

$$[M]^{(L,e)}\{\ddot{u}\} + [K]^{(L,e)}\{u\} = \{f\} \tag{8.37}$$

where

[M] is the mass matrix for a 2D axial member,

$$[M]^{(L,e)} = \frac{\rho A L}{6} \begin{bmatrix} 2 & 1 \\ 1 & 2 \end{bmatrix},$$

[K] is the stiffness matrix for a 2D axial member,

$$[K]^{(L,e)} = \frac{EA}{L} \begin{bmatrix} 1 & -1 \\ -1 & 1 \end{bmatrix}, \text{ and}$$

{f}, is the vector of external loads.

8.3.2 **2D beam member in LCS**

A 2D beam member is its LCS can be described in Fig. 8.39. Each node in a beam member consists of two displacements, that is, y-axis displacement (u_y) and z-axis rotational displacement (θ_z). Correspondingly, the velocities on these two displacement directions are defined.

The interpolations in a 2D beam member for displacements and velocities are performed separately as,

$$u = [S]\{u\} = \begin{bmatrix} 1 - \dfrac{3x^2}{L^2} + \dfrac{2x^3}{L^3} & x - \dfrac{2x^2}{L} + \dfrac{x^3}{L^2} & \dfrac{3x^2}{L^2} - \dfrac{2x^3}{L^3} & -\dfrac{x^2}{L} + \dfrac{x^3}{L^2} \end{bmatrix} \begin{Bmatrix} u_{iy} \\ \theta_{iz} \\ u_{jy} \\ \theta_{jz} \end{Bmatrix}$$

(8.38)

$$\dot{u} = [S]\{\dot{u}\} = \begin{bmatrix} 1 - \dfrac{3x^2}{L^2} + \dfrac{2x^3}{L^3} & x - \dfrac{2x^2}{L} + \dfrac{x^3}{L^2} & \dfrac{3x^2}{L^2} - \dfrac{2x^3}{L^3} & -\dfrac{x^2}{L} + \dfrac{x^3}{L^2} \end{bmatrix} \begin{Bmatrix} \dot{u}_{iy} \\ \dot{\theta}_{iz} \\ \dot{u}_{jy} \\ \dot{\theta}_{jz} \end{Bmatrix}$$

(8.39)

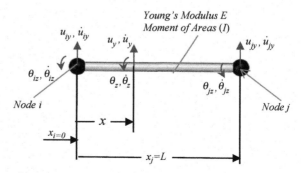

■ **FIGURE 8.39** Two-dimensional beam member in LCS for modal analysis.

The strain of a beam member can be found from Eq. (8.38) as,

$$\{\boldsymbol{\varepsilon}\} = -y\frac{\partial^2 u}{\partial x^2} = -y\left[\frac{\partial^2 S}{\partial x^2}\right]\{\boldsymbol{u}\}$$

$$= -y\left[-\frac{6}{L^2}+\frac{12x}{L^3} \quad -\frac{4}{L}+\frac{6x}{L^2} \quad \frac{6}{L^2}-\frac{12x}{L^3} \quad -\frac{2}{L}+\frac{6x}{L^2}\right]\begin{Bmatrix} u_{iy} \\ \theta_{iz} \\ u_{jy} \\ \theta_{jz} \end{Bmatrix} \quad (8.40)$$

The strain energy of a beam number is calculated as,

$$\Lambda = \int_V \frac{1}{2}E\left(-y\frac{d^2u}{dx^2}\right)^2 dV = \frac{EI}{2}\int_L \{\boldsymbol{u}\}^T\left[\frac{\partial^2\{\boldsymbol{S}\}}{\partial x^2}\right]^T\left[\frac{\partial^2\{\boldsymbol{S}\}}{\partial x^2}\right]\{\boldsymbol{u}\}dl \quad (8.41)$$

And the kinetic energy is calculated as,

$$T = \int_V \frac{1}{2}\rho(\dot{u})^2 dV = \frac{\rho A}{2}\int_L \{\dot{\boldsymbol{u}}\}^T[S]^T[S]\{\dot{\boldsymbol{u}}\}dl \quad (8.42)$$

Thus, substituting Eqs. (8.41) and (8.42) into Eq. (8.34) to obtain the conditions for the minimized potential energy as,

$$[\boldsymbol{M}]^{(L,e)}\{\ddot{\boldsymbol{u}}\} + [\boldsymbol{K}]^{(L,e)}\{\boldsymbol{u}\} = \{\boldsymbol{f}\} \quad (8.43)$$

where

$[\boldsymbol{M}]^{(L,e)}$ is the mass matrix for a 2D beam element,

$$[\boldsymbol{M}]^{(L,e)} = \frac{\rho AL}{420}\begin{bmatrix} 156 & 22L & 54 & -13L \\ 22L & 4L^2 & 13L & -3L^2 \\ 54 & 13L & 156 & -22L \\ -13L & -3L^2 & -22L & 4L^2 \end{bmatrix}$$

$[\boldsymbol{K}]^{(L,e)}$ is the stiffness matrix for a 2D beam element,

$$[\boldsymbol{K}]^{(L,e)} = \frac{EI}{L^3}\begin{bmatrix} 12 & 6L & -12 & 6L \\ 6L & 4L^2 & -6L & 2L^2 \\ -12 & -6L & -12 & -6L \\ 6L & 2L^2 & -6L & 4L^2 \end{bmatrix}$$

8.3.3 Modeling of 2D frame element

As shown in Fig. 8.40, a 2D beam member consists of *x*- and *y*-displacements and *z*-rotational displacement. Under the assumption of the small deformation, the deformations under a variety of loads can be summed

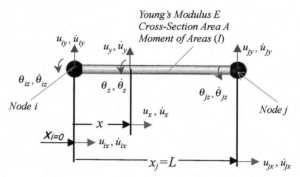

■ **FIGURE 8.40** 2D frame member in LCS for modal analysis.

linearly. Accordingly, the mass matrix and stiffness matrix are obtained by assembling Eqs. (8.37) and (8.43) as,

$$[M]^{(L,e)}\{\ddot{u}\} + [K]^{(L,e)}\{u\} = \{f\} \tag{8.44}$$

where

$[M]^{(L,e)}$ is the mass matrix for a 2D frame element,

$$[M]^{(L,e)} = \frac{\rho A L}{420} \begin{bmatrix} 140 & 0 & 0 & 70 & 0 & 0 \\ 0 & 156 & 22L & 0 & 54 & -13L \\ 0 & 22L & 4L^2 & 0 & 13L & -3L^2 \\ 70 & 0 & 0 & 140 & 0 & 0 \\ 0 & 54 & 13L & 0 & 156 & -22L \\ 0 & -13L & -3L^2 & 0 & -22L & 4L^2 \end{bmatrix}$$

$[K]^{(L,e)}$ is the stiffness matrix of a 2D frame element,

$$[K]^{(L,e)} = \begin{bmatrix} \dfrac{EA}{L} & 0 & 0 & -\dfrac{EA}{L} & 0 & 0 \\ 0 & \dfrac{12EI}{L^3} & \dfrac{6EI}{L^2} & 0 & -\dfrac{12EI}{L^3} & \dfrac{6EI}{L^2} \\ 0 & \dfrac{6EI}{L^2} & \dfrac{4EI}{L} & 0 & -\dfrac{6EI}{L^2} & \dfrac{2EI}{L} \\ -\dfrac{EA}{L} & 0 & 0 & \dfrac{EA}{L} & 0 & 0 \\ 0 & -\dfrac{12EI}{L^3} & -\dfrac{6EI}{L^2} & 0 & \dfrac{12EI}{L^3} & -\dfrac{6EI}{L^2} \\ 0 & \dfrac{6EI}{L^2} & \dfrac{2EI}{L} & 0 & -\dfrac{6EI}{L^2} & \dfrac{4EI}{L} \end{bmatrix}$$

8.3.4 **Modal analysis of 2D structure with frame members**

For a 2D structure consisting of frame members, all of the element models in their LCSs have to be transformed into the corresponding ones in GCS, so that they can be assembled into a system model. In the coordinate transformation, let

$$\{\ddot{U}\} = [T]\{\ddot{u}\}, \{U\} = [T]\{u\}\{F\} = [T]\{f\} \tag{8.45}$$

where the coordinate transformation T from LCS to GCS is

$$[T] = \begin{bmatrix} \cos\theta & \sin\theta & 0 & 0 & 0 & 0 \\ -\sin\theta & \cos\theta & 0 & 0 & 0 & 0 \\ 0 & 0 & 1 & 0 & 0 & 0 \\ 0 & 0 & 0 & \cos\theta & \sin\theta & 0 \\ 0 & 0 & 0 & -\sin\theta & \cos\theta & 0 \\ 0 & 0 & 0 & 0 & 0 & 1 \end{bmatrix}$$

Substituting Eq. (8.45) into Eq. (8.44) gets the model of a 2D frame element in GCS as,

$$[M]^{(G,e)}\{\ddot{U}\} + [K]^{(G,e)}\{U\} = \{F\} \tag{8.46}$$

where

$[M]^{(G,e)}$ is the mass matrix for 2D frame element in GCS,
$[M]^{(G,e)} = [T]^T[M]^{(L,e)}[T]$
$[K]^{(G,e)}$ is the stiffness matrix of a 2D frame element in GCS,
$[K]^{(G,e)} = [T]^T[K]^{(L,e)}[T]$, and
$[F]^{(G,e)}$ is the load of a 2D frame element in GCS, $\{F\} = [T]\{f\}$.

For a 2D structure consisting of 2D frame elements, the assembled system model in GCS becomes,

$$[M]^{(G)}\{\ddot{U}\}^{(G)} + [K]^{(G)}\{U\}^{(G)} = \{F\}^{(G)} \tag{8.47}$$

where

$[M]^{(G)}, [K]^{(G)}$ are the mass matrix and stiffness matrix of the 2D frame structure, respectively;
$\{U\}^{(G)}, \{F\}^{(G)}$ are the vectors of displacements and loads in the structure.

In modal analysis, the harmonic solution of Eq. (8.47) is concerned, and the solution is assumed as,

$$\left.\begin{array}{l} \{U\}^{(G)} = \{X\} \cdot \sin(\omega t + \phi) \\ \{F\}^{(G)} = 0 \end{array}\right\} \tag{8.48}$$

Substituting Eq. (8.48) into Eq. (8.47) gives,

$$\left(\omega^2[\boldsymbol{M}]^{(G)} - [\boldsymbol{K}]^{(G)}\right)\{\boldsymbol{X}\}\sin(\omega t + \phi) = 0 \qquad (8.49)$$

Eq. (8.49) specifies the conditions of natural frequencies as,

$$\left|\omega^2[\boldsymbol{M}]^{(G)} - [\boldsymbol{K}]^{(G)}\right| = 0 \qquad (8.50)$$

where $|\bullet|$ is the determinate of a matrix.

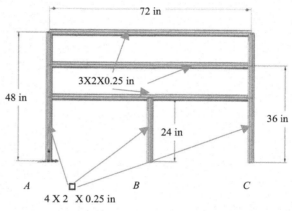

■ **FIGURE 8.41** Example of two-dimensional frame for modal analysis.

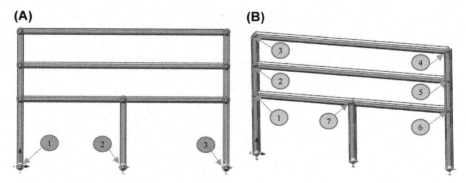

■ **FIGURE 8.42** Boundary conditions on two-dimensional frame structure: (A) fixed nodes at three supports; (B) roller nodes at other seven intersections.

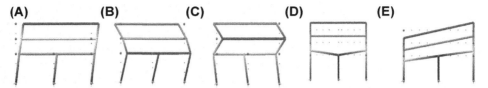

(A)   **(B)**   **(C)**   **(D)**   **(E)**

■ **FIGURE 8.43** Five vibration modes of structure: (A) first mode; (B) second mode; (C) third mode; (D) fourth mode; (E) fifth mode.

Example 8.5

Fig. 8.41 shows a 2D frame structure. Three elements in the vertical direction have a rectangle thin-tube cross-section of $4 \times 2 \times 0.25$ in, and other elements in the horizontal direction have a rectangle thin-tube cross-section of $3 \times 2 \times 0.25$ in. The structure is completely fixed at A, B, and C. The plate uses the plain carbon steel with the elastic modulus of $E = 3.046 \times 10^7$ psi, the density of $\rho = 0.282$ lb/in^3, the Poisson' ratio $\nu = 0.28$, and the yield strength $S_y = 3.20 \times 10^4$ psi. Determine the first five natural frequencies of the structure.

Solution

In the SolidWorks simulation, *2D Simplification* option is unavailable for modal analysis. Therefore, the 2D frame model is modeled as a 3D beam structure, but roller constraints are applied to all of the intersections to remove DOF off the $x–y$ plane. The modeling process consists of the following steps: (1) define a sketch using the dimensions provided in Fig. 8.41 for the frame; (2) create two groups of structural members with different cross-section areas ($4 \times 2 \times 0.25$ in and $3 \times 2 \times 0.25$ in) for vertical and horizontal frame elements; (3) fix the displacements at three support nodes, and define the roller supports for the rest of seven interactions of frame as shown in Fig. 8.42.

Solving the defined FEA generates the results of the first five natural frequencies as,

Mode	Frequencies (rad/s)	Frequencies (Hz)
1	284.64	45.302
2	931.55	148.26
3	2420.9	385.3
4	3468.0	551.95
5	2470.5	552.34

Correspondingly, the vibration modes for above natural frequencies are illustrated in Fig. 8.43.

8.4 FATIGUE ANALYSIS

Fatigue is a phenomenon of materials where the accumulative damage occurs by repetitive loads. Structures and systems in many applications are subjected to dynamic and repetitive loads, which induce fluctuating or cyclic stresses. Invisible damage on the materials caused by such loads is accumulated until it leads to a structural fracture. Fatigue causes over 90% of all of mechanical service failures (ASM International, 2008).

Fig. 8.44 shows few applications where fatigue analysis must be performed in the designs of mechanical structures or components. In these applications, the magnitude of stress causes fatigue damage could be much less than the ultimate strength of material. However, the frequency of a repetitive load is very high, and the product is generally required to run safely for a long time. This implies that the materials must endure a repetitive load with a large number of loading cycles. Taking an example of a car engine, if an average reciprocating speed is 4000 revolutions per minute and a 400-h duration test is performed, the expected fatigue life is 9.6×10^7 cycles (Shariyat et al., 2016). By all means, the experimental solution to a fatigue analysis usually takes a long time. In addition, a fatigue test often involves a high cost for the development of a testing platform and instrumentation. Moreover, it is impractical to test a large number of application scenarios for products. Therefore, FEA-based simulations have been widely adopted for the fatigue analysis of mechanical designs.

Fatigue refers to the weakening of a material caused by cyclic loads on the material (Ramachandran et al., 2005). Fatigue is the progressive and localized structure damage under a dynamic load. The progress of fatigue can be divided into three stages; i.e., *crack initiation*, *crack growth*, and *overload*

Springs

Gears

Bearings

Damping

Guideways

■ **FIGURE 8.44** Examples of products where fatigue analysis is needed.

failure. The fatigue behavior of material not only relates to the properties of material, but also relates to many other factors such as the application environment, characteristics of loads, temperature, and surface conditions (Nanninga, 2008). The methods to analyze the fatigue life of a machine element have been discussed extensively (Hamrock et al., 1999; Budynas and Nisbett, 2015), and three major methods are the *strain-life method*, *the linear-elastic fracture mechanics method*, and *the stress-life method*.

8.4.1 **Strain-life method**

The strain-life method is the best approach yet advanced to explain the nature of fatigue failure (Budynas and Nisbett, 2015). However, it was based on some idealizations and assumptions, which brings the uncertainties in predicting fatigue damages. A fatigue failure is assumed to begin at a local discontinuity (e.g., corner, notch, crack, or other stress concentration). When the stress at the discontinuity exceeds the elastic limit, the plastic strain occurs, and a fatigue fracture corresponds to an accumulation of cyclic plastic strains. Correspondingly, the total strain at the critical area can be quantified as

$$\frac{\Delta \varepsilon}{2} = \frac{\sigma'_F}{E}(2N)^b + \varepsilon'_F(2N)^c \tag{8.51}$$

where

$\Delta \varepsilon$, is the total strain,
E, is the elastic modulus of material,
N, is the number of reversals of cyclic loading,
b and c, are the slope of the elastic and plastic strain lines, respectively,
σ'_F, is the true stress corresponding to fracture in one reversal, and
ε'_F, is the true strain corresponding to fracture in one reversal.

Eq. (8.51) is also referred to as the Mason—Coffin relation of fatigue life and total strains. The coefficients applied in Eq. (8.51) can be found in SAE J1099 Standards (SAE, 2014). However, the strain-life method is rarely used in practice due to two main reasons: (1) it is unclear how to determine the total strain at the discontinuity, and (2) no data are available for strain concentration factors.

8.4.2 **Linear elastic fracture mechanics method**

In the linear elastic fracture mechanics method, the fatigue damage is measured by crack sizes. Fatigue cracks nucleate and grow when the stress varies. As shown in Fig. 8.45A, where the size of an initial crack is denoted

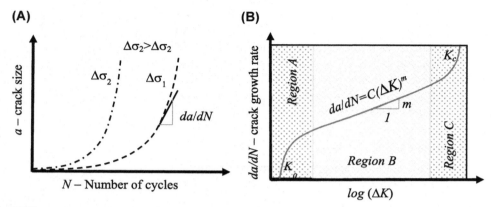

■ **FIGURE 8.45** Linear elastic fracture mechanics method: (A) crack growth with the number of cycles; (B) three phases of crack growth.

as a_i, a higher range of the stress change corresponds to a quicker increase of the stress intensity. Thus, the rate of crack size growth with respect to the loading cycle is given by

$$\frac{da}{dN} = C(\Delta K_I)^m = C(\beta\Delta\sigma\sqrt{\pi a})^m \qquad (8.52)$$

where

a is the crack size,
β is the tress intensity modification factor,
N is the number of reversals of cyclic loading,
C and m, are the empirical material constants,
ΔK_I is the change of stress intensity, and
$\Delta\sigma$ is the change of stress.

Upon integrating Eq. (8.52), the number of cycles N_f corresponding to a fatigue failure is found as,

$$N_f = \frac{1}{C}\int_{a_i}^{a_f} \frac{da}{(\beta\Delta\sigma\sqrt{\pi a})^m} \qquad (8.53)$$

where a_i and a_f are the crack size at the beginning and the fracture state, respectively.

Because the stress intensity modification factor actually changes with the crack size, the linear elastic fracture mechanics method was adopted in only in a few of numerical tools (NASA/FLAGRO, 2014) in some special areas. Fatigue analysis based on the linear elastic fracture mechanics needs some essential information such as crack parameters and loading schedule, which is unavailable in most of applications.

8.4.3 **Stress-life method**

In a stress-life method, the fatigue damage is measured by *fatigue strength*. The fatigue strength is defined in terms of the number of cycles. In particular, the fatigue strength is called an *endurance limit* when the number of loading cycles exceeds the required number of loading cycles. A product has an infinite fatigue life if its fatigue strength is higher than the endurance limit. The basic relation of the fatigue strength and the number of loading cycles is commonly known as an *S−N curve*; i.e.,

$$S_f' = \frac{(fS_{ut})^2}{S_e} N^{\left(-\frac{1}{3}\log\left(\frac{fS_{ut}}{S_e}\right)\right)} \tag{8.54}$$

where

S_f', is the fatigue strength,
N, is the number of fully reversed cycles,
f, is the fatigue strength fraction for 10^3 loading cycles,
S_{ut}, is the ultimate tensile strength, and
S_e, is the endurance limit.

Fatigue strength S_f' and endurance limit S_e in Eq. (8.54) are applicable only to a fully reversed load under standardized testing conditions. For the real-world applications, modification factors are introduced to take into account of the difference between actual and testing loading conditions. For example, if the load is not fully reversed, the design criteria in Fig. 8.46 should be applied to consider the impact of the mean stress on the fatigue behavior of material. In the case when the magnitude of a dynamic load varies, the Minor's Rule is used to calculate the accumulated damage in the given loading period. The stress-life method works well when the material deforms in its elastic range (Unigovski et al., 2013).

8.4.4 **Selection of fatigue analysis methods**

There are some reasons why three fatigue analysis methods have coexisted for a long time. The advantages and disadvantages of each method are relative and depend on where and when the fatigue analysis is needed. Table 8.1 provided some general tips for users in selecting an appropriate method for fatigue analysis (Aparcio, 2013).

8.4.5 **Case study of fatigue analysis**

One company manufactures composite parts for aerial-lift truck manufacturers. Parts are made of fiberglass (0 degree/90 degrees e-glass). As shown in Fig. 8.47A, these parts are called as tie rods, and they are used to support

(A)

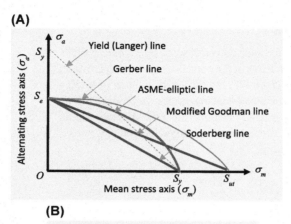

(B)

Design Criterion	Equation
Goodman line	$\dfrac{\sigma_a}{S_e} + \dfrac{\sigma_m}{S_{ut}} = \dfrac{1}{n}$
Soderberg line	$\dfrac{\sigma_a}{S_e} + \dfrac{\sigma_m}{S_y} = \dfrac{1}{n}$
Gerber line	$\dfrac{n\sigma_a}{S_e} + \left(\dfrac{n\sigma_m}{S_{ut}}\right)^2 = 1$
ASME-elliptic line	$\left(\dfrac{n\sigma_a}{S_e}\right)^2 + \left(\dfrac{n\sigma_m}{S_y}\right)^2 = 1$
Langer static yield	$\sigma_a + \sigma_m = \dfrac{S_y}{n}$

■ **FIGURE 8.46** Stress-life method for fatigue analysis: (A) design diagram; (B) design formula.

Table 8.1 Guides for Selection of Fatigue Analysis Methods

Strain-Life Method	Linear Elastic Fracture Method	Stress-Life Method
■ Mostly defect free, metallic structures or components. ■ Components where the crack initiation is the important failure criterion. ■ Locating the point(s) where cracks may initiate, and hence the growth of a crack should be considered. ■ Evaluating the effect of alternative materials and different surface conditions. ■ Components which are made from metallic, isotropic ductile materials which have symmetric cyclic stress-strain behaviour. ■ Components that experience short lives—low cycle fatigue—where plasticity is dominant.	■ Precracked structures or structures which must be presumed to be already cracked when manufactured such as welds. ■ Prediction of test programs to avoid testing components where cracks will not grow. ■ Planning inspection programs to ensure checks are carried out with the correct frequency. ■ To simply determine the amount of life left after crack initiation. ■ Components which are made from metallic, isotropic ductile materials which have symmetric cyclic stress—strain behaviors.	■ Long life or high cycle fatigue problems, where there is little plasticity since the S—N method is based on nominal stress. ■ Components where a crack initiation or crack growth modeling is inappropriate, e.g., composites, welds, plastics, and other nonferrous materials. ■ Situations where large amount of preexisting S—N data exist. ■ Components, which are required by a control body to be designed for fatigue using standard data. ■ Spot weld analysis and random vibration induced fatigue problems.

(A) **(B)**

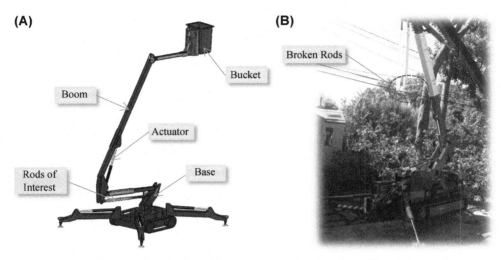

■ **FIGURE 8.47** Product fatigue failure in case study: (A) product structure; (B) part failure.

the working platform attached on the boom. The main purpose of using composite materials for tie rods is to avoid an electric shortcut from the workplace to the ground. The original design of tie rods was based on static analysis. It turned out that the impact of dynamic loads was significantly underestimated. The prototype caused a severe damage at the testing phase. As shown in Fig. 8.47B, the fracture occurred to the connection between the tie rod and lifting base. It happened around 100 h of service. The failure diagnosis in this incident was used as a case study in this section (Bi and Mueller, 2016). FEA is used to (1) determine if the fatigue failure is the actual cause of failure, (2) choose an appropriate size of fiberglass sleeve so that no fatigue failure occurs to tie rods, and (3) ensure that one rod is strong enough to support the boom for a period of over 10 min in case that the other tie rod breaks.

To conduct fatigue analysis by FEA, the first step is to collect raw data of the original design problem and specify the inputs of the FEA model accordingly.

8.4.5.1 Preparation of Model data

Tie rods are made of 0 degree/90 degrees e-glass, which are one of mostly used composite materials. Its material properties are available at a number of sources over the Internet. However, one caution should be paid to determine what material properties are used in the FEA. The properties of a composite material relate to how the materials are made, which can vary from one place

Table 8.2 Basic Material Properties of Tie Rod (Performance Composites, 2014)

	Young's Modulus	Poisson Ratio	Tensile Strength	Yield Strength	Compression Strength	Density
50% fabric, dry, room temperature, e-glass	24 GPa	0.2	440 MPa	400 MPa	425 MPa	1.900 kg/m³

to another. Given the fact that the client used e-glass, 50% of fiberglass, and 0 degree/90 degrees configuration; the basic properties in Table 8.2 were applied (Performance Composites, 2014).

In the fatigue analysis, the curve of *strength—number of cycles* ($S-N$ curve) is essential. The data by Sutherland and Mandell (2004) were used in this case study. The corresponding $S-N$ data and curve are included in Fig. 8.48.

Another important input for fatigue analysis is the characteristics of dynamic loads. As shown in Fig. 8.49, mean load (σ_m) and alternating load (σ_a) must be quantified based on the given working conditions. Generally, a commercial machinery follows operation standards to specify the loading conditions. UNI-EN-280 for mobile elevating work platforms (SAI Global, 2013) is used to determine the loads in Table 8.3.

To analysis the stress distribution in the tie rod, the force applied on the product must be modeled to characterize the loads on all components including tie rods. The motion simulation is performed based on the working condition in Table 8.3 (Bi and Mueller, 2016), and the results about the dynamic load on the tie rod are illustrated in Fig. 8.49.

(A)

Sample No.	N	$S_f(R=-1)$
1	1	400MPa
2	100	325MPa
3	1000	275MPa
4	10000	210MPa
5	100000	149MPa
6	1000000	110MPa

S-N Data

(B)

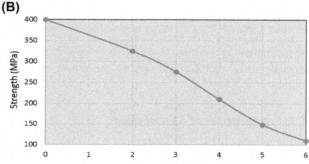

S-N Curve

■ **FIGURE 8.48** The $S-N$ curve in finite element analysis (Bi and Mueller, 2016): (A) $S-N$ data; (B) $S-N$ curve.

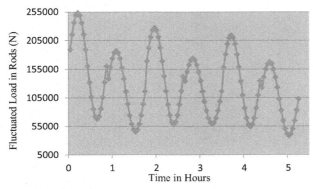

■ FIGURE 8.49 Characteristics of dynamic load on tie rod.

Table 8.3 Characteristics of Dynamic Loads

Nominal load:	Two persons + tools = 2×80 kg + 40 kg = 200 kg. In fatigue analysis, the nominal load is increased of a factor $f_1 = 1.15$, it leads to a total load of 230 kg.
Inertia loads:	The inertial loads are assumed as $0.1 \times mp$, where mp is the mass of the part in movement. The direction of inertia load is the direction of movement.
Operation forces:	A force of 20 kg $\times$ 1.1 = 22 kg (for each person in cage) is considered acting on cage hand rail. These forces are considered acting in the worst possible direction. The manual forces are not considered combined with inertia forces.
Wind forces:	The wind forces in the operating condition are considered acting with the horizontal direction for a reference pressure of 100 N/m^2 $\times$ 1.1 = 110 N/m^2.

The rod is symmetric about its transverse plane in the middle, the parametric representation of the rod geometry at one end is given in Fig. 8.50, and the dimensions for these parameters in the original design are listed in Table 8.4.

8.4.5.2 FEA simulation on original design

FEA has been conducted in the SolidWorks. Due to the symmetric nature of a tie rod, a half of the tie rod is taken into consideration to reduce the computation. The FEA model for fatigue analysis requires the simulation result from a static analysis where a nominal load is applied in the model as a reference. In the static analysis, the inputs of material properties are given by Table 8.2. In defining loading condition, axial loads are applied

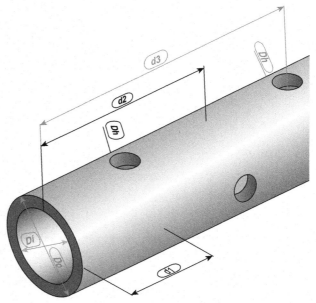

■ **FIGURE 8.50** Parametric representation of tie rod.

Table 8.4 Dimensions of Rod in Original Design

Label	Description	Value in Original Design (m)
L	Total length of rod	0.8763
D_o	The outside diameter of rod	0.051
D_i	The inside diameter of rod	0.038
D_h	The size of bolt hole	0.013
d_1, d_2, d_3	The distances from holes of fastening to the rod end	0.04762, 0.09525, 0.14287

on the hole-walls of six fasteners evenly. In defining boundary conditions, roller/slider boundary conditions are applied to the symmetric plane of the rod in the middle. In the meshing process, default 3D tetrahedral elements are used, and the resulted mesh is illustrated in Fig. 8.51A. It included total 326,280 nodes and 208,730 elements. The FEA simulation in Fig. 8.51B and C shows that the maximum stress occurs to the middle edges of the third set of bolted holes; the corresponding safety factors at these critical positions is 2.02 against yield strength.

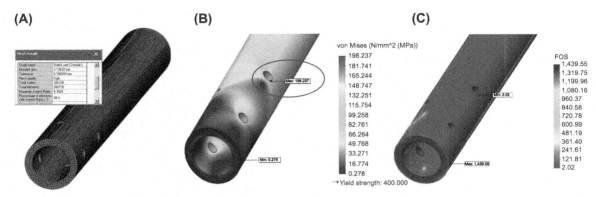

■ **FIGURE 8.51** Simulation result of finite element analysis static analysis: (A) meshed part; (B) von Mises stress; (C) safety factor.

In fatigue analysis, the $S-N$ curve in Fig. 8.48 was input as the fatigue properties of the rod, the loading curve in Fig. 8.49 was used to define the fluctuated load, and the Goodman criterion is used to take into account of the mean stress. The modification factor for the fatigue strength is given as,

$$K_f = K_R \times K_c \times K_d = 0.59024 \tag{8.55}$$

where

$K_R = 0.868$ is the modification factor for the reliability of 95%,
$K_c = 0.85$ is the modification factor for axial load, and
$K_d = 0.80$ is the distribution factor considering the variant of distribution over three bolts.

The predicted life cycles over the tie rod from fatigue analysis are shown in Fig. 8.52. The position corresponding to the maximized static stress has the shortest life cycles of 14.82 blocks. Accordingly, the number of operating hours before a fatigue fracture is 77.90 h.

8.4.5.3 FEA simulation on new design

To address unsatisfactory fatigue life of rod, a parametric study is defined by setting the outside diameter of rod as the design variable. The simulation from the parametric study has shown that the diameter of a tie rod should be larger than 0.07905 m to meet the expectation of the fatigue life. The distributions of the von Mises stress, safety factor, and fatigue life block for the rod with the new dimensions in Fig. 8.53A–C, respectively. Accordingly, the expected number of years of the tie rod with new dimension is 15.35 (years). It indicates that the tie rod with new dimension meets the expectation of over 10-year fatigue life.

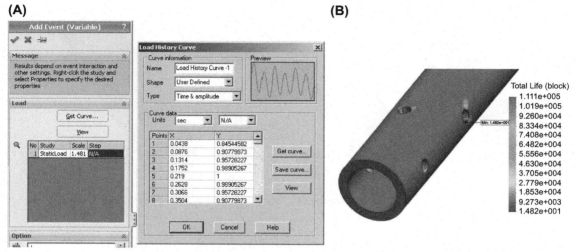

■ **FIGURE 8.52** Simulation result of fatigue analysis: (A) input of dynamic loads; (B) fatigue life.

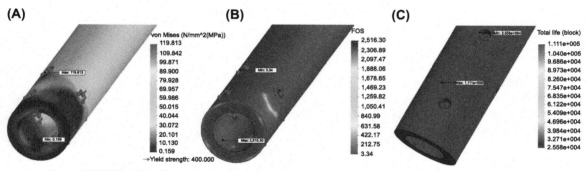

■ **FIGURE 8.53** Simulation results of rod with new dimensions: (A) von Mises stress; (B) safety factor; (C) fatigue life.

8.4.5.4 Verification of FEA simulation

One critical step in the postprocessing of FEA is the verification. It is especially important when FEA is applied to support the design of actual products. The verification can be performed based on a simplified analytical model or experimental data. Because the part in this case study can be simplified as a binary element, a simplified model can be used to estimate the fatigue life of the part using an analytical approach.

For the static analysis, each rod can be treated as a two-force element in Fig. 8.54. Once the pulling force F_R is found, the corresponding stress within the tie rod can be calculated can be estimated; it is mainly determined by the area of the cross-section shown in Fig. 8.55. The minimal cross-section includes two holes of bolts.

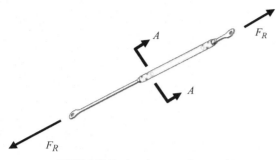

■ **FIGURE 8.54** Tie the rod as a two-force member.

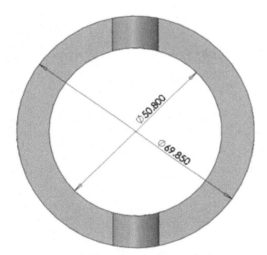

■ **FIGURE 8.55** A-A cross-section of rod.

The stress concentration has to be taken into consideration at the locations of bolts. Therefore, an individual FEA has been conducted to obtain the stress concentration factor under an axial loading condition. The stress concentration factor is found as 2.35. The tie rods are connected to the base by bolts, and four modes of failure are tensile failure, bearing failure, tear-out failure, and shear failure (Budynas and Nisbett, 2015). The corresponding safety factors against these failures are evaluated in Table 8.5, these equations are programed in the Matlab. Table 8.6 shows the obtained safety factors against four failure modes subjected to the maximized load of 253,665 N.

Table 8.5 Failure Modes, Stresses, and Safety Factors

Failure Mode	Stress Equation	Safety Factor
Tensile Failure	$\sigma_{average} = \dfrac{F_R/2}{\pi\left[(D_0/2)^2-(D_i/2)^2\right]-Dh\times(D_0-D_i)_Y}$	$n_{f,t} = \dfrac{S_Y}{K_t \cdot \sigma_{average}}$
Bearing Failure	$\sigma_{compression} = \dfrac{(F_R/(\text{number of rods}))/(\text{number of walls of a hole})}{(\text{Diameter of Bolt})\times(\text{thickness of wall})\times 2}$	$n_{f,c} = \dfrac{S_C}{K_t \cdot \sigma_{compression}}$
Tear-out Failure		
Shear Failure	$\tau_{shear} = \dfrac{(F_R/(\text{number of rods}))/(\text{number of walls of a hole})}{2\times(\text{thickness of wall})\times(\text{the shortest distance from a hole to the end of rod})}$	$n_{f,s} = \dfrac{S_Y/2}{K_t \cdot \tau_{shear}}$

Table 8.6 Safety Factors against Four Failure Modes under Maximized Static Load

Failure Mode	Max Stress (MPa)	Strength	Safety Factor
Tensile failure	192.64	S_Y	2.08
Bearing failure	386.84	S_C	1.10
Tear-out failure			
Shear failure	54.78	$S_Y/2$	3.65

To predict the fatigue life analytically, the Minor's Rule is used to take into account of the fluctuated load, i.e.,

$$\frac{n_1}{N_1} + \frac{n_2}{N_2} + \frac{n_3}{N_3} + \cdots \leq \frac{1}{Df_{\text{fatigue}}} \tag{8.56}$$

n_i is the number of cycles at stress level σ_i

N_i is the number of cycles to failure at stress level σ_i

Df_{fatigue} is the safety factor under varying amplitude loads

The Goodman's equation is applied to find the equivalent reverse stress $S'_{f,i}$ at a stress level $(\sigma_{m,i}, \sigma_{a,i})$, i.e.,

$$\frac{\sigma_{a,i}}{\sigma_{f,i}} + \frac{\sigma_{m,i}}{S_{ut}} = 1 \tag{8.57}$$

where

$\sigma_{m,i}$, $\sigma_{a,i}$ are the mean and alternating stress at scenario i, respectively.

S_{ut} is the ultimate strength, and

$\sigma_{f,i}$ is the equivalent reverse stress.

The equivalent reverse stress in Eq. (8.57) for the dynamic load in Fig. 8.49 is calculated by

$$\sigma_{f,i} = \frac{\sigma_{a,i}}{1 - \sigma_{m,i}/S_{ut}} \tag{8.58}$$

Thus, the number of cycles at the stress level $(\sigma_{m,i}, \sigma_{a,i})$ is,

$$N_i = \left(\frac{\sigma_{f,i}}{a}\right)^{1/b} \tag{8.59}$$

The coefficients a and b in Eq. (8.59) are defined based the $S-N$ curve in Fig. 8.48 as,

$$a = \frac{\left((S_f^i)_{10^3}\right)^2}{(S_f^i)_{10^6}} = 687.5 \text{ MPa} \tag{8.60}$$

$$b = -\frac{1}{3}\log\frac{(S_f^i)_{10^3}}{(S_f^i)_{10^6}} = -0.13265 \tag{8.61}$$

where $(S_f^i)_{10^3} = 275$ MPa, $\quad (S_f^i)_{10^6} = 110$ MPa

The dynamic load curve in Fig. 8.49 was derived based on six application scenarios (Bi and Mueller, 2016). The mean and alternating stresses can be calculated for these scenarios, the equivalent fully reversely stresses are then estimated by Eq. (8.58), and the Minor's Rule in Eq. (8.56) is used to predict the fatigue life. The result from the analytical model is given in Table 8.7. Recalling the predicted life 77.90 h from the FEA in Fig. 8.53, the discrepancy of the simulation and analytic results is less than 10%.

To suggest the new dimension of the tie rod for the expected fatigue life of 10 years, the optimized diameter of the tie rod from the FEA simulation is

Table 8.7 Verification of Finite Element Analysis Simulation on Original Design

Scenario	Mean Force (N)	Alternating Force (N)	Mean Stress (MPa)	Alternating Stress (MPa)	Equivalent Reverse Stress (MPa)	Predicted Life
1	160,630	93,035	121.98	70.65	477.26	13.46 (blocks)
2	116,148	71,114	88.20	54.00	240.22	0.0081 (years)
3	142,837	84,267	108.47	63.99	358.01	70.77 (h)
4	116,276	57,705	88.30	43.82	195.12	
5	133,941	79,882	101.72	60.66	312.53	
6	102,803	64,537	78.07	49.01	197.75	

Table 8.8 Verification of Finite Element Analysis Simulation on New Design

Scenario	Mean Stress (MPa)	Alternating Stress (MPa)	Equivalent Reverse Stress (MPa)	Predicted Life
1	75.54	43.75	172.53	22,806.4 (blocks)
2	54.62	33.44	111.06	13.68 (years)
3	67.17	39.63	145.37	119,870 (h)
4	54.68	27.14	90.16	
5	62.99	37.57	133.16	
6	48.35	30.35	96.23	

$D_o = 0.07905$ m. The analytical model is used to calculate the fatigue life for the rod, and the result is shown in Table 8.8. Generally, the predicted lives of the rod from the analytical and FEA models agree with each other. The fatigue life (13.68 years) from the simplified model is 10% less than that of the FEA simulation (15.35 years).

8.5 SUMMARY

FEA was developed originally for numerical solutions of complex problems in solid mechanics. FEA is by far the most widely used and versatile technique for simulating deformable solids. This chapter gives an overview of the physical and mathematic background required to understand the FEA implementation for solid mechanics' problems. The physical behaviors of mechanical structures or systems are analyzed, and the minimum potential energy principle is used to develop element models. The procedures for FEA modeling are discussed for a few of classic solid mechanics' problems such as truss structure, plane stress, plane strain, modal analysis, as well as fatigue analysis.

REFERENCES

Aparcio, C., 2013. What Is Fatigue Analysis? http://simulatemore.mscsoftware.com/what-is-fatigue-analysis-msc-nastran/.

ASM International, 2008. Elements of Metallurgy and Engineering Alloys (#05224G): Chapter 14 Fatigue. http://www.asminternational.org/documents/10192/1849770/05224G_Chapter14.pdf.

Bi, Z.M., Mueller, D., 2016. Finite element analysis for diagnosis of fatigue failure of composite materials in product development. International Journal of Advanced Manufacturing Technology 87 (5), 2245–2257.

Budynas, R., Nisbett, J.K., 2015. Shigley's Mechanical Engineering Design, tenth ed. McGraw Hill, ISBN 978-0073398204.

Costa, M.Y.P., Voorwald, H.J.C., Pigatin, W.L., Guimaraes, V.A., Cioffi, M.O.H., 2006. Evaluation of shot peening on the fatigue strength of anodized Ti-6AI-4V alloy. Materials Research 9 (1), 107–109.

Hamrock, B.J., Jacobson, B., Schmid, S.R., 1999. Fundamental of Machine Elements. WCB/McGraw-Hill, ISBN 0-256-19069-0.

Mott, R.L., 2014. Machine Elements in Mechanical Design, fifth ed. Pearson, Inc., Upper Saddle River, New Jersey, U.S.A., ISBN 0-13-507793-1

Nanninga, N.E., 2008. High Cycle Fatigue of AA6082 and AA6063 Aluminum Extrusions (Doctorial thesis). Michigan Technological University. http://digitalcommons.mtu.edu/etds/18.

NASA/FLAGRO, 2014. Fatigue User's Guide Crack Growth. NASA/FLAGRO. http://www.mscsoftware.com/training_videos/patran/Reverb_help/index.html#page/Fatigue%20Users%20Guide/fat_growth.0.html.

Performance Composites, 2014. Material Properties for Carbon Fibers. http://www.performance-composites.com/carbonfibre/mechanicalproperties_2.asp.

Ramachandran, V., Raghuram, A.C., Krishnan, R.V., Bhaumik, S.K., 2005. Failure Analysis of Engineering Structures: Methodology and Case Histories. http://nusretmeydanlik.trakya.edu.tr/MLZM%20MKN/Failure_Analysis_of_Engineering_Structures__Methodology_and_Case_Histories.pdf.

SAE, 2014. J1099, SAE Standards: Technical Report on Low Cycle Fatigue Properties Ferrous and Non-ferrous of Materials. http://standards.sae.org/j1099_197502/.

SAI Global, 2013. UNI-EN-280: Mobile Elevating Work Platforms - Design Calculations - Stability Criteria - Construction - Safety - Examinations and Tests. http://infostore.saiglobal.com/store/Details.aspx?productID=1691031.

Shariyat, M., Sola, J.F., Jazayeri, S.A., 2016. Experimentally validated combustion and piston fatigue life evaluation procedures for the bi-fuel engines, using an integral-type fatigue criterion. Latin American Journal of Solids and Structures 13 (6), 1030–1053.

Sutherland, H.J., Mandell, J.F., 2004. Updated Goodman Diagrams for Fiberglass Composite Materials Using the DOE/MSU Fatigue Database. windpower.sandia.gov/other/Global04_18983_Sutherland_Final.pdf.

Unigovski, Y.B., Grinberg, A., Gerafi, E., Gutman, E.M., Moisa, S., 2013. Low-cycle fatigue of an aluminum alloy plated with multi-layer deposits. Journal of Optoelectronics and Advanced Materials 15 (7–8), 863–868.

■ PROBLEMS

8.1. Fig. 8.56 shows a truss structure in two-dimensional space with the specified loads $F = 100$ lbf at two nodes (C and D); Nodes A and B are fixed. All truss members use the same material – Gray Case Iron, which has the Young's modulus of $E = 9.598 \times 10^6$ psi and the Poisson' ratio of 0.27. The yield strength is $S_y = 2.20 \times 10^4$ psi. The cross-section areas of all truss members are given as $A = 1$ in^2. Use

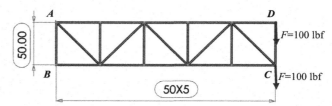

■ FIGURE 8.56 A truss structure for problem 8.1.

    the SolidWorks simulation to predict the maximal stress and deflection of the truss structure.

8.2. Fig. 8.57 shows a truss structure in three-dimensional space with the specified loads $F = (0, 50, -100)$ lbf at two nodes (G and I) and $F = (0, 0, -500)$ lbf at node H. The nodes at the ground (A, B, C, D, E, F) are all fixed. All truss members use the same material – Gray Case Iron, which has the Young's modulus of $E = 9.598 \times 10^6$ psi and the Poisson' ratio of 0.27. The yield strength is $S_y = 2.20 \times 10^4$ psi. The cross-section areas of all truss members are given as $A = 1$ in^2. Use the SolidWorks simulation to predict the maximal stress and deflection of the truss structure.

8.3. Fig. 8.58 shows a thin plate with a thickness of 0.2-in. The plate includes three discontinuities of geometries, a 1.00×2.00 rectangle, a hexagon in a circle of 1.5-in diameter, and a shoulder with a round of 0.25-in. The plate uses Aluminum 1060 alloy with the elastic modulus of $E = 1.0 \times 10^7$ psi, the Poisson' ratio is 0.33, yield strength $S_y = 3.999 \times 10^3$ psi. Determine stress concentration factors at three discontinuity sections under a bending load.

8.4. Fig. 8.59 shows a thin aluminum part to transfer the load in a composite structure. The part has a thickness of 5-mm, and the material

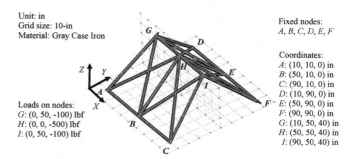

■ FIGURE 8.57 A Truss Structure for Problem 8.2

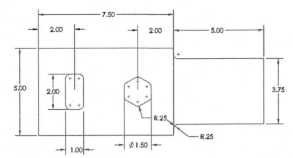

■ **FIGURE 8.58** A Plane Stress Plate with a Thickness of 0.2-in (aluminum 1060 alloy).

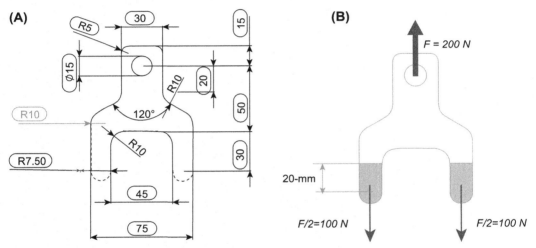

■ **FIGURE 8.59** Aluminum Part Subjected to Tensile Load (A) Dimensions (unit in mm and thickness 5-mm) (B) Free body diagram.

is AISI 304 with the elastic modulus of 1.90×10^{11} N/m^2, the yield strength of 2.068×10^8 N/m^2, the density of 8000 kg/m^3, shear Modulus of 7.5×10^{10} N/m^2, and the Poisson modules of 0.29. Assume that the minimal safety factor against the static failure is 2.0, (1) create and run an FEA model, and determine if the part can withstand a total load of 200 *N* without a static failure, (2) what is the maximized load one can apply on the part without a static failure (for the safety factor of 2.0)?

8.5. Fig. 8.60 shows a structural member made of 1023 Carbon steel with the elastic modulus of 2.05×10^{11} N/m^2, the yield strength of 2.827×10^8 N/m^2, the density of 7858 kg/m^3, and the Poisson modules of 0.29. Assume the left end is fixed, determine the first five natural frequencies of the structure member.

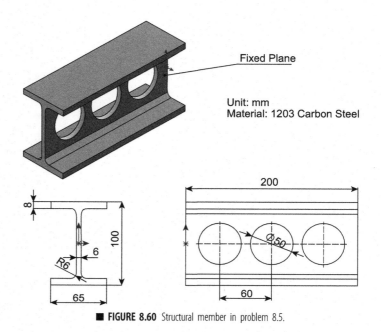

■ **FIGURE 8.60** Structural member in problem 8.5.

8.6. Costa et al. (2006) analyzed the fatigue behavior of anodized Ti−6 Al−4V alloy and the influence of shot peening pre-treatment on the experimental data. Axial fatigue tests (R = 0.1) were performed, and a significant reduction in the fatigue strength of anodized Ti−6Al−4 V was observed. A number of data sets from the experiment was given in Table 8.9. The chemical composition of Ti−6Al−4V used was 6.13% Al, 4.0% V, 0.23% Fe, 0.19% O, 0.07% Ni, 0.0124%

Table 8.9 A Number of S−N Experimental Data of Ti−6Al−4V

Number of Cycles		5.0×10^3	1.0×10^4	2.0×10^4	3.0×10^4	1.0×10^5	3.0×10^5	6.0×10^5	1.0×10^7
Fatigue strengths (MPa)	Base Ti−6Al−4V	1000	965			975	975		
	Anodized Ti−6Al−4V		980	900					800
	Shot Peened Ti−6Al−4V				980	950		935	

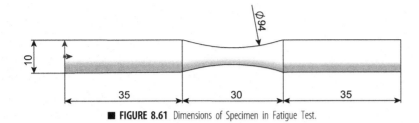

■ **FIGURE 8.61** Dimensions of Specimen in Fatigue Test.

Table 8.10 *S–N* Curve of Ti-6AL-4V from Another Source

n	Fatigue Strength (N/m^2)
35,000	792,897,090
80,000	758,423,300
100,000	744,633,790
350,000	689,475,730
900,000	620,528,160
1,000,000	606,738,640
3,000,000	551,580,580
10,000,000	537,791,070

H, 0.004% Y, 0.39% residual elements, and Ti for the reminder. Mechanical properties of this alloy are: (35−40) HRc and ultimate strength of 1270 MPa, in the annealed condition. Axial fatigue tests according to ASTM 466 were conducted using a sinusoidal load of frequency 20 Hz and ratio R = 0.1, at room temperature considering, as fatigue strength, the complete specimens fracture or 10^7 load cycles.

Create a fatigue analysis model for the specimen in Fig. 8.61 as a verification of FEA model. Assume the *S–N* curve from another source for the same material is given in Table 8.10. Choose a few of points in experimental *S–N* to verify the FEA model.

Applications—Heat Transfer Problems

9.1 INTRODUCTION

In thermodynamics, most important physical quantities are *energy*, *temperature*, and *entropy*. The relations of these quantities are governed by four laws of thermodynamics (Kittel and Kroemer, 1980; Guggenheim, 1985):

- **The zeroth law**. If two systems (*A* and *B*) are in a thermal equilibrium with a third system (*C*). These two systems (*A* and *B*) are in a thermal equilibrium with each other. This law provides the necessity of defining the notion of temperature to describe thermal equilibrium.
- **The first law**. When energy passes, as work, heat or matter, into or out from a system, the internal energy of system changes based on the energy conservation. In other words, energy cannot be created or destroyed; it can only be transferred from one form to another. This law tells that perpetual motion machines of the first kind are impossible.
- **The second law**. In a natural thermodynamic process, the sum of the entropies of the interacting thermodynamic systems increases.
- **The third law**. The entropy of a system approaches a constant value when the temperature of the system approaches absolute zero. With the exception of noncrystalline solids, the entropy of a system at absolute zero is typically close to zero, and is equal to the logarithm of the product of the quantum ground states.

Heat transfer refers to an energy flow in the form of heat where the internal energy of system is changed, which is governed by the first law of thermodynamics. Heat is transferred from a region of high temperature to another region of a lower temperature, which is governed by the second law of thermodynamics. Heat transfer will occur in a direction that increases the entropy of the collection of systems (Faghri et al., 2010). Because a mechanical system is to transfer motion and energy from one component to another; moreover, material properties relate closely to heat and temperature. Therefore, heat transfer problems have been widely investigated in a variety of

Finite Element Analysis Applications. http://dx.doi.org/10.1016/B978-0-12-809952-0.00009-1

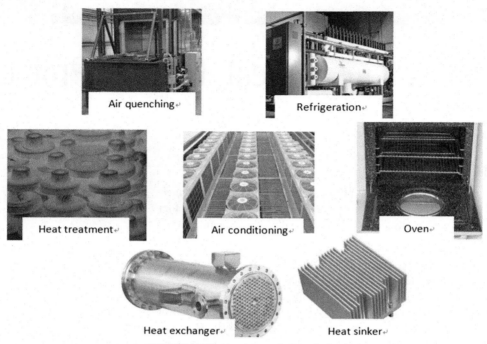

■ **FIGURE 9.1** Heat transfer examples in applications.

engineering applications. The properties of the matter vary significantly with respect to temperature. Various heat transfer processes have been developed for different products or processes, so that the temperature or the temperature change can be controlled as expected. As shown in Fig. 9.1, heat transfer processes are developed to (2) change materials properties is in air quenching and heat treatment, (2) control operation environments in air conditioning, (3) protect or process goods in refrigeration or ovens, and (4) exchange or dissipate energy in heat exchangers and heat sinkers.

9.2 GOVERNING EQUATIONS OF HEAT TRANSFER

Heat transfer is the transferring of thermal energy in a body or two physical systems. The rate of heat transfer depends on the temperature gradients and the properties of involved materials through which the heat is transferred. Three fundamental modes of heat transfer are *conduction*, *convection*, and *radiation*. The governing equations are introduced for these heat transfer models, and then the equations for the energy conservation with all of these heat transfer modes are discussed.

9.2.1 **Conduction**

Heat conduction is the transfer of internal thermal energy by the collisions of microscopic particles and movement of electrons within a body. The microscopic particles in the heat conduction can be molecules, atoms, and electrons. Internal energy includes kinematic and potential energy of microscopic particles. Heat conduction is governed by *the law of heat conduction*, which is also called the Fourier's law: the time rate of heat transfer through a material is proportional to the negative gradient in the temperature and to the area. Under a Cartesian coordinate system, the Fourier's law for the rate of heat transfer by conduction is expressed as,

$$
\left.
\begin{aligned}
q_x &= q_x' A_x = -k_x A_x \frac{\partial T}{\partial x} \\
q_y &= q_y' A_y = -k_y A_y \frac{\partial T}{\partial y} \\
q_z &= q_z' A_z = -k_z A_z \frac{\partial T}{\partial z}
\end{aligned}
\right\}
\tag{9.1}
$$

where

$(q_x, \ q_y, \ q_z)$ are the rate of heat transfer,

$\left(q_x', \ q_y', \ q_z'\right)$ are the heat flux,

k_x, k_y, and k_z are the conductivity coefficients along x-, y-, and z-axis,
$(A_x, \ A_y, \ A_z)$ are the cross-section areas, and
$\left(\dfrac{\partial T}{\partial x}, \ \dfrac{\partial T}{\partial y}, \ \dfrac{\partial T}{\partial z}\right)$ are the temperature gradients along three axes of
the Cartesian coordinate system.

9.2.2 **Heat convection**

Heat convection is a mode of heat transfer by the mass motion of a fluid such as air. Heat convection occurs to the surface of an object where the surrounding fluid of object is heated and moved energy away from the source of heat. Convective heat transfer occurs when the surface temperature differs from that of surrounding fluid. The governing equation of heat convection behaviors is the Newton's law,

$$
q_h = hA(T_s - T_f)
\tag{9.2}
$$

where

q_h is convective heat on surface,
A is cross-section of boundary surface,
T_s is the surface temperature, and
T_f is the temperature of the surrounding fluid.

9.2.3 **Heat radiation**

Heat radiation is the transfer of internal energy in the form of electromagnetic waves. It represents a conversion of thermal energy into electromagnetic energy through the emission of a spectrum of electromagnetic radiation due to an object's temperature. Any matter emits heat via radiation. Heat radiation is governed by the Stefan—Boltzmann law of thermal radiation as (Wellons, 2007),

$$q_r = \sigma \varepsilon \left(T^4 - T_0^4\right) \tag{9.3}$$

where

σ Stefan—Boltzmann constant,

ε is the surface emission coefficient,

T_0 is absolute temperature of the environment,

T is absolute temperature of the object emitting or absorbing thermal radiation, and.

q_r is the incident radiant heat flow per unit area of surface.

9.2.4 **Mathematic models**

Fig. 9.2 shows the energy state of an infinitesimal volume in a continuous domain. The heat can be conducted from and to in any direction (*x*-, *y*-, and *z*-axis). The change of thermal energy in this unit depends on (1) conducted heats in these axes, (2) the heat for temperature changes of the mass in unit, and (3) the heat generated from any source in unit. The thermal energy in all of forms is conserved.

To develop an energy conservation equation, the parts of energy in three forms should be quantified.

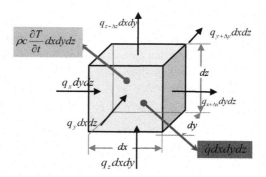

■ **FIGURE 9.2** Heat equilibrium state in a continuous domain.

1. Input energy E_{in} transferred by the heat conduction along x-, y-, and z-axis,

$$E_{in} = -k_x dydz \frac{\partial T}{\partial x} - k_y dxdz \frac{\partial T}{\partial y} - k_z dxdy \frac{\partial T}{\partial z} \qquad (9.4)$$

where

q_x, q_y, and q_z are the rates of heat energy flow per unit area defined in Eq. (9.1), and

$\frac{\partial T}{\partial x}$, $\frac{\partial T}{\partial y}$, $\frac{\partial T}{\partial z}$ are the temperature gradients along x-, y-, and z-axis, respectively.

2. Output energy E_{out} transferred by the heat conduction along x-, y-, and z-axis,

$$
\begin{aligned}
E_{out} = {} & -k_x dydz \frac{\partial T}{\partial x} - dydz \cdot \partial \left(\frac{\partial (k_x T)}{\partial x} \right) \\
& -k_y dxdz \frac{\partial T}{\partial z} - dxdz \cdot \partial \left(\frac{\partial (k_y T)}{\partial y} \right) \\
& -k_z dxdy \frac{\partial T}{\partial z} - dxdy \cdot \partial \left(\frac{\partial (k_z T)}{\partial z} \right)
\end{aligned}
\qquad (9.5)
$$

where $\frac{\partial (k_x T)}{\partial x}$, $\frac{\partial (k_y T)}{\partial y}$, $\frac{\partial (k_z T)}{\partial z}$ are the derivatives of temperature related quantities with respect to x-, y-, and z-axis, respectively.

3. Increased energy E_g generated by internal heat source,

$$E_g = \dot{q} dxdydz \qquad (9.6)$$

where $\dot{q}$ is the rate of generated heat per unit volume from the internal heat source.

4. Storage energy E_s in element

$$E_s = \rho c \frac{\partial T}{\partial t} dxdydz \qquad (9.7)$$

where

ρ and c are the density and specific heat of the matter, and

$\frac{\partial T}{\partial t}$ is the rate of temperature change with respect to time.

Applying the principle of energy conservation and substituting the Fourier law Eq. (9.1) for the conduction components yield,

$$\frac{\partial}{\partial x} \left(k_x \frac{\partial T}{\partial x} \right) + \frac{\partial}{\partial y} \left(k_y \frac{\partial T}{\partial y} \right) + \frac{\partial}{\partial z} \left(k_z \frac{\partial T}{\partial z} \right) - \rho c \frac{\partial T}{\partial t} + \dot{q} = 0 \qquad (9.8)$$

Note that Eq. (9.8) governing the heat conduction behaviours inside an object. On the surfaces of an object, heat convection and radiation occur, and the transferred energy depend on the temperatures and its gradients on boundaries. Fig. 9.3 shows four common types of boundary conditions in a heat transfer problem: (1) the fixed temperature on a boundary surface, (2) the given heat flux (or the rate of heat energy transfer) through a boundary surface, (3) the convection on a boundary surface, and (4) the radiation through a surface. In addition, if no heat transfer occurs on a boundary, such a surface is called an insulated boundary.

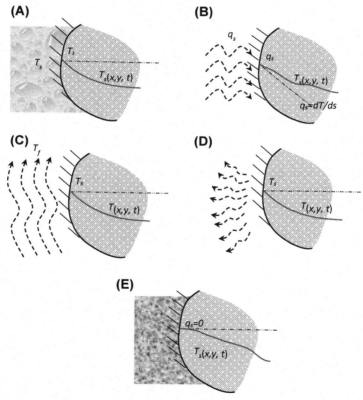

■ **FIGURE 9.3** Common boundary conditions in heat transfer problems: (A) fixed temperature on surface; (B) given heat flux on surface; (C) convection through surface; (D) radiation through surface; (E) insulated surface.

9.3 FEA MODELING OF HEAT TRANSFER PROBLEMS

In an FEA model of a heat transfer problem, the temperatures are treated as state variables. After a continuous domain is discretized into elements and nodes, the temperature of any position in an element can be interpolated by shape functions as,

$$T(x, y, z) = [S(x, y, z)]\{T(x_i, y_i, z_i)\} \tag{9.9}$$

where

$T(x, y, z)$, is the temperature at an arbitrary position (x, y, z) in the corresponding element,
$[S(x, y, z)]$, is the vector of the shape functions $S_i (x, y, z)$ where $i = 1, 2, ...N$, and N is the number of nodes in the element, and
$[T(x_i, y_i, z_i)]$, is the vector of the temperatures on N nodes and $i = 1, 2, ...N$.

The temperature gradient for the heat conduction can be obtained by taking derivative of Eq. (9.9) as,

$$\begin{Bmatrix} k_x \dfrac{\partial T}{\partial x} \\[2mm] k_y \dfrac{\partial T}{\partial y} \\[2mm] k_z \dfrac{\partial T}{\partial z} \end{Bmatrix} = [D] \cdot \{T(x_i, y_i, z_i)\} = \begin{bmatrix} k_x \dfrac{\partial S_1}{\partial x} & k_x \dfrac{\partial S_2}{\partial x} & \cdots \\[2mm] k_y \dfrac{\partial S_1}{\partial y} & k_y \dfrac{\partial S_2}{\partial y} & \cdots \\[2mm] k_z \dfrac{\partial S_1}{\partial z} & k_z \dfrac{\partial S_2}{\partial z} & \cdots \end{bmatrix} \cdot \{T(x_i, y_i, z_i)\} \tag{9.10}$$

where $[D]$ is the matrix with the derived quantities for the evaluation of the vector of temperature gradients.

To develop the elemental model governed by Eq. (9.8), the Galerkin method is used where N shape functions are used as test functions in the evaluations of weighted residuals, i.e.,

$$\int_V \left(\frac{\partial}{\partial x}\left(k_x \frac{\partial T}{\partial x}\right) + \frac{\partial}{\partial y}\left(k_y \frac{\partial T}{\partial y}\right) + \frac{\partial}{\partial z}\left(k_z \frac{\partial T}{\partial z}\right) - \rho c \frac{\partial T}{\partial t} + \dot{q} \right) \cdot S_i dV = 0 \tag{9.11}$$

Performing the integral for Eq. (9.11) for three components separately as,

$$\boxed{\text{I}} \rightarrow \int_V \left(\frac{\partial}{\partial x}\left(k_x \frac{\partial T}{\partial x}\right) + \frac{\partial}{\partial y}\left(k_y \frac{\partial T}{\partial y}\right) + \frac{\partial}{\partial z}\left(k_z \frac{\partial T}{\partial z}\right) \right) \cdot S_i dV$$

$$\boxed{\text{II}} \rightarrow -\int_V \rho c \frac{\partial T}{\partial t} \cdot S_i dV + \int_V \dot{q} dV = 0 \quad \leftarrow \boxed{\text{III}} \tag{9.12}$$

Applying the divergence theorem in the first component of Eq. (9.12) gets,

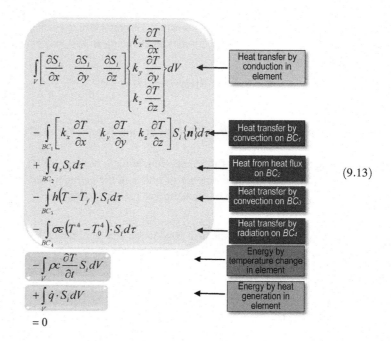

$$\int_V \begin{bmatrix} \dfrac{\partial S_i}{\partial x} & \dfrac{\partial S_i}{\partial y} & \dfrac{\partial S_i}{\partial z} \end{bmatrix} \begin{Bmatrix} k_x \dfrac{\partial T}{\partial x} \\[4pt] k_y \dfrac{\partial T}{\partial y} \\[4pt] k_z \dfrac{\partial T}{\partial z} \end{Bmatrix} dV \qquad \text{Heat transfer by conduction in element}$$

$$- \int_{BC_1} \begin{bmatrix} k_x \dfrac{\partial T}{\partial x} & k_y \dfrac{\partial T}{\partial y} & k_z \dfrac{\partial T}{\partial z} \end{bmatrix} S_i \{n\} d\tau \qquad \text{Heat transfer by convection on } BC_1$$

$$+ \int_{BC_2} q_s S_i d\tau \qquad \text{Heat from heat flux on } BC_2$$

$$- \int_{BC_3} h(T - T_f) \cdot S_i d\tau \qquad \text{Heat transfer by convection on } BC_3$$

$$- \int_{BC_4} \sigma \varepsilon (T^4 - T_0^4) \cdot S_i d\tau \qquad \text{Heat transfer by radiation on } BC_4$$

$$- \int_V \rho c \dfrac{\partial T}{\partial t} S_i dV \qquad \text{Energy by temperature change in element}$$

$$+ \int_V \dot{q} \cdot S_i dV \qquad \text{Energy by heat generation in element}$$

$$= 0$$

(9.13)

where

BC_i ($i = 1,2,3,4$), is the set of four possible boundaries if an element is on the boundary of a continuous domain,

$\{n\}$, is the normal of the boundary, i.e., $\{n\} = \{n_x, n_y, n_z\}$,

h, is the convection coefficient,

T_s, is the set of temperatures of nodes on surface to be determined,

T_f, is the given temperature of surrounding fluid, and

σ, ε, and T_0, are coefficients for heat radiation defined in Eq. (9.3).

Substituting Eq. (9.10) into Eq. (9.13) and reformatting it in a matrix gets,

$$[C] \cdot \left\{ \dfrac{\partial T}{\partial t} \right\} + ([K_C] + [K_h] + [K_r])\{T\}$$
$$= \{R_T\} + \{R_Q\} + \{R_q\} + \{R_h\} + \{R_r\}$$

(9.14)

where $\left\{\frac{\partial T}{\partial t}\right\}$ is the vector of the derivatives of nodal temperatures with respect to time.

$$[C] = \int_V \rho c[S]^T[S]dV, \qquad\qquad [K_C] = \int_V \left[\frac{\partial S}{\partial x} \ \frac{\partial S}{\partial y} \ \frac{\partial S}{\partial z}\right] \cdot [D]dV,$$

$$[K_h] = \int_{BC_3} h[S]^T[S]d\tau \qquad\qquad [K_r] = \int_{BC_4} \sigma \varepsilon T^4[S]^T d\tau,$$

$$\{R_T\} = -\int_{BC_1} [D]\{n\}[S]^T d\tau, \qquad \{R_Q\} = \int_V \dot{q}[S]^T dV$$

$$\{R_q\} = \int_{BC_2} q_s[S]^T d\tau, \qquad\qquad \{R_h\} = \int_{BC_3} hT_f[S]^T d\tau,$$

$$\{R_r\} = \int_{BC_4} \sigma \varepsilon T_0^4[S]^T d\tau$$

Note that the term $[K_r]$ by the radiative boundary condition brings the nonlinearity of the elemental model. Different heat transfer problems represented by the general model in Eq. (9.14) can be further classified into the following four types based on the time dependence and linearity.

Type I—steady linear problems:

$$([K_C] + [K_h])\{T\} = \{R_Q\} + \{R_q\} + \{R_h\} \qquad (9.15)$$

Type II—steady nonlinear problems:

$$([K_C] + [K_h] + [K_r])\{T\} = \{R_Q\} + \{R_q\} + \{R_h\} + \{R_r\} \qquad (9.16)$$

Type III—transient linear problems:

$$[C] \cdot \left\{\frac{\partial T}{\partial t}\right\} + ([K_C] + [K_h])\{T\} = \{R_Q\} + \{R_q\} + \{R_h\} \qquad (9.17)$$

Type IV—transient nonlinear problems:

$$[C] \cdot \left\{\frac{\partial T}{\partial t}\right\} + ([K_C] + [K_h] + [K_r])\{T\} = \{R_Q\} + \{R_q\} + \{R_h\} + \{R_r\}$$
$$(9.18)$$

9.4 1D STEADY HEAT TRANSFER PROBLEMS

As shown in Fig. 9.4, a wall consists of a number of layers with different materials to sustain a comfortable living environment inside a house. Due to the temperature difference, heat is transferred from inside to outside or vice versa. Our interest is to determine the temperature distribution crosswall. It can be represented by 1D steady thermal analysis model.

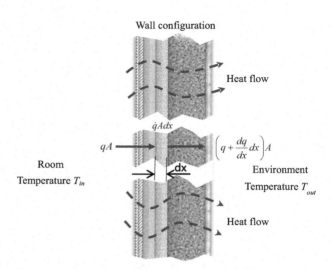

Wall configuration

Heat flow

Room
Temperature T_{in}

Environment
Temperature T_{out}

Heat flow

■ **FIGURE 9.4** Example of one-dimensional steady heat transfer problem.

Taking an element in the wall configuration as an example, the governing equation Eq. (9.8) can be written as,

$$qA + (\dot{q}A)dx + \left(q + \frac{dq}{dx}dx\right)A = 0 \qquad (9.19)$$

where

q is the heat flux,
A is the cross-section area of element,
dx is the length of element, and
$\dot{q}$ is the rate of heat flux which refers of the generated heat in the element.

According to the Fourier's law, $q = -k\frac{dT}{dx}$, substituting it into Eq. (9.19) yields,

$$\frac{d}{dx}\left(k\frac{dT}{dx}\right) + \dot{q} = 0 \qquad (9.20)$$

The temperature of an arbitrary position (X) ion in a linear element is interpolated as,

$$T(X) = \left[\frac{X_j - X}{X_j - X_i} \quad \frac{X - X_i}{X_j - X_i}\right]\begin{Bmatrix} T_i \\ T_j \end{Bmatrix} \qquad (9.21)$$

The temperature gradient can be derived from Eq. (9.10) as,

$$k\frac{\partial T}{\partial X} = [\boldsymbol{D}] \cdot \left\{ \begin{array}{c} T_i \\ T_j \end{array} \right\} = \left[\begin{array}{cc} \dfrac{-k}{X_j - X_i} & \dfrac{k}{X_j - X_i} \end{array} \right] \cdot \left\{ \begin{array}{c} T_i \\ T_j \end{array} \right\} \qquad (9.22)$$

Imposing two boundary conditions for a 1D heat transfer problem gets,

$$\left. \begin{array}{c} T_{X=0} = T_0 \\[2mm] q_{X=L} = -k\dfrac{\partial T}{\partial X} = h(T_L - T_f) \end{array} \right\} \qquad (9.23)$$

Therefore, the general Eq. (9.15) can be rewritten for a 1D heat transfer element as,

$$([\boldsymbol{K}_C] + [\boldsymbol{K}_h]) \left\{ \begin{array}{c} T_i \\ T_j \end{array} \right\} = \{\boldsymbol{R}_Q\} + \{\boldsymbol{R}_h\} \qquad (9.24)$$

where

$$[\boldsymbol{K}_C] = \int_{X_i}^{X_j} \left[\begin{array}{c} \dfrac{-1}{X_j - X_i} \\[3mm] \dfrac{1}{X_j - X_i} \end{array} \right] \cdot \left[\begin{array}{cc} \dfrac{-k}{X_j - X_i} & \dfrac{k}{X_j - X_i} \end{array} \right] dX = \dfrac{k}{X_j - X_i} \left[\begin{array}{cc} 1 & -1 \\ -1 & 1 \end{array} \right]$$

$$(9.25)$$

$$\{\boldsymbol{R}_Q\} = \int_{X_i}^{X_j} \dot{q} \left[\begin{array}{c} \dfrac{X_j - X}{X_j - X_i} \\[3mm] \dfrac{X - X_i}{X_j - X_i} \end{array} \right] dX = \dfrac{\dot{q}l}{2} \left[\begin{array}{c} 1 \\ 1 \end{array} \right] \qquad (9.26)$$

In addition, the boundary condition is applied to the last element ($X = L$) as,

$$[\boldsymbol{K}_h] = \left[\begin{array}{cc} 0 & 0 \\ 0 & h \end{array} \right] \text{ and } \{\boldsymbol{R}_h\} = hT_f \left[\begin{array}{c} 0 \\ 1 \end{array} \right] \qquad (9.27)$$

Example 9.1

The wall in Fig. 9.5 consists of two layers whose thermal properties are given. The room temperature is given as $T_0 = 25°C$, the outside temperature is given as $T_\infty = -5°C$. Find the distribution of temperature through the wall.

Solution I

As shown in Fig. 9.6, the geometries of two parts for wall layers A and B are modeled in the SolidWorks. Because the dimensions along y- and z-axes

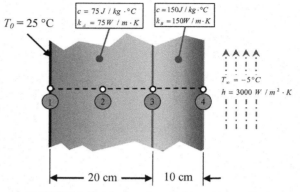

FIGURE 9.5 Example of one-dimensional heat transfer problem.

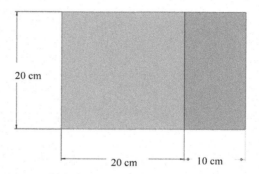

FIGURE 9.6 Assembly model for the wall construction.

can be extremely large in contrast to the dimension along x-axis, a 1D steady thermal analysis model in Fig. 9.7 has been defined to solve the problem.

Fig. 9.8 shows the defined boundary conditions: (1) the inside wall is defined as a fixed temperature 25°C and (2) the outside wall is specified as convection surface with $h = 3000$ W/m² K and $T_\infty = -5.0$°C. The solution to the given heat transfer model is shown in Fig. 9.9.

The probes are applied on the nodes and the temperatures on a few of nodes cross the wall are 25, 19, 13, and 10°C, respectively.

Solution II

To illustrate the application of the developed model in Eqs. (9.24)–(9.29), a simplified 1-D heat transfer model is developed. As shown in Table 9.1. It

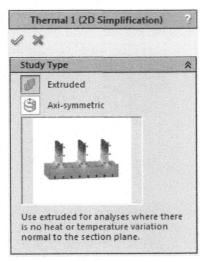

■ **FIGURE 9.7** Defining a thermal analysis model with two-dimensional simplification.

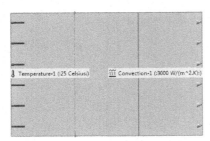

■ **FIGURE 9.8** Boundary conditions.

consists of three elements, which are connected in a sequence. The corresponding element models are determined using Eq. (9.15).

The system model can be assembled from the element models in Table 9.1 as,

$$
\begin{bmatrix}
7500 & -7500 & 0 & 0 \\
-7500 & 7500+7500 & -7500 & 0 \\
0 & -7500 & 7500+15000 & -15000 \\
0 & 0 & -15000 & 15000+3000
\end{bmatrix}
\begin{Bmatrix}
T_0 \\
T_1 \\
T_2 \\
T_3
\end{Bmatrix}
=
\begin{Bmatrix}
0 \\
0 \\
0 \\
-15000
\end{Bmatrix}
\qquad (9.28)
$$

$T_0 = 25°C$

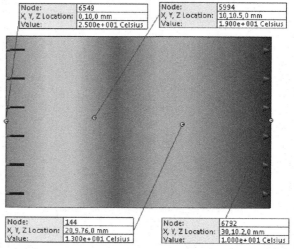

Node:	6549
X, Y, Z Location:	0,10,0 mm
Value:	2.500e+001 Celsius

Node:	5994
X, Y, Z Location:	10,10,5,0 mm
Value:	1.900e+001 Celsius

Node:	144
X, Y, Z Location:	20,9,76,0 mm
Value:	1.300e+001 Celsius

Node:	6792
X, Y, Z Location:	30,10,2,0 mm
Value:	1.000e+001 Celsius

■ **FIGURE 9.9** Temperature distribution in finite element analysis solution.

Table 9.1 Nodes and Elements in the Finite Element Analysis Model

Element	Node i	Node j	$[K_c]^{(e,i)}$	$R_q^{(e,i)} + R_h^{(e,i)}$
1	1	2	$\frac{(75)}{(0.01)}\begin{bmatrix} 1 & -1 \\ -1 & 1 \end{bmatrix}$	$\begin{Bmatrix} 0 \\ 0 \end{Bmatrix}$
2	2	3		
3	3	4	$\frac{(150)}{(0.01)}\begin{bmatrix} 1 & -1 \\ -1 & 1 \end{bmatrix} + \begin{bmatrix} 0 & 0 \\ 0 & 3000 \end{bmatrix}$	$\begin{Bmatrix} 0 \\ 0 \end{Bmatrix} + 3000(-5)\begin{Bmatrix} 0 \\ 1 \end{Bmatrix}$

Solving Eq. (9.28) gets the solution of,

$$\begin{Bmatrix} T_0 \\ T_1 \\ T_2 \\ T_3 \end{Bmatrix} = \begin{Bmatrix} 25 \\ 19 \\ 13 \\ 10 \end{Bmatrix} (°C) \qquad (9.29)$$

The above solution from the manual calculation is aligned accurately with the result of the FEA from the SolidWorks Simulation.

9.5 **2D STEADY HEAT TRANSFER PROBLEMS**

The rate of heat transfer can be increased by two ways. The first way is to increase the convection heat transfer coefficient h; it can be achieved by using highly conductive materials such as aluminum and increasing the fluid flow over the surface by pumps or fans. The second way is to increase the surface area of convection; it can be achieved by using extended surfaces called fins. Fig. 9.10 shows some examples of fins that various geometric designs have been adopted to increase the surface areas of convection for a high rate of heat transfer. Usually, surface areas for convection can be extended by attaching linear or circular pattern features on the base object.

Many heat transfer products use linear or circular patterns to extend surface areas of convection, the simplification can be made by estimating the heat transfer rate based on the results of the simulation on single pattern instance. 2D steady thermal analysis models are often used to represent heat transfer behaviors of fins.

Fig. 9.11 describes the working principle of a heat sink: the heat comes from a base part of the sink; therefore, a contact surface of a fin with the base part is given as a fixed temperature. The heat is conducted through the fin, and it is

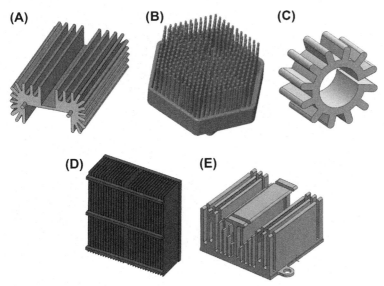

■ **FIGURE 9.10** Examples of fins for heat convections: (A) extrusions; (B) pin fins; (C) circular fins; (D) linear fins; (E) varying fins.

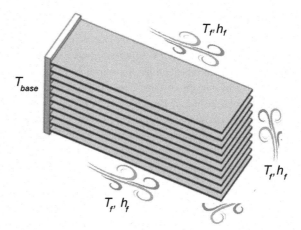

■ **FIGURE 9.11** Description of an extended heat transfer fin.

dissipated to the fluid surrounding the fin. The heat transfer coefficient h and the ambient temperature in fluids T_f have a great impact on the heat transfer rate. Fig. 9.12 shows the discretization of a 2D flat fin for a thermal analysis. Without losing the generality, 2D triangle and rectangle elements are used to represent the heat transfer domain. Each node in a triangle or rectangle element has 1° of freedom; the state variable of the node is temperature. The modeling of triangle and rectangle elements are discussed as below.

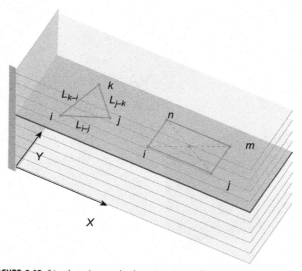

■ **FIGURE 9.12** Triangle and rectangle elements in a two-dimensional heat transfer domain.

9.5.1 **Modeling of rectangle elements**

In a rectangle element, the heat transfer behavior is represented by the temperatures $(T_i, T_j, T_m, \text{ and } T_n)$ on node $i, j, m,$ and n, and the temperature of an arbitrary position (X, Y) in this element is interpolated as,

$$T(X,Y) = [S_i(X,Y) \quad S_j(X,Y) \quad S_m(X,Y) \quad S_n(X,Y)]\begin{Bmatrix} T_i \\ T_j \\ T_m \\ T_n \end{Bmatrix} \quad (9.30)$$

The shape functions in Eq. (9.30) can be obtained by using the Lagrange interpolation formula Eq. (3.46) in the global coordinate system as,

$$\left.\begin{aligned} S_i(X,Y) &= \left(\frac{X_j - X}{l}\right)\left(\frac{Y_j - Y}{w}\right) \\ S_j(X,Y) &= \left(\frac{X - X_i}{l}\right)\left(\frac{Y_j - Y}{w}\right) \\ S_m(X,Y) &= \left(\frac{X - X_i}{l}\right)\left(\frac{Y - Y_i}{w}\right) \\ S_n(X,Y) &= \left(\frac{X_j - X}{l}\right)\left(\frac{Y - Y_i}{w}\right) \end{aligned}\right\} \quad (9.31)$$

where

$S_i(X, Y)$, $S_j(X, Y)$, $S_m(X, Y)$, and $S_n(X, Y)$ are the shape functions used in Eq. (9.30) and
$l = X_j - X_i$ and $w = Y_j - Y_i$

The temperature gradient can be derived by substituting Eq. (9.30) into Eq. (9.10) as,

$$\begin{Bmatrix} k_X\dfrac{\partial T}{\partial X} \\[2mm] k_Y\dfrac{\partial T}{\partial Y} \end{Bmatrix} = [\boldsymbol{D}]\cdot\begin{Bmatrix} T_i \\ T_j \\ T_m \\ T_n \end{Bmatrix}$$

$$= \frac{1}{lw}\begin{bmatrix} -k_X(Y_j - Y) & k_X(Y_j - Y) & k_X(Y - Y_i) & -k_X(Y - Y_i) \\ -k_Y(X_j - X) & -k_Y(X - X_i) & k_Y(X_j - X) & k_Y(X_j - X) \end{bmatrix}\cdot\begin{Bmatrix} T_i \\ T_j \\ T_m \\ T_n \end{Bmatrix}$$

$$(9.32)$$

As a consequence, Eq. (9.15) can be rewritten for the rectangle element as,

$$([\boldsymbol{K}_C] + [\boldsymbol{K}_h]) \begin{Bmatrix} T_i \\ T_j \\ T_m \\ T_n \end{Bmatrix} = \{\boldsymbol{R}_Q\} + \{\boldsymbol{R}_h\} \tag{9.33}$$

where

$$[\boldsymbol{K}_C] = \int_A \begin{bmatrix} \dfrac{\partial \boldsymbol{S}}{\partial X} & \dfrac{\partial \boldsymbol{S}}{\partial Y} \end{bmatrix} \cdot [\boldsymbol{D}] dA = \int_A \frac{1}{lw} \begin{bmatrix} -(Y_j - Y) & -(X_j - X) \\ (Y_j - Y) & -(X - X_i) \\ (Y - Y_i) & (X - X_i) \\ -(Y - Y_i) & (X_j - X) \end{bmatrix} \cdot [\boldsymbol{D}] dA$$

$$= \frac{1}{6lw} \begin{bmatrix} 2k_X w^2 + 2k_Y l^2 & -2k_X w^2 + k_Y l^2 & -k_X w^2 - k_Y l^2 & k_X w^2 - 2k_Y l^2 \\ -2k_X w^2 + k_Y l^2 & 2k_X w^2 + 2k_Y l^2 & k_X w^2 - 2k_Y l^2 & -k_X w^2 - k_Y l^2 \\ -k_X w^2 - k_Y l^2 & k_X w^2 - 2k_Y l^2 & 2k_X w^2 + 2k_Y l^2 & -2k_X w^2 + k_Y l^2 \\ k_X w^2 - 2k_Y l^2 & -k_X w^2 - k_Y l^2 & -2k_X w^2 + k_Y l^2 & 2k_X w^2 + 2k_Y l^2 \end{bmatrix}$$

$$\tag{9.34}$$

$$\{\boldsymbol{R}_Q\} = \int_A \dot{q}[\boldsymbol{S}]^T dA = \int_A \dot{q} \begin{bmatrix} S_i \\ S_j \\ S_m \\ S_n \end{bmatrix} dA = \frac{\dot{q}(lw)}{4} \begin{Bmatrix} 1 \\ 1 \\ 1 \\ 1 \end{Bmatrix} \tag{9.35}$$

In addition, $[\boldsymbol{K}_h]$ and $\{\boldsymbol{R}_h\}$ exist when one of edges (i–j, j–m, m–n, and n–i) involves convective heat transfer.

For the convection on edge i–j:

$$[\boldsymbol{K}_h] = \int_{i-j} h[\boldsymbol{S}]^T[\boldsymbol{S}] d\tau = \int_{X_i}^{X_j} h \begin{bmatrix} \dfrac{X_j - X}{l} \\ \dfrac{X - X_i}{l} \\ 0 \\ 0 \end{bmatrix} \cdot \begin{bmatrix} \dfrac{X_j - X}{l} & \dfrac{X - X_i}{l} & 0 & 0 \end{bmatrix} d\tau$$

$$= \frac{hl}{6} \begin{bmatrix} 2 & 1 & 0 & 0 \\ 1 & 2 & 0 & 0 \\ 0 & 0 & 0 & 0 \\ 0 & 0 & 0 & 0 \end{bmatrix}$$

$$\tag{9.36}$$

$$\{R_h\} = \int_{i-j} hT_f[S]^T d\tau = hT_f \int_{X_i}^{X_j} \begin{bmatrix} \dfrac{X_j - X}{l} \\ \dfrac{X - X_i}{l} \\ 0 \\ 0 \end{bmatrix} d\tau = \dfrac{hT_f l}{2} \begin{Bmatrix} 1 \\ 1 \\ 0 \\ 0 \end{Bmatrix} \quad (9.37)$$

For the convection on edge $j-m$:

$$[K_h] = \int_{j-m} h[S]^T[S] d\tau = \int_{Y_i}^{Y_j} h \begin{bmatrix} 0 \\ \dfrac{Y_j - Y}{w} \\ \dfrac{Y - Y_i}{w} \\ 0 \end{bmatrix} \cdot \begin{bmatrix} 0 & \dfrac{Y_j - Y}{w} & \dfrac{Y - Y_i}{w} & 0 \end{bmatrix} d\tau$$

$$= \dfrac{hw}{6} \begin{bmatrix} 0 & 0 & 0 & 0 \\ 0 & 2 & 1 & 0 \\ 0 & 1 & 2 & 0 \\ 0 & 0 & 0 & 0 \end{bmatrix}$$

$$(9.38)$$

$$\{R_h\} = \int_{j-m} hT_f[S]^T d\tau = hT_f \int_{Y_i}^{Y_j} \begin{bmatrix} 0 \\ \dfrac{Y_j - Y}{w} \\ \dfrac{Y - Y_i}{w} \\ 0 \end{bmatrix} d\tau = \dfrac{hT_f w}{2} \begin{Bmatrix} 0 \\ 1 \\ 1 \\ 0 \end{Bmatrix} \quad (9.39)$$

For the convection on edge $m-n$:

$$[K_h] = \int_{m-n} h[S]^T[S] d\tau = \int_{X_j}^{X_i} h \begin{bmatrix} 0 \\ 0 \\ \dfrac{X - X_i}{l} \\ \dfrac{X_j - X}{l} \end{bmatrix} \cdot \begin{bmatrix} 0 & 0 & \dfrac{X - X_i}{l} & \dfrac{X_j - X}{l} \end{bmatrix} d\tau$$

$$= \dfrac{hl}{6} \begin{bmatrix} 0 & 0 & 0 & 0 \\ 0 & 0 & 0 & 0 \\ 0 & 0 & 2 & 1 \\ 0 & 0 & 2 & 1 \end{bmatrix}$$

$$(9.40)$$

$$\{R_h\} = \int_{m-n} hT_f[S]^T d\tau = hT_f \int_{X_j}^{X_i} \begin{bmatrix} 0 \\ 0 \\ \dfrac{X - X_i}{l} \\ \dfrac{X_j - X}{l} \end{bmatrix} d\tau = \frac{hT_f l}{2} \begin{Bmatrix} 0 \\ 0 \\ 1 \\ 1 \end{Bmatrix} \qquad (9.41)$$

For the convection on edge $n-i$:

$$[K_h] = \int_{n-i} h[S]^T[S] d\tau = \int_{Y_j}^{Y_i} h \begin{bmatrix} \dfrac{Y_j - Y}{w} \\ 0 \\ 0 \\ \dfrac{Y - Y_i}{w} \end{bmatrix} \cdot \begin{bmatrix} \dfrac{Y_j - Y}{w} & 0 & 0 & \dfrac{Y - Y_i}{w} \end{bmatrix} d\tau$$

$$= \frac{hw}{6} \begin{bmatrix} 2 & 0 & 0 & 1 \\ 0 & 0 & 0 & 0 \\ 0 & 0 & 0 & 0 \\ 1 & 0 & 0 & 2 \end{bmatrix}$$

$$(9.42)$$

$$\{R_h\} = \int_{n-i} hT_f[S]^T d\tau = hT_f \int_{Y_j}^{Y_i} \begin{bmatrix} \dfrac{Y_j - Y}{w} \\ 0 \\ 0 \\ \dfrac{Y - Y_i}{w} \end{bmatrix} d\tau = \frac{hT_f w}{2} \begin{Bmatrix} 1 \\ 0 \\ 0 \\ 1 \end{Bmatrix} \qquad (9.43)$$

9.5.2 **Modeling of triangular elements**

In a triangular element, the heat transfer behavior is represented by the temperatures (T_i, T_j, and T_k) on node i, j, and k, and the temperature of an arbitrary position (X, Y) in this element is interpolated as,

$$T(X, Y) = [S_i(X, Y) \quad S_j(X, Y) \quad S_k(X, Y)] \begin{bmatrix} T_i \\ T_j \\ T_k \end{bmatrix} \qquad (9.44)$$

The shape functions in Eq. (9.44) are given Eq. (3.73) and rewritten here as

$$\left.\begin{aligned}
S_i(x,y) &= \frac{\alpha_i + \beta_i \cdot x + \delta_i \cdot y}{|\Delta|} \\
S_j(x,y) &= \frac{\alpha_j + \beta_j \cdot x + \delta_j \cdot y}{|\Delta|} \\
S_k(x,y) &= \frac{\alpha_k + \beta_k \cdot x + \delta_k \cdot y}{|\Delta|}
\end{aligned}\right\} \tag{9.45}$$

where all of the coefficients (α_i, β_i, δ_i, and $|\Delta|$) are determined by the geometry of the triangular element in Eq. (3.71).

The temperature gradient can be derived by substituting Eq. (9.44) into Eq. (9.10) as,

$$\begin{aligned}
\left\{\begin{array}{c} k_X \dfrac{\partial T}{\partial X} \\[2mm] k_Y \dfrac{\partial T}{\partial Y} \end{array}\right\} &= [D] \cdot \left\{\begin{array}{c} T_i \\ T_j \\ T_k \end{array}\right\} \\[4mm]
&= \frac{1}{|\Delta|}\begin{bmatrix} k_X\beta_i & k_X\beta_j & k_X\beta_k \\ k_Y\delta_i & k_Y\delta_j & k_Y\delta_k \end{bmatrix} \cdot \left\{\begin{array}{c} T_i \\ T_j \\ T_k \end{array}\right\}
\end{aligned} \tag{9.46}$$

As a consequence, Eq. (9.15) can be rewritten for the triangular element as,

$$([K_C] + [K_h])\left\{\begin{array}{c} T_i \\ T_j \\ T_k \end{array}\right\} = \{R_Q\} + \{R_h\} \tag{9.47}$$

where

$$\begin{aligned}
[K_C] &= \int_A \left[\frac{\partial S}{\partial X} \quad \frac{\partial S}{\partial Y}\right] \cdot [D]dA = \int_A \frac{1}{|\Delta|}\begin{bmatrix} \beta_i & \delta_i \\ \beta_j & \delta_j \\ \beta_k & \delta_k \end{bmatrix} \cdot [D]dA \\[4mm]
&= \frac{1}{4A}\begin{bmatrix} k_x\beta_i^2 + k_y\delta_i^2 & k_x\beta_i\beta_j + k_y\delta_i\delta_j & k_x\beta_i\beta_k + k_y\delta_i\delta_k \\ k_x\beta_i\beta_j + k_y\delta_i\delta_j & k_x\beta_j^2 + k_y\delta_j^2 & k_x\beta_j\beta_k + k_y\delta_j\delta_k \\ k_x\beta_i\beta_k + k_y\delta_i\delta_k & k_x\beta_j\beta_k + k_y\delta_j\delta_k & k_x\beta_k^2 + k_y\delta_k^2 \end{bmatrix}
\end{aligned} \tag{9.48}$$

$$\{R_Q\} = \int_A \dot{q}[S]^T dA = \int_A \dot{q} \begin{bmatrix} S_i \\ S_j \\ S_k \end{bmatrix} dA = \frac{\dot{q}A}{3} \begin{Bmatrix} 1 \\ 1 \\ 1 \end{Bmatrix} \qquad (9.49)$$

In addition, $[K_h]$ and $\{R_h\}$ exist when one of the edges ($i-j$, $j-k$, and $k-i$) involves convective heat transfer.

For the convection on edge $i-j$ (L_{i-j} is the length of edge $i-j$):

$$[K_h] = \int_{i-j} h[S]^T[S] d\tau = = \frac{hL_{i-j}}{6} \begin{bmatrix} 2 & 1 & 0 \\ 1 & 2 & 0 \\ 0 & 0 & 0 \end{bmatrix} \qquad (9.50)$$

$$\{R_h\} = \int_{i-j} hT_f[S]^T d\tau = \frac{hT_fL_{i-j}}{2} \begin{Bmatrix} 1 \\ 1 \\ 0 \end{Bmatrix} \qquad (9.51)$$

For the convection on edge $j-k$ (L_{j-k} is the length of edge $j-k$):

$$[K_h] = \int_{j-k} h[S]^T[S] d\tau = \frac{hL_{j-k}}{6} \begin{bmatrix} 0 & 0 & 0 \\ 0 & 2 & 1 \\ 0 & 1 & 2 \end{bmatrix} \qquad (9.52)$$

$$\{R_h\} = \int_{j-k} hT_f[S]^T d\tau = \frac{hT_fL_{j-k}}{2} \begin{Bmatrix} 0 \\ 1 \\ 1 \end{Bmatrix} \qquad (9.53)$$

For the convection on edge $k-i$ (L_{k-i} is the length of edge $k-i$):

$$[K_h] = \int_{k-i} h[S]^T[S] d\tau = \frac{hL_{k-i}}{6} \begin{bmatrix} 2 & 0 & 1 \\ 0 & 0 & 0 \\ 1 & 0 & 2 \end{bmatrix} \qquad (9.54)$$

$$\{R_h\} = \int_{k-i} hT_f[S]^T d\tau = \frac{hT_fL_{k-i}}{2} \begin{Bmatrix} 1 \\ 0 \\ 1 \end{Bmatrix} \qquad (9.55)$$

9.5.3 2D steady heat transfer example

In this section, the 2D steady heat transfer models for rectangle elements from Section 9.5.2 are used to determine the heat transfer performance of a flat plate fin.

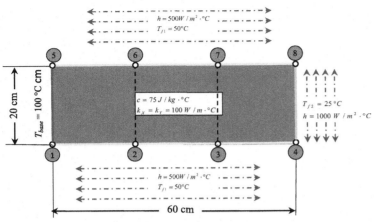

■ **FIGURE 9.13** Example of two-dimensional steady heat transfer problem.

Example 9.2

Fig. 9.13 gives the geometry of a fin from a heat sink. The base temperature on the left edge is given as $T_{base} = 100°C$. The parameters for the convection on lateral surfaces are $h_1 = 500$ W/m^2 °C and $T_{f1} = 50°C$. The parameters for the convection on tip are $h_2 = 1000$ W/m^2 °C and $T_{f2} = 25°C$. Determine the temperature on tip of fin using a 2D steady heat transfer model.

Solution I

The model with Eqs. (9.33)–(9.43) in Section 9.5.2 are used in element modeling. The fin is divided into three rectangle elements and eight nodes as shown in Fig. 9.13. The relations of elements and nodes are listed in Table 9.2. The conduction matrices and load vectors in three element models are calculated using the equations in the model, and the results are also included in Table 9.2.

The system model is then assembled from three element models in Table 9.2 as,

$$
\begin{bmatrix}
100 & 0 & 0 & 0 & -16.67 & -33.33 & 0 & 0 \\
0 & 200 & 0 & 0 & -33.33 & -33.33 & -33.33 & 0 \\
0 & 0 & 200 & 0 & 0 & -33.33 & -33.33 & -33.33 \\
0 & 0 & 0 & 166.67 & 0 & 0 & -33.33 & 16.67 \\
-16.67 & -33.33 & 0 & 0 & 100 & 0 & 0 & 0 \\
-33.33 & -33.33 & -33.33 & 0 & 0 & 200 & 0 & 0 \\
0 & -33.33 & -33.33 & -33.33 & 0 & 0 & 200 & 0 \\
0 & 0 & -33.33 & 16.67 & 0 & 0 & 0 & 166.67
\end{bmatrix}
\begin{Bmatrix}
T_1 \\ T_2 \\ T_3 \\ T_4 \\ T_5 \\ T_6 \\ T_7 \\ T_8
\end{Bmatrix}
=
\begin{Bmatrix}
2500 \\ 5000 \\ 5000 \\ 5000 \\ 2500 \\ 5000 \\ 5000 \\ 5000
\end{Bmatrix}
$$

$$(9.56)$$

Table 9.2 Nodes and Elements in Two-Dimensional Heat Transfer Finite Element Analysis Model

Nodes i, j, m, n	$[K_c]^{(e,i)}$	$R_q^{(e,i)} + R_h^{(e,i)}$
① ② ⑥ ⑤	$\dfrac{(100)}{6}\begin{bmatrix} 4 & -1 & -2 & -1 \\ -1 & 4 & -1 & -2 \\ -2 & -1 & 4 & -1 \\ -1 & -2 & -1 & 4 \end{bmatrix}$	$\begin{Bmatrix} 0 \\ 0 \\ 0 \\ 0 \end{Bmatrix} + \dfrac{(500)(50)(0.2)}{2}\begin{Bmatrix} 1 \\ 1 \\ 0 \\ 0 \end{Bmatrix}$
② ③ ⑦ ⑥	$+\dfrac{(500)(0.2)}{6}\begin{bmatrix} 2 & 1 & 0 & 0 \\ 1 & 2 & 0 & 0 \\ 0 & 0 & 0 & 0 \\ 0 & 0 & 0 & 0 \end{bmatrix} + \dfrac{(500)(0.2)}{6}\begin{bmatrix} 0 & 0 & 0 & 0 \\ 0 & 0 & 0 & 0 \\ 0 & 0 & 2 & 1 \\ 0 & 0 & 2 & 1 \end{bmatrix}$	$+\dfrac{(500)(50)(0.2)}{2}\begin{Bmatrix} 0 \\ 0 \\ 1 \\ 1 \end{Bmatrix}$
③ ④ ⑧ ⑦	$\dfrac{(100)}{6}\begin{bmatrix} 4 & -1 & -2 & -1 \\ -1 & 4 & -1 & -2 \\ -2 & -1 & 4 & -1 \\ -1 & -2 & -1 & 4 \end{bmatrix} + \dfrac{(500)(0.2)}{6}\begin{bmatrix} 2 & 1 & 0 & 0 \\ 1 & 2 & 0 & 0 \\ 0 & 0 & 0 & 0 \\ 0 & 0 & 0 & 0 \end{bmatrix}$ $+\dfrac{(1000)(0.2)}{6}\begin{bmatrix} 0 & 0 & 0 & 0 \\ 0 & 2 & 1 & 0 \\ 0 & 1 & 2 & 0 \\ 0 & 0 & 0 & 0 \end{bmatrix} + \dfrac{(500)(0.2)}{6}\begin{bmatrix} 0 & 0 & 0 & 0 \\ 0 & 0 & 0 & 0 \\ 0 & 0 & 2 & 1 \\ 0 & 0 & 2 & 1 \end{bmatrix}$	$\begin{Bmatrix} 0 \\ 0 \\ 0 \\ 0 \end{Bmatrix} + \dfrac{(500)(50)(0.2)}{2}\begin{Bmatrix} 1 \\ 1 \\ 0 \\ 0 \end{Bmatrix}$ $+\dfrac{(1000)(25)(0.2)}{2}\begin{Bmatrix} 0 \\ 1 \\ 1 \\ 0 \end{Bmatrix} + \dfrac{(500)(50)(0.2)}{2}\begin{Bmatrix} 0 \\ 0 \\ 1 \\ 1 \end{Bmatrix}$

(Element)

Applying Boundary Conditions (BCs) on nodes 1 and 5 with a fixed temperature 100°C gets the modified system model as,

$$\begin{bmatrix} 200 & 0 & 0 & -33.33 & -33.33 & 0 \\ 0 & 200 & 0 & -33.33 & -33.33 & -33.33 \\ 0 & 0 & 166.67 & 0 & -33.33 & 16.67 \\ -33.33 & -33.33 & 0 & 200 & 0 & 0 \\ -33.33 & -33.33 & -33.33 & 0 & 200 & 0 \\ 0 & -33.33 & 16.67 & 0 & 0 & 166.67 \end{bmatrix} \begin{Bmatrix} T_2 \\ T_3 \\ T_4 \\ T_6 \\ T_7 \\ T_8 \end{Bmatrix}$$

$$= \begin{Bmatrix} 8333.33 \\ 5000 \\ 5000 \\ 8333.33 \\ 5000 \\ 5000 \end{Bmatrix}$$

(9.57)

Solving Eq. (9.57) gets the complete solution of nodal temperatures as,

$$\{T\} = \{ 100, \quad 59.8 \quad 49.2 \quad 36.2 \quad 100 \quad 59.8 \quad 49.2 \quad 36.2 \}(°C)$$

(9.58)

The temperature at the tip is 36.2°C.

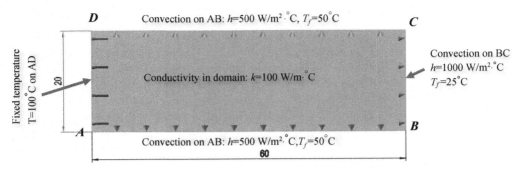

■ **FIGURE 9.14** Defining a two-dimensional finite element analysis model for heat transfer fin (unit: cm).

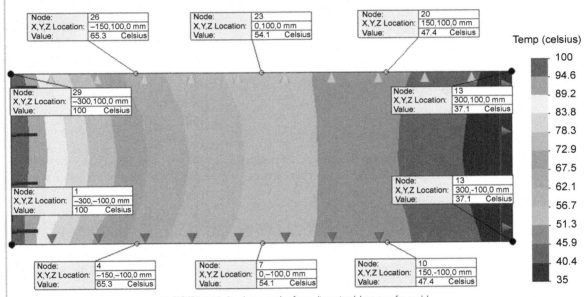

■ **FIGURE 9.15** Simulation result of two-dimensional heat transfer model.

Solution II

Fig. 9.14 shows a 2D steady heat transfer model created in the SolidWorks Simulation. The boundary conditions and loads are defined on four edges based on the information in Fig. 9.13. The size of the mesh is set as 10 mm.

The result from the simulation of the developed FEA model is illustrated in Fig. 9.15. The minimized temperature obtained on the fin tip is 37.1°C, which is ~2.4% of the discrepancy in comparison to the results from the manual calculation.

9.6 TRANSIENT HEAT TRANSFER PROBLEMS

In general, the temperature of a body in a heat transfer problem varies with spatial position as well as time. There are often circumstances where the transient response to heat transfer is critical. To find the solution to a transient heat transfer problem, the time domain is discretized, the time-dependent solution is made by a time integration procedure. Therefore, the time step in the discretization is critical to both of stability and accuracy of results. If the time step is too small, a spurious oscillation may occur which leads to meaningless results. Otherwise, if the time step is too large, temperature gradients cannot be evaluated appropriately.

Biot number $\left(Bi = \frac{h\Delta x}{k_{\text{solid}}}\right)$ and Fourier number $\left(Fo = \frac{\alpha\Delta t}{(\Delta x)^2}\right)$ can be used to determine a reasonable time step, where Δx denotes the mean edge length of an element and Δt is the time step.

If the Biot number Bi is less than 1 ($Bi < 1$) in an FEA model, the time step should be determined based on the condition of convergence $0.1 \leq Fo = b = \frac{\alpha\Delta t}{(\Delta x)^2} \leq 0.5$,

$$\Delta t = \frac{(\Delta x)^2 Fo}{\alpha} \tag{9.59}$$

where $Fo \in [0.1, 0.5]$

If the Biot number Bi is larger than 1 ($Bi > 1$) in an FEA model, the time step should be determined based on the product of Biot and Fourier numbers $(Fo)(Bi) = b$,

$$\Delta t = \frac{(\Delta x)k_{\text{solid}}}{\alpha h}b = \frac{(\Delta x)\rho c}{h}b \tag{9.60}$$

We have provided the generic governing models for linear and nonlinear transient heat transfer problems in Eqs. (9.17) and (9.18). In this section, the following simplified equations in a matrix form are used to replace Eqs. (9.17) and (9.18).

$$[C] \cdot \left\{\frac{\partial T}{\partial t}\right\} + [K] \cdot \{T\} = \{Q\} \tag{9.61}$$

where

$$[K] = [K_C] + [K_h] \tag{9.62}$$

$$\{Q\} = \left\{ \begin{array}{ll} \{R_Q\} + \{R_q\} + \{R_h\} & \text{for a linear transient problem} \\ \{R_Q\} + \{R_q\} + \{R_h\} + \{R_r\} & \text{for a nonlinear transient problem} \end{array} \right\} \tag{9.63}$$

Other matrix and vector components in Eq. (9.61) are defined in Eq. (9.14).

Using the Taylor's expansion formula over the time domain gets,

$$\{T\}^{(p+1)} \approx \{T\}^{(p)} + \Delta t \{\dot{T}\}^{(p)} + \frac{\Delta t^2}{2} \{\ddot{T}\}^{(p)} \tag{9.64}$$

In order to provide the flexibility of the approximation in dealing with the weights of $\{\dot{T}\}^{(p)}$ and $\{\dot{T}\}^{(p+1)}$, we introduce the Euler parameter $\theta \in (0, 1)$ to replace 1/2 for the adjustment in Eq. (9.64),

$$\{T\}^{(p+1)} \approx \{T\}^{(p)} + \Delta t \{\dot{T}\}^{(p)} + \theta \Delta t^2 \{\ddot{T}\}^{(p)} \tag{9.65}$$

Different values of the Euler parameter refer to different schemes of the approximation, i.e.,

$$\theta = \begin{cases} 0 & \text{Forward difference (Euler)} \\ 1/2 & \text{Crank} - \text{Nicolson difference} \\ 1 & \text{Backward difference} \end{cases} \tag{9.66}$$

The second-order derivative in Eq. (9.65) can be approximated by the first-order derivatives as,

$$\{\ddot{T}\}^{(p)} = \frac{\{\dot{T}\}^{(p+1)} - \{\dot{T}\}^{(p)}}{\Delta t} \tag{9.67}$$

Substituting Eq. (9.67) into (9.65) yields,

$$\{T\}^{(p+1)} \approx \{T\}^{(p)} + (1 - \theta)\Delta t \cdot \{\dot{T}\}^{(p)} + \theta \cdot \Delta t \cdot \{\dot{T}\}^{(p+1)} \tag{9.68}$$

Eq. (9.68) shows that $\{T\}^{(p+1)}$ and $\{\dot{T}\}^{(p+1)}$ are dependent and both of them are unknowns to be determined. $\{\dot{T}\}^{(p+1)}$ can be calculated based on given $\{T\}^{(p+1)}$ as,

$$\{\dot{T}\}^{(p+1)} = \frac{(\{T\}^{(p+1)} - \{T\}^{(p)})}{\theta \cdot \Delta t} - \left(\frac{1}{\theta} - 1\right)\{\dot{T}\}^{(p)} \tag{9.69}$$

Assume that both of $\{T\}^{(p)}$ and $\{\dot{T}\}^{(p)}$ are known at the previous time step p. Finally, $\{T\}^{(p+1)}$ can be found by substituting $\{\dot{T}\}^{(p+1)}$ in Eq. (9.69) into Eq. (9.61) as,

$$\left(\frac{[C]}{\theta \cdot \Delta t} + [K]\right) \cdot \{T\}^{(p+1)} = \{Q\} + \frac{[C]}{\theta \cdot \Delta t} \cdot \{T\}^{(p)} + \left(\frac{1}{\theta} - 1\right)[C]\{\dot{T}\}^{(p)} \tag{9.70}$$

where

$[K_m] = \frac{[C]}{\theta \cdot \Delta t} + [K]$ is modified conductive matrix, and.

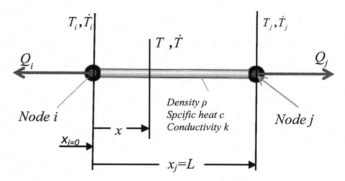

■ FIGURE 9.16 One-dimensional transient heat transfer element.

$$\{Q_m\} = \{Q\} + \frac{[C]}{\theta \cdot \Delta t} \cdot \{T\}^{(p)} + \left(\frac{1}{\theta} - 1\right)[C]\{\dot{T}\}^{(p)} \text{ is a modified load}$$

vector.

In the following, the model of 1D transient heat transfer element is discussed to demonstrate the element modeling and the procedure for solving a transient heat transfer problem. 1D transient heat element is described in Fig. 9.16.

The shape functions of a 1D element under a local coordinate system are $S_i(x) = 1 - x/L$ and $S_j(x) = x/L$. The formula in Eq. (9.14) are used to define $[C]$ and $[K]$ as,

$$[C] = \int_V \rho c [S]^T [S] dV = \rho c A \int \begin{bmatrix} S_i^2 & S_i S_j \\ S_i S_j & S_j^2 \end{bmatrix} dl = \frac{\rho c A L}{6} \begin{bmatrix} 2 & 1 \\ 1 & 2 \end{bmatrix} \quad (9.71)$$

$$[K_C] = \int_V \left[\frac{\partial S}{\partial x}\right] \cdot [D] dV = A \int_L \begin{bmatrix} \frac{\partial S_i}{\partial x} \\ \frac{\partial S_j}{\partial x} \end{bmatrix} \begin{bmatrix} k\frac{\partial S_i}{\partial x} & k\frac{\partial S_j}{\partial x} \end{bmatrix} dl = \frac{kA}{L} \begin{bmatrix} 1 & -1 \\ -1 & 1 \end{bmatrix}$$

$$(9.72)$$

where

ρ and c are the density and specific heat of materials, respectively;

k is the conductive coefficient;

A and L are the cross-section area and length of element.

Therefore, the system model with Eqs. (9.61)—(9.63) can be tailored for a 1D transient heat transfer element as,

$$[K_M]_{2\times 2} \cdot \begin{Bmatrix} T_i \\ T_j \end{Bmatrix}^{(p+1)} = \{Q_M\}_{2\times 1} \quad (9.73)$$

where $[K_M]$ is the modified conductive matrix as

$$[K_M] = \left(\frac{\rho c A L}{6\theta \Delta t} \begin{bmatrix} 2 & 1 \\ 1 & 2 \end{bmatrix} + \frac{kA}{L} \begin{bmatrix} 1 & -1 \\ -1 & 1 \end{bmatrix} \right) \quad (9.74)$$

$[Q_M]$ is the modified load vector determined by the state of the precedent step as,

$$\{Q_M\} = \left\{ \begin{matrix} Q_i \\ Q_j \end{matrix} \right\} + \frac{\rho c A L}{6\theta \Delta t} \begin{bmatrix} 2 & 1 \\ 1 & 2 \end{bmatrix} \cdot \left\{ \begin{matrix} T_i \\ T_j \end{matrix} \right\}^{(p)}$$
$$+ \left(\frac{1}{\theta} - 1 \right) \frac{\rho c A L}{6} \begin{bmatrix} 2 & 1 \\ 1 & 2 \end{bmatrix} \left\{ \begin{matrix} \dot{T}_i \\ \dot{T}_j \end{matrix} \right\}^{(p)} \quad (9.75)$$

After the nodal temperatures at $(p + 1)$ have been found, the temperature gradients can be derived by,

$$\left\{ \begin{matrix} \dot{T}_i \\ \dot{T}_j \end{matrix} \right\}^{(p+1)} = \frac{1}{\theta \Delta t} \left\{ \begin{matrix} T_i^{(p+1)} - T_i^{(p)} \\ T_j^{(p+1)} - T_j^{(p)} \end{matrix} \right\} - \left(\frac{1}{\theta} - 1 \right) \left\{ \begin{matrix} \dot{T}_i \\ \dot{T}_j \end{matrix} \right\}^{(p)} \quad (9.76)$$

Example 9.3

A 2-in thick plate of 1060 alloy is heated to 300°C uniformly, it is then air-quenched in the forced air. It is assumed that the temperature of the plate surface is suddenly set as 25°C. The materials density $\rho = 0.0975437$ lb/in³, the conductivity $k = 0.002675$ Btu/(sin °C), and specific heat $c = 0.214961$ Btu/(lb °C). Determine the changing curve of temperature at the neutral plane of the plate with respect to time in 5 s.

Solution I

Because the plate is symmetric about its neutral axis, a 1D FEA model is developed for half of the plate thickness. As shown in Fig. 9.17; it is assumed to have five elements and six nodes. Every element ha an even length of 0.2-in.

Let the Fourier number be $F_0 = 0.5$, the step time Δt is found from Eq. (9.59) as,

$$\Delta t = \frac{(\Delta x)^2 F_0}{\alpha} = \frac{(0.2)^2(0.5)}{(0.002675/(0.0975437 * .214961))} = 0.1568(s)$$

If the history of the temperature change in 5 s is concerned, the total number N of the time steps is,

$$N = \frac{(5)}{\Delta t} = 32(steps)$$

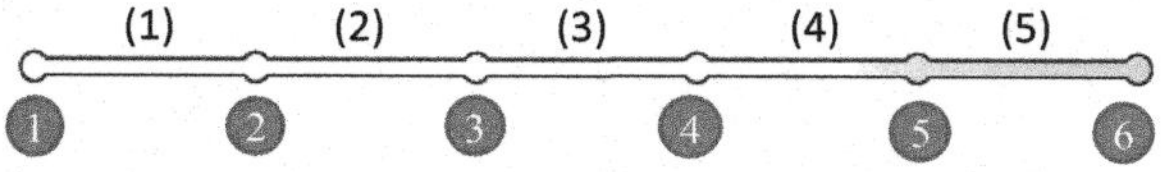

$$T_1^{(0)} = T_2^{(0)} = T_3^{(0)} = T_4^{(0)} = T_5^{(0)} = T_6^{(0)} = 300°C$$
$$\dot{T}_1^{(0)} = \dot{T}_2^{(0)} = \dot{T}_3^{(0)} = \dot{T}_4^{(0)} = \dot{T}_5^{(0)} = \dot{T}_6^{(0)} = 0$$
$$T_6^{(p)} = 25°C \qquad p > 0$$

■ **FIGURE 9.17** Example of one-dimensional transient heat transfer model.

Let the Euler parameter $\theta = 0.5$ in Eq. (9.66) where the Crank–Nicolson difference is used in the approximation. At each time step p from 1 to N, the system model can be assembled from five element models expressed in Eqs. (9.74) and (9.75) as,

$$[K]^{(G)} \begin{Bmatrix} T_1 \\ T_2 \\ T_3 \\ T_4 \\ T_5 \\ T_6 \end{Bmatrix}^{(p+1)} = \{Q\}^{(p)}$$

$$= \begin{Bmatrix} 2c_1 T_1^{(p)} + c_1 T_2^{(p)} + 2c_2 \dot{T}_1^{(p)} + c_2 \dot{T}_2^{(p)} \\ c_1 T_1^{(p)} + 4c_1 T_2^{(p)} + c_2 \dot{T}_1^{(p)} + 4c_2 \dot{T}_2^{(p)} + c_1 T_3^{(p)} + c_2 \dot{T}_3^{(p)} \\ c_1 T_2^{(p)} + 4c_1 T_3^{(p)} + c_2 \dot{T}_2^{(p)} + 4c_2 \dot{T}_3^{(p)} + c_1 T_4^{(p)} + c_2 \dot{T}_4^{(p)} \\ c_1 T_3^{(p)} + 4c_1 T_4^{(p)} + c_2 \dot{T}_3^{(p)} + 4c_2 \dot{T}_4^{(p)} + c_1 T_5^{(p)} + c_2 \dot{T}_5^{(p)} \\ c_1 T_4^{(p)} + 4c_1 T_5^{(p)} + c_2 \dot{T}_4^{(p)} + 4c_2 \dot{T}_5^{(p)} + c_1 T_6^{(p)} + c_2 \dot{T}_6^{(p)} \\ c_1 T_5^{(p)} + 2c_1 T_6^{(p)} + c_2 \dot{T}_5^{(p)} + 2c_2 \dot{T}_6^{(p)} \end{Bmatrix}$$

$$(9.77)$$

where

$$[K]^{(G)} = \begin{bmatrix} 0.0312 & -0.0045 & 0 & 0 & 0 & 0 \\ -0.0045 & 0.0624 & -0.0045 & 0 & 0 & 0 \\ 0 & -0.0045 & 0.0624 & -0.0045 & 0 & 0 \\ 0 & 0 & -0.0045 & 0.0624 & -0.0045 & 0 \\ 0 & 0 & 0 & -0.0045 & 0.0624 & -0.0045 \\ 0 & 0 & 0 & 0 & -0.0045 & 0.0312 \end{bmatrix}$$

$$(9.78)$$

c_1 and c_2 are two constants used in an element model Eq. (9.76) as,

$$\left.\begin{array}{l} c_1 = \dfrac{\rho cAL}{6\theta\Delta t} = 0.0089 \\[3mm] c_2 = \left(1 - \dfrac{1}{\theta}\right)\dfrac{\rho cAL}{6} = -0.00069893 \end{array}\right\} \qquad (9.79)$$

At each step p, after $\{T\}^{(p+1)}$ is found from Eq. (9.77), temperature gradients on nodes can be updated based on Eq. (9.75) as,

$$\begin{Bmatrix} \dot{T}_1 \\ \dot{T}_2 \\ \dot{T}_3 \\ \dot{T}_4 \\ \dot{T}_5 \\ \dot{T}_6 \end{Bmatrix}^{(p+1)} = \begin{Bmatrix} \dfrac{1}{\theta\Delta t}\left(T_1^{(p+1)} - T_1^{(p)}\right) - \left(\dfrac{1}{\theta} - 1\right)\dot{T}_1^{(p)} \\[3mm] \dfrac{1}{\theta\Delta t}\left(T_2^{(p+1)} - T_2^{(p)}\right) - \left(\dfrac{1}{\theta} - 1\right)\dot{T}_2^{(p)} \\[3mm] \dfrac{1}{\theta\Delta t}\left(T_3^{(p+1)} - T_3^{(p)}\right) - \left(\dfrac{1}{\theta} - 1\right)\dot{T}_3^{(p)} \\[3mm] \dfrac{1}{\theta\Delta t}\left(T_4^{(p+1)} - T_4^{(p)}\right) - \left(\dfrac{1}{\theta} - 1\right)\dot{T}_4^{(p)} \\[3mm] \dfrac{1}{\theta\Delta t}\left(T_5^{(p+1)} - T_5^{(p)}\right) - \left(\dfrac{1}{\theta} - 1\right)\dot{T}_5^{(p)} \\[3mm] \dfrac{1}{\theta\Delta t}\left(25 - T_6^{(p+1)}\right) - \left(\dfrac{1}{\theta} - 1\right)\dot{T}_6^{(p)} \end{Bmatrix} \qquad (9.80)$$

Note that the calculation for the temperature gradient of the boundary node 6 is different since the temperature at this node is fixed as 25°C. Accordingly, when the calculation result at step p is moved to next step $p + 1$, the temperature at node 6 must be set back to 25°C as the boundary constraint. The temperature changes on nodes with respect to time are obtained in Table 9.3.

The plots of temperature and temperature on node 1 (i.e., the center of plate) are illustrated in Figs. 9.18 and 9.19, respectively.

Solution II

A transient heat transfer model in defined for the given problem. As shown in Fig. 9.20, 2D simplification is activated, and 1060 alloyed is specified as the materials of the plate. Two boundary conditions are (1) an initial temperature of 300°C in the whole domain, and (2) a fixed temperature of 25°C for the edge AB. The meshing size is set as 0.2 in.

Fig. 9.21 shows the change of temperature at the central axis with respect to time, which is aligned with the result from manual calculation

Table 9.3 The History of Nodal Temperatures in 5 s

Step		Temperature (°C)					
No.	Time (s)	Node 1	Node 2	Node 3	Node 4	Node 5	Node 6
0	0	300	300	300	300	300	25
1	0.1568	299.9945	299.9618	299.4711	292.6341	197.4066	25
2	0.3136	299.8771	299.3362	293.5781	248.8355	120.4615	25
3	0.4704	298.8845	295.6335	274.0357	212.8518	138.0203	25
4	0.6272	294.7467	286.1748	255.8324	204.9803	127.9079	25
5	0.784	285.8694	275.4648	245.3924	192.5891	122.1926	25
6	0.9408	275.5284	265.5057	234.3123	183.8132	115.9687	25
7	1.0976	265.4401	255.0242	224.6841	175.3348	110.9712	25
8	1.2544	255.091	245.0755	215.2958	167.9028	106.3968	25
9	1.4112	245.1051	235.2671	206.5492	160.9382	102.3165	25
10	1.568	235.3315	225.8797	198.1785	154.4945	98.5393	25
11	1.7248	225.9358	216.8209	190.2477	148.4187	95.0359	25
12	1.8816	216.8846	208.1518	182.6812	142.6915	91.7441	25
13	2.0384	208.2126	199.8444	175.4773	137.2591	88.6407	25
14	2.1952	199.9055	191.9035	168.6055	132.101	85.7005	25
15	2.352	191.9622	184.3126	162.0528	127.1924	82.9092	25
16	2.5088	184.3696	177.0622	155.8008	122.5178	80.254	25
17	2.6656	177.1169	170.1378	149.8358	118.0622	77.7256	25
18	2.8224	170.1904	163.5267	144.1437	113.8137	75.3161	25
19	2.9792	163.577	157.2152	138.7118	109.7612	73.0187	25
20	3.136	157.2633	151.1904	133.5279	105.895	70.8274	25
21	3.2928	151.2364	145.4396	128.5806	102.206	68.737	25
22	3.4496	145.4835	139.9505	123.859	98.6858	66.7424	25
23	3.6064	139.9925	134.7113	119.3527	95.3265	64.8391	25
24	3.7632	134.7514	129.7108	115.0518	92.1205	63.0228	25
25	3.92	129.7491	124.9382	110.9471	89.0608	61.2894	25
26	4.0768	124.9747	120.383	107.0294	86.1406	59.6352	25
27	4.2336	120.4178	116.0353	103.2904	83.3536	58.0563	25
28	4.3904	116.0686	111.8859	99.7218	80.6937	56.5495	25
29	4.5472	111.9177	107.9255	96.3159	78.1551	55.1114	25
30	4.704	107.9559	104.1457	93.0652	75.7322	53.7389	25
31	4.8608	104.1746	100.5382	89.9627	73.4197	52.4289	25
32	5.0176	100.5658	97.095	87.0016	71.2127	51.1787	25

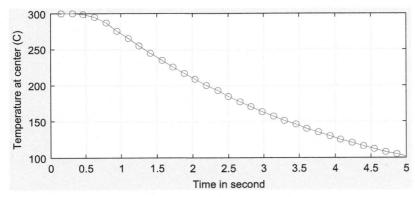

■ **FIGURE 9.18** Temperature at center of plate with respect to time.

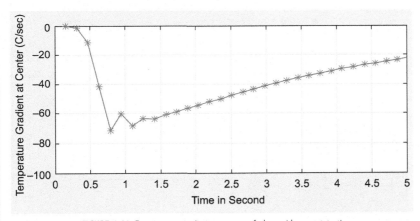

■ **FIGURE 9.19** Temperature gradient at center of plate with respect to time.

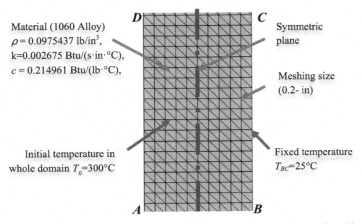

■ **FIGURE 9.20** A finite element analysis model for two-dimensional transient heat transfer problem.

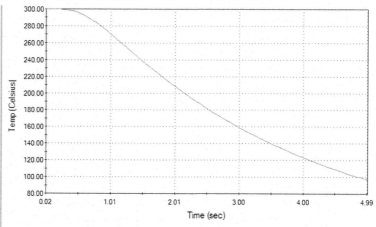

■ **FIGURE 9.21** Change of temperature at center with respect to time.

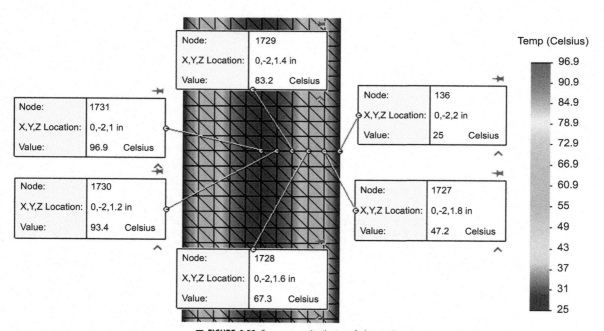

■ **FIGURE 9.22** Temperature distribution of plate in 5 s.

in Fig. 9.18. The temperature distribution of the plate in 5 seconds is illustrated in Fig. 9.22. It shows that the temperature at the center axis of plate is 96.9°C, which is 4% discrepancy. Therefore, both of the estimations from manual calculation and computer simulation are accepted.

9.7 **SUMMARY**

In this article, the relations of heat transfer problems and thermodynamics are discussed; the four laws of thermodynamics are reviewed to explain the physical behaviors in heat transfer problems. Heat transfer behaviors are classified into heat conduction, heat convection, and heat radiation. The corresponding governing equations are presented. The models for 2D heat transfer elements are developed based on the energy conservation. The procedures of FEA modeling for both of steady and transient heat transfer problems are covered. A few of 1D and 2D examples are provided to illustrate computation implementation and FEA modeling for solving steady or transient heat transfer problems.

REFERENCES

Faghri, A., Zhang, Y., Howell, J., 2010. Advanced Heat and Mass Transfer. Global Digital Press, Columbia, MO, ISBN 978-0-9842760-0-4.

Guggenheim, E.A., 1985. Thermodynamics. An Advanced Treatment for Chemists and Physicists, seventh ed. North Holland, Amsterdam, ISBN 0-444-86951-4.

Kittel, C., Kroemer, H., 1980. Thermal Physics, second ed. W. H. Freeman, San Francisco, ISBN 0-7167-1088-9.

Wellons, M., 2007. The Stefan-Boltzmann Law. http://physics.wooster.edu/JrIS/Files/Wellons_Web_Article.pdf.

■ **PROBLEMS**

9.1 Fig. 9.23 shows a fin to transfer the heat from base by convection; the dimensions of lengths are in inch. The temperature of fluid flow is 20°C and the heat transfer coefficient $h = 0.1$ Btu/in^2 s °C. The coefficient of conduction of fin is $k = 3.0$ Btu/in s °C, determine the temperature distribution in the fin.

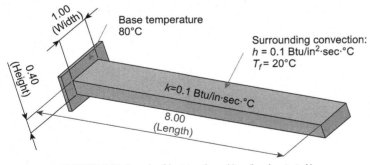

■ **FIGURE 9.23** Example of heat transfer problem (length unit: inch).

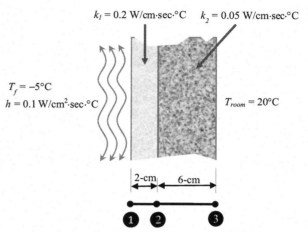

$k_1 = 0.2$ W/cm·sec·°C $k_2 = 0.05$ W/cm·sec·°C

$T_f = -5°C$
$h = 0.1$ W/cm²·sec·°C

$T_{room} = 20°C$

2-cm   6-cm

■ **FIGURE 9.24** One-dimensional heat transfer through wall.

9.2 Fig. 9.24 shows a wall consisting of two materials, i.e., the first layer is 2 cm with the conductivity of $k_1 = 0.2$ W/cm s °C and the second later is 6 cm with the conductivity of $k_2 = 0.05$ W/cm s °C. Other parameters related to the boundary conditions are shown in Fig. 9.24. Use a 2-element simplified model to estimate the distribution of temperature in the wall.

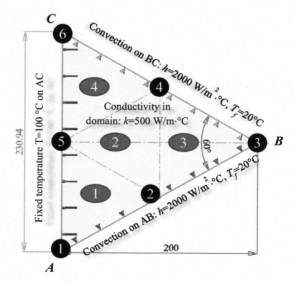

■ **FIGURE 9.25** Triangle-shape heat transfer fin (unit: mm).

9.3 Fig. 9.25 gives the geometry of a fin from a heat sink. The conductivity coefficient of the material $k = 500$ W/m °C. The base temperature on the left edge is given as $T_{base} = 100°$C. The parameters for the convection on lateral surfaces are $h = 2000$ W/m^2 °C and $T_f = 20°$C: (1) use a 2-D steady heat transfer model with four triangle elements and six nodes to estimate the temperature on the tip of the fin; (2) create a thermal analysis model in Solidworks to verify the result of manual calculation.

9.4 A 10-mm thick plate of AISI 304 is heated to 400°C. It is then quenched in water with a temperature of 25°C. Assume that plate surface temperature is fixed as 25°C during quenching. The materials density $\rho = 8000$ kg/m^3, the conductivity $k = 16$ W/(m °C), specific heat $c = 500$ J/(kg °C), Determine the changing curve of temperature at the neutral plane of the plate with respect to time in 5 s.

Applications—Fluid Mechanics

10.1 INTRODUCTION

Fluid mechanics studies the systems with fluid such as liquid or gas under static and dynamics loads. Fluid mechanics is a branch of *continuous mechanics*, in which the kinematics and mechanical behavior of materials are modeled as a continuous mass rather than as discrete particles. The relation of fluid mechanics and continuous mechanics has been discussed by Bar-Meir (2008). In fluid mechanics, the continuous domain does not hold certain shapes and geometry like solids, and in many applications, the density of fluid varies with time and position. Fig. 10.1 has shown some common problems involved in fluid mechanics. In this chapter, we will discuss the FEA modeling of some basic problems including pipe flow, laminate flow, and groundwater flow.

Fluid differs from solid by its reaction to shear stress: the fluid is continuously and permanently deformed under shear stress, whereas a solid object exhibits a small deformation that does not change with time. Liquid cannot return to its original state after deformation. The fluid falls into two categories, i.e., liquid and gas. Gas will occupy the whole volume, whereas liquid usually have a nearly fixed volume. A fluid cannot resist any deformation force, therefore, fluid moves or flows under the action of the force. Its shape will change continuously when the external force is applied. Computer simulation on fluid flows began in the early 1960s with potential flows (Pironneau, 1988). A potential flow describes a velocity field as the gradient of a scalar function and it is characterized by an irrotational velocity field for incompressible fluids. It was then expanded to compressible fluid (Ritchmeyer and Morton, 1967). A fluid flow transfers mass, momentum, and energy in the flow. Fluid movement is described by the conservation equations for mass, momentum, and energy; these equations are commonly referred as the Navier—Stokes equations. The first FEA implementation for the Navier—Stokes equations was introduced in the 1970s (Temam, 1977; Chung, 2002). Nowadays, computational fluid dynamics has been widely applied for resolving different fluid mechanics problems in aerospace, automobile, heat, ventilation, and air-conditioning, oil and gas industry, manufacturing processes, and hydrodynamics.

Finite Element Analysis Applications. http://dx.doi.org/10.1016/B978-0-12-809952-0.00010-8

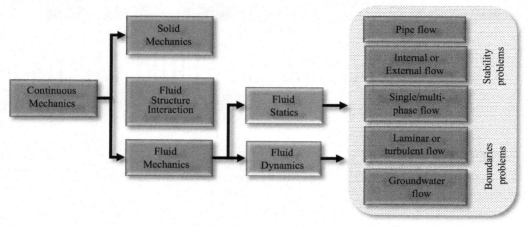

■ **FIGURE 10.1** Fluid mechanics as a branch of continuous mechanics.

10.2 **MATHEMATICAL MODELS**

Fluid flow has its characteristic for its inability to sustain shear stresses at the rest state. Only the hydrostatic stress is possible to be sustained. Therefore, the numerical simulation on fluids focuses on the flow motion, which differs from the displacements of solids in solid mechanics. In other words, the primary important variables for a fluid flow are the distributions of flow velocities v in a continuous domain. At a given position of a three-dimensional (3D) continuous space, the velocity of a fluid flow is expressed as (Zienkiewicz and Taylor, 2000),

$$v = u \cdot \boldsymbol{i} + v \cdot \boldsymbol{j} + w \cdot \boldsymbol{k} \tag{10.1}$$

where

v is the vector of the fluid flow velocity,
u, v, and w are the components of velocity along x-, y-, and z-axes, respectively.

The rates of strain are the main sources of the general stresses σ_{ij}. Similar to the definition of shear stress in solid mechanics, the rates of strain are defined based on given fluid velocities as,

$$\dot{\varepsilon}_{xy} = \frac{1}{2}\left(\frac{\partial u}{\partial y} + \frac{\partial v}{\partial x}\right) \quad \dot{\varepsilon}_{yz} = \frac{1}{2}\left(\frac{\partial u}{\partial z} + \frac{\partial w}{\partial y}\right) \quad \dot{\varepsilon}_{xz} = \frac{1}{2}\left(\frac{\partial u}{\partial z} + \frac{\partial w}{\partial x}\right) \tag{10.2}$$

where u, v, and w are given in Eq. (10.1).

The relation in Eq. (10.2) is known as tonsorial definition of strain rates. Conventionally, Eq. (10.2) can be rewritten into a matrix form as,

$$\dot{\boldsymbol{\varepsilon}}^T = [\begin{matrix}\dot{\varepsilon}_{11} & \dot{\varepsilon}_{22} & \dot{\varepsilon}_{33} & 2\dot{\varepsilon}_{12} & 2\dot{\varepsilon}_{23} & 2\dot{\varepsilon}_{31}\end{matrix}]^T = [S][v] \qquad (10.3)$$

where

$[v] = [\begin{matrix}u & v & w\end{matrix}]^T$ is given in Eq. (10.1)

the subscripts 1, 2, and 3 represent x, y, and z axes, respectively, and the matrix $[S]$ is the strain rate operator and determined by Eq. (10.2) as,

$$[S] = \begin{bmatrix} \frac{\partial}{\partial x} & 0 & 0 \\ 0 & \frac{\partial}{\partial y} & 0 \\ 0 & 0 & \frac{\partial}{\partial z} \\ \frac{1}{2}\frac{\partial}{\partial y} & \frac{1}{2}\frac{\partial}{\partial x} & 0 \\ 0 & \frac{1}{2}\frac{\partial}{\partial z} & \frac{1}{2}\frac{\partial}{\partial y} \\ \frac{1}{2}\frac{\partial}{\partial z} & 0 & \frac{1}{2}\frac{\partial}{\partial x} \end{bmatrix} \qquad (10.4)$$

To calculate the stress from the state of strain rates in a linear isotropic fluid, two constants should be defined. *The first one* is called deviatoric stresses T_{ij}, which relates to the deviatoric strain rates as,

$$\tau_{ij} = \sigma_{ij} - \delta_{ij}\frac{\sigma_{kk}}{3} = 2\mu\left(\dot{\varepsilon}_{ij} - \delta_{ij}\frac{\dot{\varepsilon}_{kk}}{3}\right) \qquad (10.5)$$

where

the term $\left(\dot{\varepsilon}_{ij} - \delta_{ij}\frac{\dot{\varepsilon}_{kk}}{3}\right)$ in Eq. (10.5) is called *deviatoric stress*.

δ_{ij} is the *Kronecker delta*,

the subscripts 1, 2, and 3 represent x, y, and z axes, respectively,

the coefficient μ refers to the dynamic viscosity or simple viscosity, and

a repeated index represents a summation as,

$$\left.\begin{matrix}\sigma_{ii} = \sigma_{11} + \sigma_{22} + \sigma_{33} \\ \dot{\varepsilon}_{ii} = \dot{\varepsilon}_{11} + \dot{\varepsilon}_{22} + \dot{\varepsilon}_{33}\end{matrix}\right\} \qquad (10.6)$$

The relation of the strain rate and stress is equivariant to the relation of strain and stress by shear module G in the linear elasticity.

The second one is the pressure in a fluid flow, which relates to the strain rate. This pressure or the mean stress can be calculated from the volumetric strain rates as,

$$p = \frac{\sigma_{ii}}{3} = -\kappa \dot{\varepsilon}_{ii} + p_0 \tag{10.7}$$

where

κ is a volumetric viscosity coefficient, which is a quantity equivalent to the bulk modulus κ in the linear elasticity;

p_0 is the initial hydrostatic pressure and it is independent of the strain rate.

The constitutive model for the relation of stress and strain can be derived from Eqs. (10.5) and (10.7) as,

$$\sigma_{ij} = \tau_{ij} - \delta_{ij}p$$
$$= 2\mu \left(\dot{\varepsilon}_{ij} - \delta_{ij} \frac{\dot{\varepsilon}_{kk}}{3} \right) + \delta_{ij} \frac{\sigma_{kk}}{3} \tag{10.8}$$
$$= 2\mu \dot{\varepsilon}_{ij} + \delta_{ij} \left(\kappa - \frac{2}{3}\mu \right) \dot{\varepsilon}_{ii} + \delta_{ij}p_0$$

For a further simplification, little evidence has been found on the existence of volumetric viscosity. Therefore, we can assume that $\kappa \dot{\varepsilon}_{ii} = 0$ and the constitutive model for the fluid flow can be summarized as,

$$\tau_{ij} = 2\mu \left(\dot{\varepsilon}_{ij} - \delta_{ij} \frac{\dot{\varepsilon}_{kk}}{3} \right) = \mu \left(\left(\frac{\partial v_i}{\partial x_j} + \frac{\partial v_j}{\partial x_i} \right) - \delta_{ij} \frac{\partial v_k}{\partial x_k} \right) \tag{10.9}$$

and

$$\sigma_{ij} = \tau_{ij} - \delta_{ij}p \tag{10.10}$$

The constitutive model for the fluid flow with Eqs. (10.9) and (10.10) has shown the similarity to those of a solid object. Note that the aforementioned model treats the coefficient μ as constant; a fluid flow with such a characteristic is called a *Newtonian fluid flow*. If the coefficient μ varies and depends on the strain rate, the corresponding fluid flow is called as *nonlinear* or *non-Newtonian* fluid flow.

10.2.1 **Mass conservation equation**

The matter is conserved; it can be not created or vanished. However, fluids and solid objects respond to external force differently. The mass of a solid object is sustained in given shapes and volume of body, whereas a fluid does not have the shape and it deforms permanently under an external force.

Rather than to track the positions of individual particles of fluid, it is more convenient to track the amount of fluid mass by the flow rate passing a given section of flow. Therefore, the continuity equation of fluids usually deals with the mass flux passing a section of the flow. As shown in Fig. 10.2, if a fluid flow is internally in a pipe, the rate of mass flow at two cross-sections must be the same, i.e.,

$$\dot{m}_1 = \dot{m}_2 = \rho_1 v_1 A_1 = \rho_2 v_2 A_2 \tag{10.11}$$

where

$\dot{m}_1$ and $\dot{m}_2$ are the mass rates,
ρ_1 and ρ_2 are the densities,
v_1 and v_2 are the velocities of flow, and
A_1 and A_2 are the areas of two cross-sections, respectively.

In an open space, the state of the mass flow at a given position is described in Fig. 10.3. An infinitesimal control volume with $dx \times dy \times dz$ is considered, the velocity of flow is represented by (u, v, w), the density of flow is denoted by ρ, $d\rho/dt$ is the rate of density change, and a source of mass at the given position is $\dot{S}_m$.

The condition of the mass conservation is written as,

$$\dot{m}_{\text{in}} - \dot{m}_{\text{out}} = \frac{dm}{dt} \tag{10.12}$$

Expanding Eq. (10.12) for the infinitesimal volume in Fig. 10.3 yields,

$$\rho u dy dz + \rho v dx dz + \rho w dx dy -$$
$$\left(\left(\rho u + \frac{\partial(\rho u)}{\partial x} dx \right) dy dz + \left(\rho v + \frac{\partial(\rho v)}{\partial y} dy \right) dx dz + \left(\rho w + \frac{\partial(\rho w)}{\partial z} dz \right) dx dy \right)$$
$$+ \frac{d\rho}{dt} dx dy dz = \dot{S}_m dx dy dz$$

$$\tag{10.13}$$

■ **FIGURE 10.2** Mass conservation in an internal fluid flow.

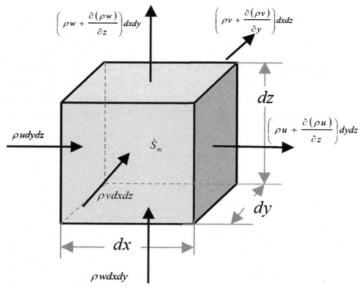

■ **FIGURE 10.3** The state of fluid flow where u, v, and w represent the velocity components on x-, y-, and z- axes.

Eq. (10.13) can be simplified as,

$$\nabla \cdot (\rho v) + \frac{d\rho}{dt} = \dot{S}_m \qquad (10.14)$$

where

$\nabla \cdot (\rho v) = \frac{\partial(\rho u)}{\partial x} + \frac{\partial(\rho v)}{\partial y} + \frac{\partial(\rho w)}{\partial z}$ is the divergence of a vector quantity, which corresponds to an inner product between ∇ and a vector, and $\dot{S}_m$ is the mass generate rate per volume.

In fluid mechanics, the equations for the mass conservation are referred to as the continuity equations, which have been described in a variety of forms.

10.2.2 **Momentum conservation equation**

Energy cannot be created or vanished, it can only be converted from one form to another, and the momentum is only changed by force acting through time. In the fluid flow, energy exists in three different forms (Acharya, 2016).

The first form is *the potential energy* (PE) related to the fluid height,

$$PE = W \cdot H \qquad (10.15)$$

where

PE is the amount of potential energy,

W and H are the weight and height of fluid, respectively.

The second form is *the kinetic energy* (KE) relates to the energy for fluid movement,

$$KE = W \frac{v^2}{2g} \tag{10.16}$$

The third form of energy relates to flow pressure (FP),

$$FP = \frac{W \cdot p}{\gamma} = \frac{W \cdot p}{\rho \cdot g} \tag{10.17}$$

where γ is the weight of per unit volume.

Adding three forms of energy in Eqs. (10.15−10.17) for the energy conservation at the given position yields,

$$E \equiv PE + KE + FP \equiv W \cdot \left(H + \frac{v^2}{2g} + \frac{p}{\gamma} \right) \tag{10.18}$$

If the flow is in a steady state, the difference of the summed energy at any two points of a streamline will be the energy to overcome the friction flowing from one position to another. This leads to the well-known Bernoulli's equation,

$$gH_1 + \frac{v_1^2}{2g} + \frac{p_1}{\gamma} = gH_2 + \frac{v_2^2}{2g} + \frac{p_2}{\gamma} + h_L \tag{10.19}$$

where

H_1 and H_2 are the elevation heads,

p_1 and p_2 are the pressure heads, and

$\frac{v_1^2}{2g}$ and $\frac{v_2^2}{2g}$ are velocity heads at two points of the streamline, respectively.

In Eq. (10.19), h_L is the head loss by friction, fittings, bends, and valves, which can be evaluated as (Acharya, 2016),

$$\left. \begin{array}{l} h_L = k \dfrac{v^2}{2g} \quad \text{(minor loss from values, fitting, and bents)} \\[2em] h_L = f \dfrac{L}{D} \dfrac{v^2}{2g} \quad \text{(major loss from friction and pump)} \end{array} \right\} \tag{10.20}$$

where

f is friction factor,

L is pipe length,

v is average velocity at cross-section, and

g is gravity acceleration.

In general, for a fluid follow with a controlled volume, the momentum in the jth direction is balanced, i.e., the terms of $(\rho v_j)v_i$ leaving or entering a controlled volume are in equilibrium with the stresses σ_{ij} and body forces ρf_i,

$$\frac{\partial(\rho v_j)}{\partial t} + \frac{\partial[(\rho v_j)v_i]}{\partial x_j} - \frac{\partial[\sigma_{ij}]}{\partial x_j} - \rho f_j = 0 \tag{10.21}$$

Substituting Eq. (10.10) into (10.21) yields,

$$\frac{\partial(\rho v_j)}{\partial t} + \frac{\partial[(\rho v_j)v_i]}{\partial x_j} - \frac{\partial[\tau_{ij}]}{\partial x_j} + \frac{\partial p}{\partial x_j} - \rho f_j = 0 \tag{10.22}$$

where

$\frac{\partial[\tau_{ij}]}{\partial x_j}$ is the force caused by the viscosity, and

τ_{ij} is defined in Eq. (10.9).

Using u, v, and w to replace v_i ($i = 1,2,3$) in Eq. (10.22) gets three equations for the momentum balances in three axes as,

$$\left.\begin{array}{l}\dfrac{\partial(\rho u)}{\partial t} + \dfrac{\partial(\rho u^2)}{\partial x} + \dfrac{\partial(\rho uv)}{\partial y} + \dfrac{\partial(\rho uw)}{\partial z} - \dfrac{\partial\tau_{xx}}{\partial x} - \dfrac{\partial\tau_{xy}}{\partial y} - \dfrac{\partial\tau_{xz}}{\partial z} + \dfrac{\partial p}{\partial x} - \rho f_x = 0 \\[2ex] \dfrac{\partial(\rho v)}{\partial t} + \dfrac{\partial(\rho uv)}{\partial x} + \dfrac{\partial(\rho v^2)}{\partial y} + \dfrac{\partial(\rho vw)}{\partial z} - \dfrac{\partial\tau_{xy}}{\partial x} - \dfrac{\partial\tau_{yy}}{\partial y} - \dfrac{\partial\tau_{yz}}{\partial z} + \dfrac{\partial p}{\partial y} - \rho f_y = 0 \\[2ex] \dfrac{\partial(\rho w)}{\partial t} + \dfrac{\partial(\rho uw)}{\partial x} + \dfrac{\partial(\rho vw)}{\partial y} + \dfrac{\partial(\rho w^2)}{\partial z} - \dfrac{\partial\tau_{xz}}{\partial x} - \dfrac{\partial\tau_{yz}}{\partial y} - \dfrac{\partial\tau_{zz}}{\partial z} + \dfrac{\partial p}{\partial z} - \rho f_z = 0\end{array}\right\} \tag{10.23}$$

10.3 ENERGY CONSERVATION MODEL

Eq. (10.23) works in an isothermal condition where the thermal energy is constant. If the temperature in the fluid domain varies, the corresponding change of thermal energy has to be considered in energy equilibrium.

The governing equations in Sections 10.1 and 10.2 show that three important variables in a fluid flow are *velocity* (v), *pressure* (p), and *density* (ρ). Other parameters, such as the deviatoric stresses, are dependent variables relating to the velocity, pressure, and the density. In general cases, the number of independent variables is more than the number of differential equations to find unique solutions. It is required to find another governing equation based on the energy conservation. The energy conservation equation provides additional information for the system response of fluid flow.

Generally, the density (ρ) is dependent on the pressure (p) and temperature (T) by the equation of state as,

$$\rho = \rho(p, T) \tag{10.24}$$

Taking an example of ideal gas, Eq. (10.24) can be written explicitly as,

$$\rho = \frac{p}{R \cdot T} \tag{10.25}$$

where R is the universal gas constant.

To develop an energy conservation equation, a few of energy related quantities have to be defined. The first one is called *intrinsic energy e* over per unit mass,

$$e = e(p, T) \tag{10.26}$$

where e is intrinsic energy determined by the state of the fluid flow, i.e., pressure (p) and temperature (T).

Besides intrinsic energy, the fluid also has the KE related to motion. Therefore, the total energy per unit mass (E) is a sum of two energy forms as,

$$E = e(p, T) + \frac{1}{2} v_i v_i \tag{10.27}$$

where

E is total energy per unit mass,
$e(p, T)$ is the intrinsic energy per unit mass, and
v_i ($i = 1, 2, 3$) corresponding to u, v, and w in Fig. 10.3 for the velocity of the fluid flow.

The second one is *the enthalpy*, which is defined as a thermodynamic quantity equivalent to the total heat content of a system. For per unit mass, it relates to internal energy, the pressure, and density as,

$$H = E + \frac{p}{\rho} \tag{10.28}$$

The third one is the substantial derivative of total energy $\frac{D(\rho E)}{Dt}$ (Tryggvason, 2016); it can be obtained as,

$$\frac{D(\rho E)}{Dt} = \frac{\partial}{\partial t}(\rho E) + \frac{\partial}{\partial x_i}(\rho v_i E) = E\left(\frac{\partial \rho}{\partial t} + \frac{\partial \rho v_i}{\partial x_i}\right) + \left(\frac{\partial \rho E}{\partial t} + \frac{\partial E v_i}{\partial x_i}\right) \tag{10.29}$$

Assume there is a mass source $\dot{S}_m = 0$ internally. Substituting Eq. (10.14) into (10.29) yields,

$$\frac{D(\rho E)}{Dt} = \frac{\partial}{\partial t}(\rho E) + \frac{\partial}{\partial x_i}(\rho v_i E) = \left(\frac{\partial \rho E}{\partial t} + \frac{\partial E v_i}{\partial x_i}\right) \tag{10.30}$$

In addition, energy can be transferred by convection, conduction, and radiation. The convection and radiation occur only on boundaries, and the conduction heat flux q_i for an isotropic material can be defined as,

$$q_i = -k\frac{\partial T}{\partial x_i} \tag{10.31}$$

where

    k is the thermal conductivity, and
    x_i ($i = 1, 2, 3$) correspond to x, y, z axis, respectively.

The energy is dissipated due to internal stresses. This amount of energy can be quantified by using the quantities in Eqs. (10.9) and (10.10) as,

$$\frac{\partial}{\partial x_i}(\sigma_{ij}v_j) = \frac{\partial}{\partial x_i}(\tau_{ij}v_j) - \frac{\partial}{\partial x_j}(pv_j) \tag{10.32}$$

Therefore, the energy conservation in an infinitesimal control volume can be written as,

$$\frac{\partial(\rho E)}{\partial t} + \frac{\partial}{\partial x_i}(\rho v_i E) - \frac{\partial}{\partial x_i}\left(k\frac{\partial T}{\partial x_i}\right) + \frac{\partial}{\partial x_i}(pv_i) - \frac{\partial}{\partial x_i}(\tau_{ij}v_j) - \rho g_i v_i - q_H = 0$$

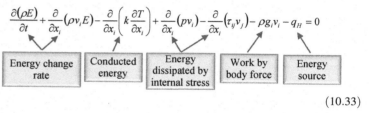

| Energy change rate | Conducted energy | Energy dissipated by internal stress | Work by body force | Energy source |

$$\tag{10.33}$$

Using the concept of the enthalpy (H) in Eqs. (10.28) and (10.33) can be simplified as,

$$\frac{\partial(\rho E)}{\partial t} + \frac{\partial}{\partial x_i}(\rho v_i H) - \frac{\partial}{\partial x_i}\left(k\frac{\partial T}{\partial x_i}\right) - \frac{\partial}{\partial x_i}(\tau_{ij}v_j) - \rho g_i v_i - q_H = 0 \tag{10.34}$$

In several special cases, not all of mass, momentum, and energy conservations have to be solved simultaneously to find the solutions to fluid mechanics problems. In the following sections, we will introduce three of these cases where either the incompressible fluids is considered (the density becomes constant) or there is some dependence between pressure, velocity, and pressure such as isothermal flow with limited compressibility.

10.4 PIPE NETWORK

A fluid flow in a pipe can be *laminar*, *transient*, or *turbulent*. The flow type depends on the Reynolds number, which is define as,

$$R_e = \frac{\text{Interntial forces}}{\text{Viscous forces}} = \frac{v_{avg}D}{v} = \frac{\rho v_{avg}D}{\mu} \tag{10.35}$$

where

v_{avg} is the average flow velocity (m/s),

D is the characteristic length of pipe geometry (m),

$v = \mu/\rho$ is the kinematic viscosity of the fluid (m²/s). It also called as *viscous diffusivity* or *diffusivity for momentum*, and

R_e is the dimensionless Reynolds number.

In most practical applications, the flow types relate to the quantities of their Reynold's numbers as,

$$\left. \begin{array}{ll} R_e \leq 2300 & \text{laminar flow} \\ 2300 < R_e \leq 4000 & \text{transient flow} \\ 4000 < R_e & \text{turbulant flow} \end{array} \right\} \tag{10.36}$$

A pipe element with a steady laminar flow is shown in Fig. 10.4. The pipe element has the given diameter D and length L, and it is at the same elevation, i.e., $H_1 = H_2$ in Eq. (10.19), and the mass conservation Eq. (10.11) leads to $v_1 = v_2$. Therefore, the moment conservation Eq. (10.19) can be simplified as,

$$\frac{p_1 - p_2}{\gamma} = \frac{p_1 - p_2}{\rho g} = h_L \tag{10.37}$$

For a fully developed laminar flow in a circular pipe, the friction coefficient can be obtained by,

$$f = \frac{64}{R_e} = \frac{64\mu}{\rho D v} \tag{10.38}$$

Taking the major friction loss in Eq. (10.20) and substituting Eq. (10.38) into (10.37) gets,

$$p_1 - p_2 = f\rho \frac{Lv^2}{2D} = \frac{32\mu Lv}{D^2} = \frac{128\mu LQ}{\pi D^4} \tag{10.39}$$

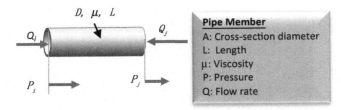

■ **FIGURE 10.4** One-dimensional element for pipe flow.

Based on Eq. (10.39), the volume rate of flow can be calculated from pressure loss as,

$$Q = \frac{\pi D^4}{128\mu L}(p_1 - p_2) \tag{10.40}$$

In developing an FEA model for a pipe network with steady fluid flows, the direct formulation method can be applied. By mapping the fluid volume rate Q as force, and the flow pressure P as displacement, the model of a pipe element can be derived directly by treating it as an equivalent spring element as,

$$\begin{bmatrix} C & -C \\ -C & C \end{bmatrix} \begin{Bmatrix} p_i \\ p_j \end{Bmatrix} = \begin{Bmatrix} Q_i \\ Q_j \end{Bmatrix} \tag{10.41}$$

where C is an equivariant stiffness coefficient from Eq. (10.40), i.e.,

$$C = \frac{\pi D^4}{128\mu L} \tag{10.42}$$

Example 10.1

A water supply system is shown in Fig. 10.5. It consists of one inlet and two outlets; the inlet has the volume rate of 1.0×10^{-3} (m³/s), and both the outlets are under the atmosphere pressure. $L_1 = 10$ m, $L_2 = 50$ m, $D_1 = 0.1$ m, $D_2 = 0.075$ m, and $D_3 = 0.06$ m. The viscosity of fluid is known as 0.3. Find the pressure at node 1 and the volume rates at nodes 5 and 7.

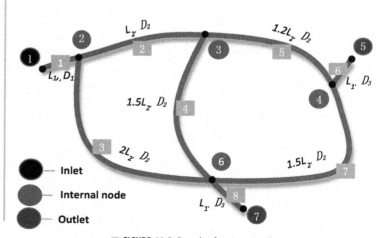

■ **FIGURE 10.5** Example of a pipe network.

Solution

The FEA model includes eight elements, and their corresponding nodes, system parameters, and boundary conditions (BCs) are given in Table 10.1.

Table 10.1 The Elements and Nodes in Finite Element Analysis Model

Element	Node i	Node j	L_i (m)	D_i (m)	Q_i (m³/s)	P_i (Pa)
1	1	2	10	0.1	$1.0\,e^{-3}$	—
2	2	3	50	0.075	—	—
3	2	6	100	0.075	—	—
4	3	6	75	0.075	—	—
5	3	4	60	0.075	—	10,325
6	4	5	10	0.06	—	—
7	4	6	75	0.075	—	10,325
8	6	7	10	0.06	—	—

The stiffness matrices for these elements are defined as follows,

$$[K^{(1)}] = \begin{bmatrix} 8.1812e\text{-}07 & -8.1812e\text{-}07 \\ -8.1812e\text{-}07 & 8.1812e\text{-}07 \end{bmatrix} \begin{matrix} 1 \\ 2 \end{matrix}$$

$$[K^{(5)}] = \begin{bmatrix} 4.3143e\text{-}08 & -4.3143e\text{-}08 \\ -4.3143e\text{-}08 & 4.3143e\text{-}08 \end{bmatrix} \begin{matrix} 3 \\ 4 \end{matrix}$$

$$[K^{(2)}] = \begin{bmatrix} 5.1772e\text{-}08 & -5.1772e\text{-}08 \\ -5.1772e\text{-}08 & 5.1772e\text{-}08 \end{bmatrix} \begin{matrix} 2 \\ 3 \end{matrix}$$

$$[K^{(6)}] = \begin{bmatrix} 1.0603e\text{-}07 & -1.0603e\text{-}07 \\ -1.0603e\text{-}07 & 1.0603e\text{-}07 \end{bmatrix} \begin{matrix} 4 \\ 5 \end{matrix}$$

$$[K^{(3)}] = \begin{bmatrix} 2.5886e\text{-}08 & -2.5886e\text{-}08 \\ -2.5886e\text{-}08 & 2.5886e\text{-}08 \end{bmatrix} \begin{matrix} 2 \\ 6 \end{matrix}$$

$$[K^{(7)}] = \begin{bmatrix} 3.4515e\text{-}08 & -3.4515e\text{-}08 \\ -3.4515e\text{-}08 & 3.4515e\text{-}08 \end{bmatrix} \begin{matrix} 4 \\ 6 \end{matrix}$$

$$[K^{(4)}] = \begin{bmatrix} 3.4515e\text{-}08 & -3.4515e\text{-}08 \\ -3.4515e\text{-}08 & 3.4515e\text{-}08 \end{bmatrix} \begin{matrix} 3 \\ 6 \end{matrix}$$

$$[K^{(8)}] = \begin{bmatrix} 1.0603e\text{-}07 & -1.0603e\text{-}07 \\ -1.0603e\text{-}07 & 1.0603e\text{-}07 \end{bmatrix} \begin{matrix} 6 \\ 7 \end{matrix}$$

The system model assembled from the above element models becomes,

$$[K] = 1.0e-6$$

$$\times \begin{bmatrix} 0.8181 & -0.8181 & 0 & 0 & 0 & 0 & 0 \\ -0.8181 & 0.8958 & -0.0518 & 0 & 0 & -0.0259 & 0 \\ 0 & -0.0518 & 0.1294 & -0.0431 & 0 & -0.0345 & 0 \\ 0 & 0 & -0.0431 & 0.1837 & -0.1060 & -0.0345 & 0 \\ 0 & 0 & 0 & -0.1060 & 0.1060 & 0 & 0 \\ 0 & -0.0259 & -0.0345 & -0.0345 & 0 & 0.2009 & -0.1060 \\ 0 & 0 & 0 & 0 & 0 & -0.1060 & 0.1060 \end{bmatrix}$$

Solving the system model obtains the volume rates and pressures at seven nodes as

$$[Q] = \begin{bmatrix} 0.001 & 0 & 0 & 0 & -0.0004 & 0 & -0.0006 \end{bmatrix}^T$$

$$[p] = 1.0e4 \begin{bmatrix} 3.4197 & 3.2975 & 2.2170 & 1.4179 & 1.0350 & 1.5952 & 1.0350 \end{bmatrix}^T$$

10.5 2D INCOMPRESSIBLE AND IRROTATIONAL FLOW

In this section, the fluid in a flow is assumed to be incompressible and the fluid density is constant. For an incompressible fluid flow ($\rho = $ constant) without a mass source $(\dot{S}_m)$, the mass conservation Eq. (10.14) can be rewritten as,

$$\nabla \cdot v = \frac{\partial u}{\partial x} + \frac{\partial v}{\partial y} + \frac{\partial w}{\partial z} = 0 \tag{10.43}$$

In 2D case, Eq. (10.43) can be reduced as,

$$\nabla \cdot v = \frac{\partial u}{\partial x} + \frac{\partial v}{\partial y} = 0 \tag{10.44}$$

For a steady flow, we introduce the *stream function* $\psi(x, y)$ and let,

$$\left. \begin{array}{c} u = \dfrac{\partial \psi(x, y)}{\partial y} \\[2mm] v = -\dfrac{\partial \psi(x, y)}{\partial x} \end{array} \right\} \tag{10.45}$$

A quick check by substituting Eq. (10.45) into (10.44) finds that the definition of the stream function satisfies the condition of the mass conservation.

Furthermore, a fluid flow is considered as an irrotational flow when it has a zero-angular velocity. When the viscosity is very small and the velocity of the fluid flow is low, the fluid flow can be simplified as an irrotational flow and the condition of zero angular velocity is given as,

$$\frac{\partial v}{\partial x} - \frac{\partial u}{\partial y} = 0 \tag{10.46}$$

Substituting Eq. (10.45) into Eq. (10.46) generates the Laplace equation of 2D incompressible irrotational flow as,

$$\frac{\partial^2 \psi}{\partial x^2} + \frac{\partial^2 \psi}{\partial y^2} = 0 \tag{10.47}$$

where $\psi(x, y)$ is the stream function of a fluid flow.

The differential Eq. (10.47) includes the stream function only. Therefore, it is completely solvable subjected to the sufficient BCs.

Other than stream function $\psi(x, y)$, another relevant concept is *potential function* $\phi(x, y)$. It is also commonly used to describe an incompressible irrotational flow, and it is defined as,

$$\left.\begin{array}{l} u = \dfrac{\partial \phi(x, y)}{\partial x} \\[3mm] v = -\dfrac{\partial \phi(x, y)}{\partial y} \end{array}\right\} \tag{10.48}$$

The definition of the potential function satisfies the condition of irrotational flow in Eq. (10.46) naturally. By substituting Eq. (10.48) into the mass conservation equation in Eq. (10.44), the governing equation for an incompressible irrotational flow is found as,

$$\frac{\partial^2 \phi}{\partial x^2} + \frac{\partial^2 \phi}{\partial y^2} = 0 \tag{10.49}$$

where $\phi(x, y)$ is the potential function defined in Eq. (10.48).

Note that the state valuable in Eq. (10.49) becomes $\phi(x, y)$ instead of the stream function $\psi(x, y)$ in Eq. (10.47). Two equations have the same differential terms because they represent for the same physical problem.

In Section 4.6, the element models for the generic expression of a Poisson equation have been developed. These models can be directly used to model an incompressible irrotational fluid flow here. The geometry and state variables of rectangle and triangular elements are shown in Fig. 10.6A and B, respectively. The stream function $\psi(x, y)$ is used as state variable in the Poisson equation.

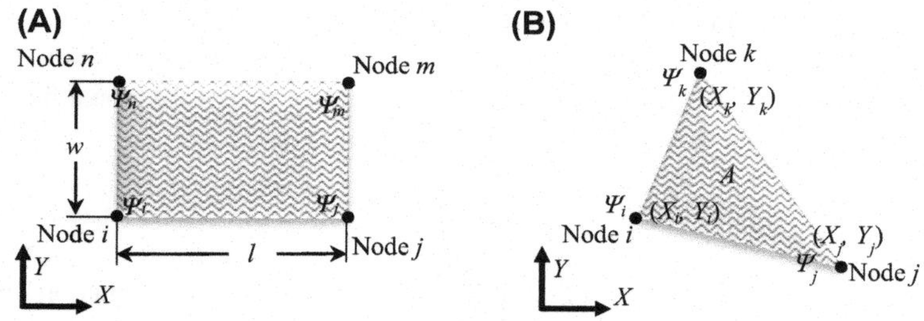

(A)

(B)

■ FIGURE 10.6 Elements in a domain of incompressible irrotational fluid flow: (A) rectangle element; (B) triangular element.

Let $c_x = c_y = 1$ in Eqs. (4.46) and (4.47), the model for a rectangle element in an incompressible irrotational fluid flow becomes,

$$[K] = [K_x] + [K_y]$$

$$= \frac{w}{6l} \begin{bmatrix} 2 & -2 & -1 & 1 \\ -2 & 2 & 1 & -1 \\ -1 & 1 & 2 & -2 \\ 1 & -1 & -2 & 2 \end{bmatrix} + \frac{l}{6w} \begin{bmatrix} 2 & 1 & -1 & -2 \\ 1 & 2 & -2 & -1 \\ -1 & -2 & 2 & 1 \\ -2 & -1 & 1 & 2 \end{bmatrix} \quad (10.50)$$

Let $c_x = c_y = 1$ in Eqs. (4.51) and (4.52), the model for a triangular element in an incompressible irrotational fluid flow becomes,

$$[K] = [K_x] + [K_y] = \frac{1}{4A} \begin{bmatrix} \beta_i^2 & \beta_i\beta_j & \beta_i\beta_k \\ \beta_j\beta_i & \beta_j^2 & \beta_j\beta_k \\ \beta_k\beta_i & \beta_k\beta_j & \beta_k^2 \end{bmatrix} + \frac{1}{4A} \begin{bmatrix} \delta_i^2 & \delta_i\delta_j & \delta_i\delta_k \\ \delta_j\delta_i & \delta_j^2 & \delta_j\delta_k \\ \delta_k\delta_i & \delta_k\delta_j & \delta_k^2 \end{bmatrix}$$

$$(10.51)$$

If the potential function is used as a state variable, the BC of an incompressible irrotational flow are known as *free-stream velocity*. At the solid surface boundary, the fluid cannot have velocity normal to the surface. This is the boundary can be modeled as ideal wall condition.

Example 10.2

A 2D domain with an incompressible irrotational fluid flow is shown in Fig. 10.7. The length is 2 m and the height is 1 m, and there is an obstacle in the middle with a diameter of 0.5 m. Assume that the reference atmosphere pressure at the outlet is and the fluid speed at the inlet is 1.0 m/s. Determine the distribution of the velocity in the fluid domain.

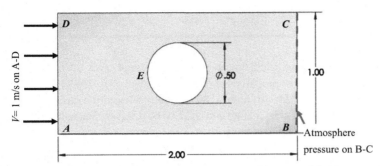

■ **FIGURE 10.7** A two-dimensional domain of an incompressible irrotational fluid flow.

Solution

The flow simulation in SolidWorks is used to model the fluid flow. As shown in Fig. 10.8, the BC for given stream function on BC is treated as fixed inlet velocity, and the flow rate on A—D is treated at given environmental pressure.

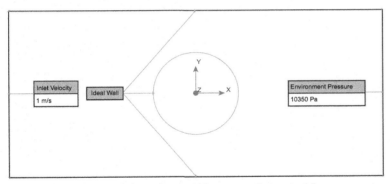

■ **FIGURE 10.8** A flow analysis model for incompressible irrotational flow.

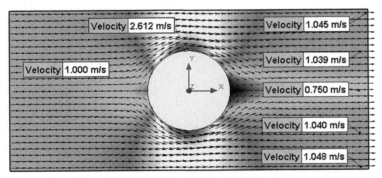

■ **FIGURE 10.9** The distribution of fluid velocity.

After the simulation, the distribution of velocity in the given continuous domain is shown in Fig. 10.9. Note that the velocity in a fluid flow is treated as the heat flux in a thermal analysis.

Example 10.3

Fig. 10.10 shows a fluid flow in a cavity, the size of cavity is $L = W = 0.1$ m, and the material properties of fluid are the density $\rho = 997.13$ kg/m³, the viscosity $\mu = 0.000891$ kg/m s. Among four sides of the cavity, only one side (i.e., the lid side) is open and is driven by a passing fluid with a constant velocity. The fluid flow in the cavity can be represented as incompressible irrotational flow governed by mass and momentum conservation equations. Assume that the Reynolds number, R_e for the fluid flow at the lid is given as 100, create an FEA model and find the distribution of velocity of fluid flow in the cavity for the problem shown in Fig. 10.10.

Solution

For the simplification, the flow is assumed to be two-dimensional, laminar, incompressible, and Newtonian. As shown in Fig. 10.11, the fluid velocity U at the lid is one BC; it can be calculated based on the given Reynolds number (R_e) 100.

Reynold's number (R_e) is the ratio of inertial to viscous forces. It influences the fluid flow in the cavity, and it relates to the flow velocity at the lid as (Omari, 2013),

$$R_e = \frac{\rho U L}{\mu} \tag{10.52}$$

Based on the given properties of fluids, the fluid velocity at the lid can be calculated from Eq. (10.52) as

$$U = \frac{R_e \mu}{\rho L} = \frac{(100)(0.000891)}{(997.13)(0.1)} = 8.9356 \times 10^{-4} \text{ m/s}$$

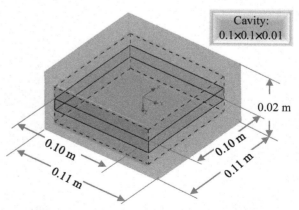

■ **FIGURE 10.10** Geometric models for simulation of a cavity flow.

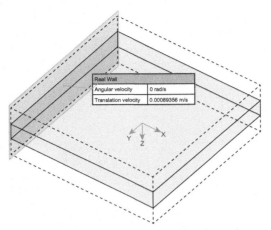

■ **FIGURE 10.11** Computational domain and boundary condition.

An FEA model is created in the flow simulation of the SolidWorks. The computational domain is made with the given size. 2D simplification is activated to reduce the computation. The simulation has shown the flow trajectory in Fig. 10.12A and the velocity distribution in Fig. 10.12B, which are aligned well with the results in the literature (Omari, 2013).

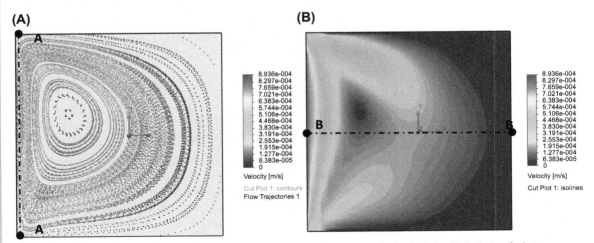

■ **FIGURE 10.12** Flow trajectory and velocity distribution of cavity. (A) Flow trajectory with colored velocity, (B) distribution of velocity.

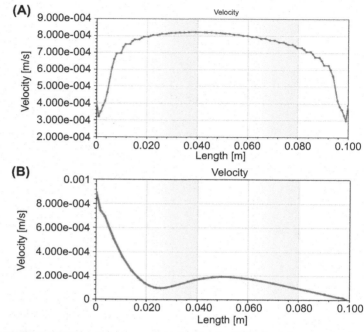

■ **FIGURE 10.13** The velocity changes along reference paths. (A) The flow velocity near the edge (A—A) with specified velocity. (B) The flow velocity in the middle plane (B—B) of the cavity flow.

The flow simulation in the SolidWorks includes many tools for users to retrieve and visualize data. For example, x–y plots can be generated to visualize the change of any quantity along a trajectory. Fig. 10.13 provides the plots for the changes of flow velocities along the paths A—A and B—B.

10.6 GROUNDWATER PROBLEMS

Groundwater is water that is found underground in cracks and spaces in the soil, sand, and rocks; an area where water fills these spaces is called a *phreatic zone* or *saturated zone*. Groundwater flow is a kind of stream flow that has infiltrated the ground. Groundwater problems related to oil-recovery technique, hydrology, solar storage, and geothermal energy (Moaveni, 2015). Groundwater flow through porous medium is governed by Darcy's law. The Darcy' law is a constitutive equation that describes the flow of a fluid through a porous medium. The equation was developed based on experiments on water flow through beds of sand. In 1D case, the velocity of fluid flow depends on the pressure drop by,

$$v = \frac{k}{\mu}\frac{dp}{dx} \tag{10.53}$$

where

v is the velocity of fluid flow,
k is the permeability of the porous medium,
μ is viscosity of fluid, and
dp/dx is the pressure gradient. For a fluid flow in 2D domain, the hydraulic head ϕ can be used to define the fluid velocity.

$$\left.\begin{aligned} v_x &= -k_x\frac{d\phi}{dx} \\ v_y &= -k_y\frac{d\phi}{dy} \end{aligned}\right\} \tag{10.54}$$

where

v_x and v_y are the components of velocity along x- and y-axis, and
k_x and k_y are the permeability of the fluid along x- and y-axis, respectively. The fluid flow in a 2D domain is governed by,

$$k_x\frac{d^2\phi}{dx^2} + k_y\frac{d^2\phi}{dy^2} = 0 \tag{10.55}$$

Eq. (10.55) is also a type of Poisson equations, which has been discussed in Section 4.6 extensively. Let $c_x = k_x$, and $c_y = k_y$ in Eqs. (4. 46) and (4.47), the model for a rectangle element in a 2D groundwater problem becomes,

$$[K] = [K_x] + [K_y]$$
$$= \frac{wk_x}{6l}\begin{bmatrix} 2 & -2 & -1 & 1 \\ -2 & 2 & 1 & -1 \\ -1 & 1 & 2 & -2 \\ 1 & -1 & -2 & 2 \end{bmatrix} + \frac{lk_y}{6w}\begin{bmatrix} 2 & 1 & -1 & -2 \\ 1 & 2 & -2 & -1 \\ -1 & -2 & 2 & 1 \\ -2 & -1 & 1 & 2 \end{bmatrix} \tag{10.56}$$

Let $c_x = k_x$, and $c_y = k_y$ in Eqs. (4. 51) and (4.52), the model for a triangular element in a 2D groundwater problem becomes,

$$[K] = [K_x] + [K_y] = \frac{k_x}{4A}\begin{bmatrix} \beta_i^2 & \beta_i\beta_j & \beta_i\beta_k \\ \beta_j\beta_i & \beta_j^2 & \beta_j\beta_k \\ \beta_k\beta_i & \beta_k\beta_j & \beta_k^2 \end{bmatrix} + \frac{k_y}{4A}\begin{bmatrix} \delta_i^2 & \delta_i\delta_j & \delta_i\delta_k \\ \delta_j\delta_i & \delta_j^2 & \delta_j\delta_k \\ \delta_k\delta_i & \delta_k\delta_j & \delta_k^2 \end{bmatrix}$$
$$\tag{10.57}$$

Example 10.4

Fig. 10.14 shows a seepage water flow under a concrete dam. The permeability of the porous soil in the bed underneath the dam is homogenous, which is $k_x = k_y = 10$ m/day. The water levels inside and outside the dam are 20 m and 1 m, respectively. Determine the velocity distribution in the soil.

Solution

Due to the similarity of the governing equations for groundwater problems and heat transfer problems, the solver for heat transfer problems can be applied to solve seepage problems. In SolidWorks Simulation, the given seepage problem is defined as a thermal analysis model (Fig. 10.15):

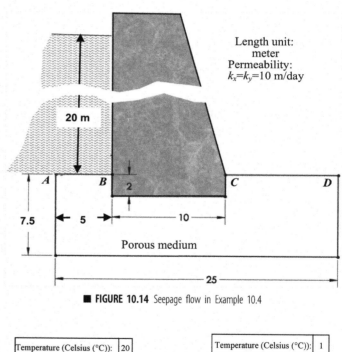

■ **FIGURE 10.14** Seepage flow in Example 10.4

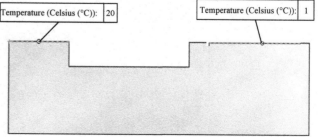

■ **FIGURE 10.15** Seepage flow modeled as a thermal analysis problem in SolidWorks.

■ **FIGURE 10.16** Hydro head distribution in porous bed.

a custom material is defined with the permeability of 10 and a 2D domain is created using the dimensions in Fig. 10.14. The BCs about the water levels are treated as the given temperatures on edges of A–B and C–D, respectively. After the simulation, the distribution of hydro head is illustrated in Fig. 10.16, the velocity distribution in the porous bed is shown in Fig. 10.17, and the seepage velocity on a cross-section M_1–M_2 is plotted in Fig. 10.18.

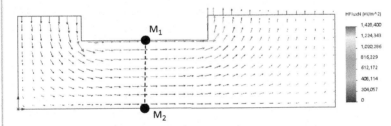

■ **FIGURE 10.17** Velocity distribution in porous bed.

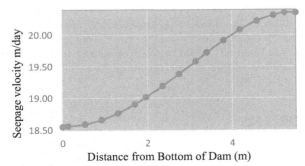

■ **FIGURE 10.18** Velocity distribution on cross-section underneath dam.

10.7 **SUMMARY**

Fluid mechanics is a branch of continuous mechanics. It differs from solid mechanics in the sense that fluid does not sustain geometry, and the deformation subjected to external force is plastic and permanent. Three primary variables in a fluid flow are velocity, pressure, and density; these variables are governed by the equations of mass conservation, momentum conservation, and energy conservation. A few of basic fluid flows can be solved without the consideration of energy conservation. In such cases, additional constraints on three primary variables must be taken into consideration. The problems of pipe network, incompressible irrotational fluid flows in 2D domains, and groundwater seepage have been discussed and modeled. Due to the similarity of mathematical representation, a pipe network system can be modeled and solved as a generic spring system and ideal fluid flows or groundwater systems can be modeled and solved as steady thermal analysis in the SolidWorks Simulation.

REFERENCES

Acharya, S., 2016. Analysis and FEM Simulation of Flow of Fluids in Pipes. Arcada University of Applied Sciences.

Bar-Meir, G., 2008. Basics of Fluid Mechanics. http://ufdcimages.uflib.ufl.edu/AA/00/01/17/21/00001/fluidMechanics.pdf.

Chung, T.J., 2002. Computational Fluid Dynamics. Cambridge University Press, New York, NY, USA, ISBN 0-521-59416-2.

Moaveni, S., 2015. Finite Element Analysis Theory and Application with ANSYS. fourth ed. Peason. ISBN-10: 0-13-384080-8.

Omari, R., 2013. CFD simulation of lid driven cavity flow at moderate Reynolds number. European Scientific Journal 9 (15), 22−35.

Pironneau, O., 1988. Finite Element Methods for Fluids. https://www.ljll.math.upmc.fr/pironneau/publi/publications/OPfemInFluids.pdf.

Ritchmeyer, R., Morton, K., 1967. Difference Methods for Initial Value Problems. Wiley.

Temam, R., 1977. Theory and Numerical Analysis of the Navier-Stokes Equations. North Holland.

Tryggvason, G., 2016. Equations of Fluid Dynamics. http://www3.nd.edu/~gtryggva/CFD-Course/NSEquations.pdf.

Zienkiewicz, O.C., Taylor, R.L., 2000. The finite element method In: Fluid Dynamics, fifth ed., vol. 3. Butterworth-Heinemann, Linacre House, Jordan Hill, Oxford, ISBN 0 7506 5050 8.

■ **PROBLEMS**

10.1. Neglecting minor losses in the pipes determine the flows in the pipes and the hydro heads at nodes in Fig. 10.19. The pressure head at $A = 98,070$ Pa (10 m), and the viscosity of water are set as 8.9×10^{-4} Pa/s. Table 10.2 shows the dimensions of pipe elements. Extend or write a program for a generic spring system and find the pressure head at all nodes ($B \sim F$).

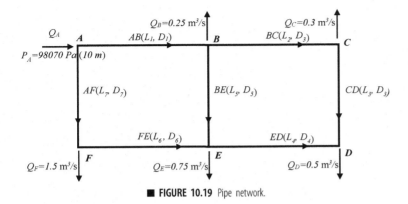

■ **FIGURE 10.19** Pipe network.

Table 10.2 Dimensions of Pipe Elements

Element	AB	BC	CD	DE	EF	AF	BE
Length (m)	600	600	200	600	600	200	200
Diameter (mm)	250	150	100	150	150	200	100

10.2. A 2D domain with an incompressible irrotational fluid flow is shown in Fig. 10.20. The length is 2 in and the height is 1 in, and there is the obstacle in the middle with four circles of ϕ 0.2-in on a circle of ϕ 0.5-in. Assume that the reference atmosphere pressure at the outlet is and the fluid speed at the inlet is 1.0 BUT/s/in^2. Create an FEA model and determine the distribution of the velocity in the fluid domain.

10.3. Fig. 10.21 shows a fluid flow in a cavity, the size of cavity is $L = W = 0.2$ m, and the fluid is water, which has the density of

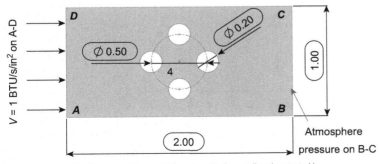

■ **FIGURE 10.20** Ideal fluid flow in a 2D domain (length unit: inch).

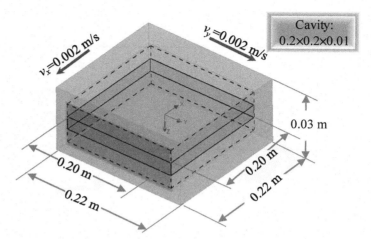

■ **FIGURE 10.21** Geometric models for simulation of a cavity flow.

$\rho = 997.13$ kg/m^3 and the viscosity of $\mu = 0.000{,}891$ kg/m s. Among four sides of cavity, two sides are open and are driven by the passing fluids with constant velocity 0.002 m/s. The fluid flow in the cavity can be represented as incompressible irrotational flow governed by mass and momentum conservation equations. Create an FEA model and find the distribution of velocity of fluid flow in the cavity for the problem.

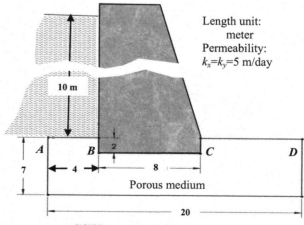

■ **FIGURE 10.22** Seepage flow in problem 10.3.

10.4. Fig. 10.22 shows a seepage water flow under a concrete dam. The permeability of the porous soil in the bed underneath the dam is homogenous, which is $k_x = k_y = 5$ m/day. The water levels inside and outside the dam are 10 m and 1 m, respectively. Determine the velocity distribution in the soil.

Applications—Multiphysics Systems

11.1 INTRODUCTION

In many applications, state variables in two or more physical subsystems are coupled with each other. State variables are subjected to extra constraints caused by the coupling of multidisciplinary behaviors. For example, structural and fluid domains interact with each other in many applications such as biomechanical systems, micromechanical systems, brake systems, or pipes (Rugonyi and Bathe, 2001). State variables such as fluid pressure at interfaces affect the behaviors of solid and fluid domains. Neither the structure nor the fluid system can be solved independently without the consideration of the other. Modeling of a coherent multiphysics system must take into the consideration of governing equations of subsystems simultaneously (Zienkiewicz and Taylor, 2013). Navier—Stokes or Euler equations are often used to represent the interactions of structural and fluid domains. Moreover, the analysis of a multiphysics system is often required to understand the behaviors of complex systems at first.

11.2 DESCRIPTION OF MATHEMATIC MODELS

In general, when modeling a multiphysics system, each subsystem can be represented by a set of mathematic equations in its own domain. The governing equations in individual domains can be partial differential equations, ordinary differential equations, or, more generally, differential algebraic equations (DAEs) (Gasmi et al., 2013).

Pawlowski et al. (2011) gave the general description of mathematic models for single-physics and multiphysics systems. Assume that a single-physics system involves n_x number of unknown variables, then its mathematic model is defined as,

$$f\left(\dot{x}, x, \{p_l\}, t\right) = 0 \qquad (11.1)$$

Finite Element Analysis Applications. http://dx.doi.org/10.1016/B978-0-12-809952-0.00011-X

where

$\{x\}$ is the vector of the state variables in n_x-dimension space $\Re^{n_x}$,

$\{\dot{x}\} = \{\partial \dot{x}/\partial t\}$ is the vector of the derivatives of state variables with respect to time,

$\{p_l\} = \{p_1, p_2, \cdots, p_{N_p}\}$ is the set of N_p subvectors and the number of system parameters in each subvector is denoted as n_{p_l}, and

$t \in [t_0, t_f]$ is the time variable from initial time t_0 to final time t_f.

The solving process for a single-physics system can be described as a mapping from the problem space to the solution space as,

$$f\left(\dot{x}, x, \{p_l\}, t\right): \Re^{\left(2n_x + \sum_{l=1}^{N_p} n_{p_l} + 1\right)} \rightarrow \Re^{n_x} \qquad (11.2)$$

To solve the system model in Eq. (11.2) in the numerical simulation, the user must define all of the system parameters used in the N_p set of subvectors. After the system model Eq. (11.2) is solved, the solution is postprocessed to compute one or more dependent quantities of interest. For example, if a conjugate heat transfer is modeled, the total heat loss through surface convection is usually an important quantity to be evaluated. The dependence of such a quantity on state variables is often referred to as *a response function*, which relates to state variables and time as,

$$g_i\left(\dot{x}, x, \{p_l\}, t\right) = 0 \quad i = 1, \ldots, N_g \qquad (11.3)$$

where N_g is the number of response functions of the system.

A response function is evaluated by a mapping from a problem space to the solution space after state variables have been solved from the system model, i.e.,

$$g_i\left(\dot{x}, x, \{p_l\}, t\right): \Re^{\left(2n_x + \sum_{l=1}^{N_p} n_{p_l} + 1\right)} \rightarrow \Re^{n_{g_i}} \qquad (11.4)$$

In a single-physics system, the response functions of system are evaluated at the phase of postprocessing. However, it may be different when a multiphysics system is modeled. The response functions in a multiphysics system may be treated as additional state variables and be computed as a part of the system solution in the solving process.

A multiphysics system involves a set of N_f coupled single-physical subsystems. The coupling of two subsystems refers to the phenomenon that the system parameters of one system depend on state variables and system parameters of another. To generalize a mathematic model of a multiphysics system, the coupled parameters $\{c_{k,i}\}$ in the i-th subsystem are isolated from the rest of uncoupled system parameters $\{p_{l,i}\}$, and a single-physics subsystem is then represented by,

$$f_i\left(\dot{x}_i, x_i, \{c_{k,i}\}, \{p_{l,i}\}, t\right) = 0, \quad for\ i = 1, 2 \cdots N_f \qquad (11.5)$$

where

$\{x_i\}$ is the vector of state variables in the n_{xi}-dimensional space for the i-th subsystem,

$\{\dot{x}_i\} = \{\partial \dot{x}_i / \partial t\}$ is the vector of the derivatives of state variables for the i-th subsystem with respect to time,

$\{c_{i,k}\} = \{c_{i,1}, c_{i,2}, \dots c_{i,N_{c,i}}\}$ is the set of $N_{c,i}$ subvectors of coupling parameters for the i-th subsystem,

$\{p_{i,l}\} = \{p_{i,1}, p_{i,2}, \dots, p_{i,N_p}\}$ is the set of N_p subvectors, the number of system parameters in each subvector is denoted as n_{p_l}, and

$t \in [t_0, t_f]$ is the time variables from initial time t_0 to final time t_f.

To solve a multiphysics system model, an iterative solution to the i-th subsystem is needed and it can be described as a mapping from a problem space to a solution space as,

$$f\left(\dot{x}_i, x_i, \{c_{i,k}\}, \{p_{i,l}\}, t\right) : \Re^{\left(2n_{x_i} + \sum_{k=1}^{N_{c,i}} n_{c_i,k} + \sum_{l=1}^{N_{p_i}} n_{p_{i,l}} + 1\right)} \rightarrow \Re^{n_{x_i}} \qquad (11.6)$$

The coupling of two or more subsystems brings an extra set of the constraints to exchange the information of $\{c_{i,k}\}$ among subsystems. For a weakly coupled system, $\{c_{i,k}\}$ can be evaluated at the postprocessing phase. Although for a strongly coupled system, $\{c_{i,k}\}$ has to be defined as transfer functions $r_{i,k}$ for the i-th subsystem as,

$$c_{i,k} = r_{i,k}\left(\{x_w\}, \{p_{w,n}\}\right), \quad for\ i = 1, 2, \cdots N_f; k = 1, 2 \cdots N_{c,i} \qquad (11.7)$$

where

$\{c_{i,k}\} = \{c_{i,1}, c_{i,2}, \dots c_{i,N_{c,i}}\}$ is the vector of $N_{c,i}$ subvectors of coupling parameters for the i-th subsystem,

$\{x_w\} = \{x_1, x_2, \dots x_{N_f}\}$ is the vector of state variables in the whole multiphysics system,

$\{p_{w,n}\} = \left\{p_{1,1}, \dots, p_{i,N_{p_1}}, \cdots p_{N_f,1}, \dots, p_{N_f,N_{p_1}}\right\}$ is the set of all subvectors of coupling parameters in the multiphysics system, and

$r_{i,k}\left(\{x_w\}, \{p_{w,n}\}\right) : \Re^{\left(\sum_{w=1}^{N_f}\left(n_{x_w} + \sum_{n=1}^{N_{p_w}} n_{p_w,n}\right)\right)} \rightarrow \Re^{n_{c_{i,k}}}$ is the transfer function of the i-th subsystem for the k-th of coupling parameter.

11.3 SIMULATION MODELS OF MULTIPHYSICS SYSTEMS

Gasmi et al. (2013) classified simulation models based on time dependence of corresponding multiphysics systems. Usually, a multiphysics system is time dependent. As shown in Fig. 11.1, both monolithic and partitioned approaches are used to model multiphysics systems.

In a *monolithic approach*, the system is modeled by a single set of differential or DAEs, all of the state variables from different subsystems are naturally and inherently coupled. A single solver is deployed to take care of the time domain and all of the state variables are updated simultaneously at each iteration of the solving process. In a *partitioned approach*, the model of a multiphysics system is divided into a set of the models of subsystems in their respective physics domains. Each subsystem is represented by a partition that is subsequently coupled with other at the system level. System coupling is reflected by input—output relations of subsystems.

A partitioned system can be either of a strongly coupled system or a weakly coupled system. A *strongly coupled partitioned system*, such as a system of differential and algebraic equations, uses a single integrator to deal with the time integration. In a weakly coupled system, the time integration is performed sequentially in a staggered procedure. Due to the flexibility in modeling each subsystem, a weakly coupled system seems more attractive; unless it runs into the difficulties to find a converged solution within a reasonable cost of computation. A weakly coupled system offers the compatibility to a variety of *time integrators*, *time increments*, and the *spatial discretization*. In other words, it allows to use existing finite element analysis (FEA) codes such as the sophisticated fluid-dynamics and structural-dynamics analysis tools to take over the computation for critical analysis.

The coupling of two systems can also be modeled *explicitly* or *implicitly*. It depends on whether or not the state variables at a current time step depend

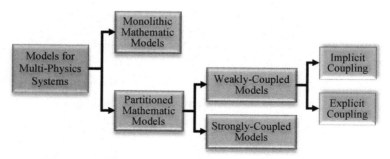

■ **FIGURE 11.1** Classification of multiphysics systems.

completely on the information at precedent steps. If the information from the precedent steps is sufficient to determine state variables at a current time, it is an explicit model. More details about the classification and formulation of multiphysics models can be found in the literature (Gasmi et al., 2013).

To analyze a multiphysics system, appropriate assumptions about the real-world applications are critical for a good balance of computation and accuracy (Tordini, 2014). In the FEA theory, system couplings are classified into two basic types, i.e., *strong coupling* and *weak coupling*. Different formulations are deployed to develop FEA models. Assume $\{X_1\}$ and $\{X_2\}$ to be the vectors of unknown variables in two respective domains, their relations to the load vectors $\{Fs\}$ are different (ANSYS, 2013).

1. For a strongly coupled problem, the matrix form of system is,

$$
\begin{bmatrix} [K_{11}] & [K_{12}] \\ [K_{21}] & [K_{22}] \end{bmatrix} \begin{Bmatrix} \{X_1\} \\ \{X_2\} \end{Bmatrix} = \begin{Bmatrix} \{F_1\} \\ \{F_2\} \end{Bmatrix} \tag{11.8}
$$

   As we introduced in Chapter 2 for system decomposition, the coupling of two subsystems is reflected by the off-diagonal submatrices $[K_{12}]$ and $[K_{21}]$. An iterative solving process will be used to update these submatrices until the constraints by coupling of two disciplines are satisfied.

2. For a weakly coupled problem, the matrix form of system becomes,

$$
\begin{bmatrix} [K_{11}(\{X_1\}, \{X_2\})] & [0] \\ [0] & [K_{22}(\{X_1\}, \{X_2\})] \end{bmatrix} \begin{Bmatrix} \{X_1\} \\ \{X_2\} \end{Bmatrix}
$$
$$
= \begin{Bmatrix} \{F_1(\{X_1\}, \{X_2\})\} \\ \{F_2(\{X_1\}, \{X_2\})\} \end{Bmatrix} \tag{11.9}
$$

The coupling is reflected by the dependency of $[K_{11}]$ and $\{F_1\}$ on $\{X_2\}$ and the dependency of $[K_{22}]$ and $\{F_2\}$ on $\{X_1\}$. Similarly, the system model has to be solved iteratively to satisfy the constraints caused by coupling. However, a submodel at each iteration does not need the full scale of relational matrix.

As shown in Table 11.1, the couplings in a design problem must be clarified before the problem can be formulated and modeled in FEA (ANSYS, 2013). The coupling nature has its impact on the selection of analysis types in simulation. For example, a magneto structural analysis in Table 11.1 is classified as a weakly coupled problem. Therefore, it will be solved sequentially by two simulations, i.e., electromagnetic analysis and structural analysis.

In the following sections, a few of multiphysics problems are discussed to modeled to illustrate the procedure of FEA modeling for coupled multiphysics systems.

Table 11.1 Exemplifying Analysis Types in ANSYS Workbench (ANSYS, 2013)

Analysis Category	Coupling Method	Example Applications
Thermal structural analysis	S, W	High temperature turbine
Magneto structural analysis	W	Solenoid, high energy magnets
Electromagnetic analysis	S	Current fed massive conductors
Piezoresistive analysis	W	Pressure and force sensors
Acoustic structural analysis	S, W	Acoustics
Thermo pressure analysis	S, W	Piping networks
Magnetic thermal analysis	W	Power interrupts, sure protection
Structural diffusion analysis	S, W	Hygroscopic swelling of polymers in electronics pacing
Structural thermal diffusion analysis	S, W	Sodium expansion in aluminum reduction cells

S, strongly coupled; W, weakly coupled.

11.4 STRUCTURAL-THERMAL ANALYSIS

FEA has been widely adopted for thermal-stress and thermal-deflection analyses. The coupling of thermal and structure systems are driven by the strains caused by thermal loads in structures (Anderson, 1994). As shown in Fig. 11.2, such physical phenomena are commonly seen in many products and systems such as rocket engines, pressure vessels, electronics, composite curing mandrels, generators, and satellites (Laird, 2015). In this section, FEA modeling for thermal-stress analyses will be discussed.

11.4.1 Structural analysis under thermal load

For any position in a three-dimensional (3D) continuous domain, if it is subjected to the temperature change, its thermal behavior induces the strain of structure,

$$\{\boldsymbol{\varepsilon}_T\} = \begin{Bmatrix} \varepsilon_x \\ \varepsilon_y \\ \varepsilon_z \\ \gamma_{xy} \\ \gamma_{yz} \\ \gamma_{xz} \end{Bmatrix} = \begin{Bmatrix} \alpha_x \Delta T \\ \alpha_y \Delta T \\ \alpha_z \Delta T \\ 0 \\ 0 \\ 0 \end{Bmatrix} \tag{11.10}$$

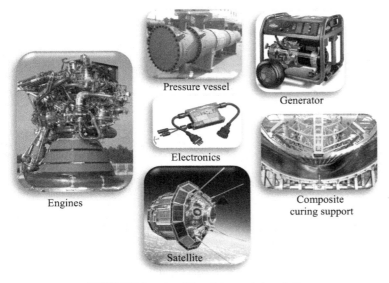

■ FIGURE 11.2 Examples of thermal-stress analysis applications.

where

$\{\boldsymbol{\varepsilon}_T\}$ is the temperature-induced strain,

α_x, α_y, and α_z are the thermal expansion coefficients along x, y, and z axes, respectively.

ΔT is temperature change.

With the presence of the temperature-induced strain, the relation of stresses and strains becomes,

$$
\begin{Bmatrix} \varepsilon_x \\ \varepsilon_y \\ \varepsilon_z \\ \gamma_{xy} \\ \gamma_{yz} \\ \gamma_{xz} \end{Bmatrix} = \begin{bmatrix} \dfrac{1}{E} & \dfrac{-v}{E} & \dfrac{-v}{E} & 0 & 0 & 0 \\[2mm] \dfrac{-v}{E} & \dfrac{1}{E} & \dfrac{-v}{E} & 0 & 0 & 0 \\[2mm] \dfrac{-v}{E} & \dfrac{-v}{E} & \dfrac{1}{E} & 0 & 0 & 0 \\[2mm] 0 & 0 & 0 & \dfrac{1}{G} & 0 & 0 \\[2mm] 0 & 0 & 0 & 0 & \dfrac{1}{G} & 0 \\[2mm] 0 & 0 & 0 & 0 & 0 & \dfrac{1}{G} \end{bmatrix} \begin{Bmatrix} \sigma_x \\ \sigma_y \\ \sigma_z \\ \sigma_{xy} \\ \sigma_{yz} \\ \sigma_{xz} \end{Bmatrix} + \begin{Bmatrix} \alpha_x \Delta T \\ \alpha_y \Delta T \\ \alpha_z \Delta T \\ 0 \\ 0 \\ 0 \end{Bmatrix} \quad (11.11)
$$

The constitutive equation in Eq. (8.7) should be modified as,

$$\{\sigma\} = [D]\{\varepsilon - \varepsilon_T\} \tag{11.12}$$

where

$$[D] = \frac{E}{(1-2v)(1+v)}
\begin{bmatrix}
1-v & v & v & 0 & 0 & 0 \\
v & 1-v & v & 0 & 0 & 0 \\
v & v & 1-v & 0 & 0 & 0 \\
0 & 0 & 0 & \dfrac{1-2v}{2} & 0 & 0 \\
0 & 0 & 0 & 0 & \dfrac{1-2v}{2} & 0 \\
0 & 0 & 0 & 0 & 0 & \dfrac{1-2v}{2}
\end{bmatrix} \tag{11.13}$$

In an FEA model, strains are dependent variables that are calculated from the state variables on nodes. Assume that a 3D element has n nodes, and the corresponding shape functions for three displacements (u_{ix}, u_{iy}, u_{iz}) are (S_{ix}, S_{iy}, S_{iz}), the relations of $\{\varepsilon\}$ and $\{U\}$ can be found as (Felippa, 2017),

$$\{\varepsilon\} = [B] \cdot \{U\} \tag{11.14}$$

where

$$\{\varepsilon\}^T = \left\{ \frac{\partial u_x}{\partial x}, \ \frac{\partial u_y}{\partial y}, \ \frac{\partial u_z}{\partial z}, \ \frac{\partial u_x}{\partial y} + \frac{\partial u_y}{\partial x}, \ \frac{\partial u_y}{\partial z} + \frac{\partial u_z}{\partial y}, \ \frac{\partial u_x}{\partial z} + \frac{\partial u_z}{\partial x} \right\}^T$$

$$\{U\}^T = \left\{ u_{1,x} \quad u_{1,y} \quad u_{1,z} \quad \cdots \quad u_{n,x} \quad u_{n,y} \quad u_{n,z} \right\}^T$$

$$[B] =
\begin{bmatrix}
\dfrac{\partial S_1}{\partial x} & 0 & 0 & \cdots & \dfrac{\partial S_n}{\partial x} & 0 & 0 \\
0 & \dfrac{\partial S_1}{\partial y} & 0 & \cdots & 0 & \dfrac{\partial S_n}{\partial y} & 0 \\
0 & 0 & \dfrac{\partial S_1}{\partial z} & \cdots & 0 & 0 & \dfrac{\partial S_n}{\partial z} \\
\dfrac{\partial S_1}{\partial y} & \dfrac{\partial S_1}{\partial x} & 0 & \cdots & \dfrac{\partial S_n}{\partial y} & \dfrac{\partial S_n}{\partial x} & 0 \\
0 & \dfrac{\partial S_1}{\partial z} & \dfrac{\partial S_1}{\partial y} & \cdots & 0 & \dfrac{\partial S_n}{\partial z} & \dfrac{\partial S_n}{\partial y} \\
\dfrac{\partial S_1}{\partial z} & 0 & \dfrac{\partial S_1}{\partial x} & \cdots & \dfrac{\partial S_n}{\partial z} & 0 & \dfrac{\partial S_n}{\partial x}
\end{bmatrix}$$

and $S_i(x, y, z)$ $(i = 1, 2, \ldots, n)$ are the shape functions in an element.

The potential energy of the solid domain is evaluated by,

$$\Pi = \frac{1}{2} \int_V \{[\boldsymbol{B}]\{\boldsymbol{U}\} + \{\boldsymbol{\varepsilon}_T\}\}^T \cdot [\boldsymbol{D}] \cdot \{[\boldsymbol{B}]\{\boldsymbol{U}\} + \{\boldsymbol{\varepsilon}_T\}\} dV - \sum \{\boldsymbol{U}\} \cdot \{\boldsymbol{F}\}$$

(11.15)

Applying the minimal potential energy equation $\dfrac{\partial \Pi}{\partial \boldsymbol{U}} = 0$ in Eq. (11.15) gets,

$$\left[\int_V [\boldsymbol{B}]^T [\boldsymbol{D}][\boldsymbol{B}] dV \right] \cdot \{\boldsymbol{U}\} = \{\boldsymbol{F}\} - \{\boldsymbol{F}_T\} = \{\boldsymbol{F}\} - \left[\int_V [\boldsymbol{B}]^T [\boldsymbol{D}] dV \right] \cdot \{\boldsymbol{\varepsilon}_T\}$$

(11.16)

where

[$\boldsymbol{D}$] and [$\boldsymbol{B}$] are the matrices defined in Eqs. (11.13) and (11.14), and
$\{\boldsymbol{F}_T\} = \left[\int_V [\boldsymbol{B}]^T [\boldsymbol{D}] dV \right] \cdot \{\boldsymbol{\varepsilon}_T\}$ is the temperature-induced load.

11.4.2 **Modeling of 2D frame members**

For a 2D frame member with the nodes of i and j in Fig. 8.40, the internal force, stress, and displacement, as well as their relations are summarized in Fig. 11.3.

From the description of a frame member in Fig. 11.3, its strain energy from the deformation can be calculated as (Anderson, 1994),

$$\Lambda = \frac{1}{2}\{\sigma\}^T\{\varepsilon\} = \frac{1}{2}\left[\sigma_x \varepsilon_x + \cancel{\sigma_y \varepsilon_y}^{\,0} + \cancel{\sigma_z \varepsilon_z}^{\,0} + \cancel{\tau_{xy}\gamma_{xy}}^{\,0} + \cancel{\tau_{yz}\gamma_{yz}}^{\,0} + \cancel{\tau_{xz}\gamma_{xz}}^{\,0}\right] = \frac{1}{2}\sigma_x \varepsilon_x$$

(11.17)

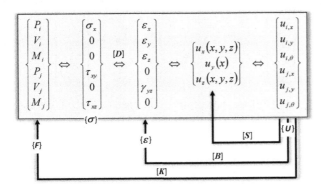

■ **FIGURE 11.3** Relations of forces, stresses, strains, and displacements in 2D frame.

Therefore, when the temperature change is considered in a 2D frame member,

$$
\left.
\begin{aligned}
&\{\boldsymbol{\varepsilon}_T\} \;=\; \{\boldsymbol{\varepsilon}_{x0}\} \;=\; \{\alpha_x \Delta T\} \\[4pt]
&[\boldsymbol{D}] \;=\; [E] \\[4pt]
&[\boldsymbol{B}]^T \;=\; \left[\dfrac{-1}{L} \quad y\!\left(\dfrac{6}{L^2} - \dfrac{12x}{L^3}\right) \quad y\!\left(\dfrac{4}{L} - \dfrac{6x}{L^2}\right) \quad \dfrac{1}{L} \quad y\!\left(\dfrac{12x}{L^3} - \dfrac{6}{L^2}\right) \quad y\!\left(\dfrac{2}{L} - \dfrac{6x}{L^2}\right) \right]
\end{aligned}
\right\}
$$

$$(11.18)$$

where $[\boldsymbol{B}]^T$ is found by combining the results from Eq. (8.9) for axial member and Eq. (8.40) for a beam member.

Substituting Eq. (11.18) into Eq. (11.16) gets the temperature-induced load as,

$$
\{\boldsymbol{F}_T\} \;=\; \left[\int_V [\boldsymbol{B}]^T [\boldsymbol{D}] dV \right] \cdot \{\boldsymbol{\varepsilon}_T\} \;=\; E\alpha_x \Delta T
\begin{Bmatrix}
-A \\
0 \\
Q \\
A \\
0 \\
-Q
\end{Bmatrix}
\qquad (11.19)
$$

where $Q = \int_{dA} y\, dA$ is the first moment of area with respect to the reference coordinate system at bonded interface.

With the thermal load, the model for a 2D frame is described as (Anderson, 1994),

$$
[\boldsymbol{K}]
\begin{Bmatrix}
u_{ix} \\
u_{iy} \\
u_{i\theta} \\
u_{jx} \\
u_{jy} \\
u_{j\theta}
\end{Bmatrix}
\;=\; \{\boldsymbol{F}\} + \{\boldsymbol{F}_T\} \;=\; \{\boldsymbol{F}\} + E\alpha_x \Delta T
\begin{Bmatrix}
-A \\
0 \\
Q \\
A \\
0 \\
-Q
\end{Bmatrix}
\qquad (11.20)
$$

where $\{\boldsymbol{F}\}$ and $\{\boldsymbol{F}_T\}$ are the vectors for mechanical force loads and thermal load, respectively, $[\boldsymbol{K}]$ is given from Eq. (8.44) as,

$$[K] = \begin{bmatrix} \dfrac{EA}{L} & 0 & 0 & -\dfrac{EA}{L} & 0 & 0 \\[2mm] 0 & \dfrac{12EI}{L^3} & \dfrac{6EI}{L^2} & 0 & -\dfrac{12EI}{L^3} & \dfrac{6EI}{L^2} \\[2mm] 0 & \dfrac{6EI}{L^2} & \dfrac{4EI}{L} & 0 & -\dfrac{6EI}{L^2} & \dfrac{2EI}{L} \\[2mm] -\dfrac{EA}{L} & 0 & 0 & \dfrac{EA}{L} & 0 & 0 \\[2mm] 0 & -\dfrac{12EI}{L^3} & -\dfrac{6EI}{L^2} & 0 & \dfrac{12EI}{L^3} & -\dfrac{6EI}{L^2} \\[2mm] 0 & \dfrac{6EI}{L^2} & \dfrac{2EI}{L} & 0 & -\dfrac{6EI}{L^2} & \dfrac{4EI}{L} \end{bmatrix}$$

Example 11.1

A composite beam consists of aluminum alloy layer and steel alloy layer with the same C-section at the opposite directions. The configuration of the composite beam is given in Fig. 11.4A and the dimensions of the C-section are illustrated in Fig. 11.4B and C. The properties of two materials are included in Table 11.2.

(A) The configuration and dimension of beam

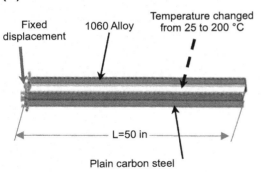

(B) Enlarged view of bonded cross-sections

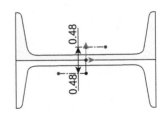

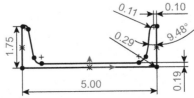

$I_y = 0.94$ in^4; $I_z = 7.48$ in^4, $Q_z = \pm 0.9456$ in^3

(C) The C-cross section

■ **FIGURE 11.4** Constitution of composite beam: (A) the configuration and dimension of beam, (B) enlarged view of bonded cross-sections, (C) the C-cross section.

Table 11.2 Basic Properties of Aluminum Alloy (1016 Alloy) and Steel (Plain Steel)

Materials	Elastic Modulus (psi)	Thermal Expansion Coefficient (1/°C)	Temperature Change (°C)
Aluminum (1060 Alloy)	$E_A = 1.001 \times 10^7$	$\alpha_A = 2.4 \times 10^{-5}$	$\Delta T = 175$
Steel (plain carbon)	$E_S = 3.046 \times 10^7$	$\alpha_S = 1.3 \times 10^{-5}$	

The composite beam is bonded at 25°C, it is fixed at one end. Find the deflection of the beam when the temperature is raised to 200°C.

Solution

As shown in Fig. 11.5, 2D frame members are used to model the composite beam. It consists of two frame members for aluminum and steel

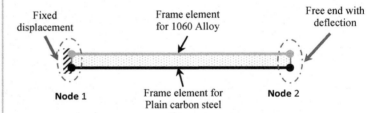

Fixed displacement

Frame element for 1060 Alloy

Free end with deflection

Node 1

Frame element for Plain carbon steel

Node 2

■ **FIGURE 11.5** Finite element analysis model of composite beam.

beams, respectively. They share the nodes at two ends due to the bonded interface. Applying Eq. (11.20) to model two elements and assembling element models together get the system model as,

$$
\begin{bmatrix}
\frac{E_{comp}A}{L} & 0 & 0 & -\frac{E_{comp}A}{L} & 0 & 0 \\
0 & \frac{12E_{comp}I}{L^3} & \frac{6E_{comp}I}{L^2} & 0 & -\frac{12E_{comp}I}{L^3} & \frac{6E_{comp}I}{L^2} \\
0 & \frac{6E_{comp}I}{L^2} & \frac{4E_{comp}I}{L} & 0 & -\frac{6E_{comp}I}{L^2} & \frac{2E_{comp}I}{L} \\
-\frac{E_{comp}A}{L} & 0 & 0 & \frac{E_{comp}A}{L} & 0 & 0 \\
0 & -\frac{12E_{comp}I}{L^3} & -\frac{6E_{comp}I}{L^2} & 0 & \frac{12E_{comp}I}{L^3} & -\frac{6E_{comp}I}{L^2} \\
0 & \frac{6E_{comp}I}{L^2} & \frac{2E_{comp}I}{L} & 0 & -\frac{6E_{comp}I}{L^2} & \frac{4E_{comp}I}{L}
\end{bmatrix}
\begin{Bmatrix}
u_{1x} \\ u_{1y} \\ u_{1\theta} \\ u_{2x} \\ u_{2y} \\ u_{2\theta}
\end{Bmatrix}
=
\begin{Bmatrix}
-(E_A\alpha_A + E_S\alpha_S)A\Delta T \\
0 \\
-(E_A\alpha_A Q_A + E_S\alpha_S Q_S)\Delta T \\
(E_A\alpha_A + E_S\alpha_S)A\Delta T \\
0 \\
-(E_A\alpha_A Q_A + E_S\alpha_S Q_S)\Delta T
\end{Bmatrix}
\tag{11.21}
$$

where $E_{comp} = E_A + E_S$

After imposing the boundary condition in the system model Eq. (11.21), the displacements at the free end can be found as,

$$
\begin{Bmatrix} u_{2x} \\ u_{2y} \\ u_{2\theta} \end{Bmatrix} =
\begin{Bmatrix}
\dfrac{(E_A\alpha_A + E_S\alpha_S)L\Delta T}{E_{comp}} \\[2ex]
-\dfrac{(E_A\alpha_A - E_S\alpha_S)QL^2\Delta T}{2E_{comp}I} \\[2ex]
-\dfrac{(E_A\alpha_A - E_S\alpha_S)QL\Delta T}{E_{comp}I}
\end{Bmatrix} =
\begin{Bmatrix}
\dfrac{(1.001 \times 2.4 + 3.046 \times 1.3) \times 10^2(50)(175)}{(4.0047 \times 10^7)} \\[2ex]
-\dfrac{(1.001 \times 2.4 - 3.046 \times 1.3) \times 10^2(0.9456)(50)^2(175)}{2 \times (4.0047 \times 10^7)(0.94)} \\[2ex]
-\dfrac{(1.001 \times 2.4 - 3.046 \times 1.3) \times 10^2(0.9456)(50)(175)}{(4.0047 \times 10^7)(0.94)}
\end{Bmatrix}
$$

$$
= \begin{Bmatrix}
0.138 (\text{in}) \\
\boldsymbol{0.854} (\text{in}) \\
0.0178 (\text{radian})
\end{Bmatrix}
$$

$$(11.22)$$

The total displacement at the free-end is $u_2 = \sqrt{u_{2x}^2 + u_{2y}^2} = 0.865$ (in). The accuracy of displacement is expected to be increased if more elements are used in the FEA model.

To verify the result in a FEA software, a 3D composite beam is modeled. As shown in Fig. 11.6, the beam is fixed on the left end, the initial temperature is set at 25°C and a thermal temperature load 200°C is applied.

(A)   **(B)**

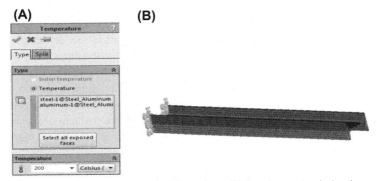

■ **FIGURE 11.6** Composite model in SolidWorks simulation: (A) thermal temperature load on beam and (B) 3D model of composite beam.

The defined FEA model of the composite beam can be solved readily in the SolidWorks Simulation and Fig. 11.7 about the displacement shows that the maximized displacement under the thermal load is 1.088 (in).

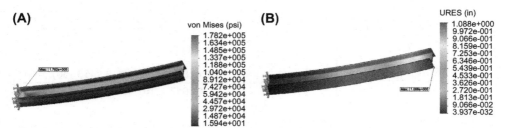

■ FIGURE 11.7 Example of temperature-induced stress in static analysis: (A) stress distribution with a max of 1.782×10^5 (psi); (B) deflection with a max of 1.088 (in).

A comparison between a simplified frame model and 3D static model has shown the discrepancy of two solvers as,

$$\varepsilon = \left| \frac{1.088 - 0.865}{1.088} \right| \approx 20.4\% \tag{11.23}$$

It shows that the analytical calculation from an FEA with two 2D frame elements gives a good estimation of the structural deflection subject to thermal loads.

Bonding parts with different materials is a very common technique for the measurement and instrumentation. For example, bimetal strips are often used in small-scale circuit breakers to break the circuit when the current-induced temperature exceeds the specified limit. The heat from the current causes the bimetal strip to bend due to thermal expansion. As long as the strip bends enough to break the circuit, no more heat will be generated and built up. To analyze such a system, multiphysics analysis capabilities are required to take into account of structural and thermal behaviors at the same time (Tordini, 2014).

11.5 **CONJUGATE HEAT TRANSFER**

Conjugate heat transfer occurs between solid and fluid flow. Heat conduction a solid region is influenced by the heat convection in an adjacent fluid. Similar physical phenomena exist in many practical applications such as in biomedical engineering, air-cooled packaging, and thermal insulation. Conjugate heat transfer is a typical multiphysics problem. Modeling of conjugate heat transfer has been proven challenging due to (1) its inherent coupling of governing equations in fluid and solid domains and (2) the nonlinear phenomenon of the convection terms is presented in both of the momentum equations and the energy equations (Malatip et al., 2009).

The governing equations for a conjugate heat transfer problem are the conservation of *mass*, *momentum*, and *energy*. Correspondingly, *the continuity equation*, *the momentum equation*, and *the energy equation* form the complete model of a conjugate heat transfer problem. Mathematic models in FEA codes for conjugate heat transfer are all based on these equations of conservation (ANSYS, 2009; Ma, 2012; Hosain and Fdhila, 2015; Sobachkin et al., 2014).

The equations for mass, momentum, and energy conservations have been introduced in Chapter 10 (Eqs. (10.14), (10.23), and (10.34)). For an ease reference, these equations are rewritten here:

Mass conservation:

$$\nabla \cdot (\rho v) + \frac{d\rho}{dt} = \dot{S}_m \tag{11.24}$$

Momentum conservation:

$$\left.\begin{array}{l} \dfrac{\partial(\rho u)}{\partial t} + \dfrac{\partial(\rho u^2)}{\partial x} + \dfrac{\partial(\rho uv)}{\partial y} + \dfrac{\partial(\rho uw)}{\partial z} - \dfrac{\partial \tau_{xx}}{\partial x} - \dfrac{\partial \tau_{xy}}{\partial y} - \dfrac{\partial \tau_{xz}}{\partial z} + \dfrac{\partial p}{\partial x} - \rho f_x = 0 \\[2mm] \dfrac{\partial(\rho v)}{\partial t} + \dfrac{\partial(\rho uv)}{\partial x} + \dfrac{\partial(\rho v^2)}{\partial y} + \dfrac{\partial(\rho vw)}{\partial z} - \dfrac{\partial \tau_{xy}}{\partial x} - \dfrac{\partial \tau_{yy}}{\partial y} - \dfrac{\partial \tau_{yz}}{\partial z} + \dfrac{\partial p}{\partial y} - \rho f_y = 0 \\[2mm] \dfrac{\partial(\rho w)}{\partial t} + \dfrac{\partial(\rho uw)}{\partial x} + \dfrac{\partial(\rho vw)}{\partial y} + \dfrac{\partial(\rho w^2)}{\partial z} - \dfrac{\partial \tau_{xz}}{\partial x} - \dfrac{\partial \tau_{yz}}{\partial y} - \dfrac{\partial \tau_{zz}}{\partial z} + \dfrac{\partial p}{\partial z} - \rho f_z = 0 \end{array}\right\} \tag{11.25}$$

Energy conservation:

$$\frac{\partial(\rho E)}{\partial t} + \frac{\partial}{\partial x_i}(\rho v_i H) - \frac{\partial}{\partial x_i}\left(k\frac{\partial T}{\partial x_i}\right) - \frac{\partial}{\partial x_i}(\tau_{ij}v_j) - \rho g_i v_i - q_H = 0 \tag{11.26}$$

However, the number of unknown variables in these equations is more than the number of available equations. Therefore, it is unsolvable without additional information.

To make the set of governing equations solvable (i.e., the number of unknown variables equals to the number of equations), the following state equations should be included in the model,

$$
\left.
\begin{aligned}
\rho &= \rho(p, T) \\
dH &= \frac{\partial H}{\partial T}\bigg|_p dT + \frac{\partial H}{\partial p}\bigg|_T dp = c_p dT + \frac{\partial H}{\partial p}\bigg|_T dp \\
c_p &= c_p(p, T)
\end{aligned}
\right\}
\tag{11.27}
$$

where

ρ is the density,
p is the pressure,
T is the temperature,
H is the enthalpy, and
c_p is specific heat coefficient.

The aforementioned equations fall in the discipline of fluid mechanics. Due to the coupling of fluid and solid mechanics, the heat transfer may occur at the fluid—solid interface. Energy in the solid domain is in the forms of *solid motion*, *conduction*, and *volumetric heat sources*, which is conserved as,

$$
\frac{\partial(\rho H)}{\partial t} + \nabla(\rho U_s H) = \nabla(k\nabla T) + q_E
\tag{11.28}
$$

where

H, ρ, k are the enthalpy, density, and the thermal conductivity of the solid.
U_s is the velocity of the solid object, and
q_E is volumetric heat source if there is.

11.5.1 Numerical solution to conjugate heat transfers

The energy conservation Eq. (11.26) is well-known as the *Navier—Stokes equations*, which is applicable to both of laminar and turbulent flow. Although many applications involve turbulent flows in some regions, a turbulent flow with a large Reynold number spans a large range of turbulent length and time scales. To solve the Navier—Stokes equation for turbulent flows, it requires a very fine mesh, which is impractical to be solved for the majority of industry-scale problems. The direct numerical simulation on fluid dynamics model needs computers in many orders of the magnitude higher than that of foreseeable ones in the future (ANSYS, 2009).

To reduce the computation in practical simulation, unsteady Navier–Stokes equations can be modified by using averaged and fluctuating quantities to produce the *Reynolds Averaged Navier–Stokes* (RANS) equations. RANS represent the mean flow quantities only; RANS is used to model turbulence behaviors without considering the fluctuation by turbulence. The models based on RANS are called as *statistical turbulence models*. To develop the solution to RANS, the fluid velocity is decomposed as,

$$U_i = \overline{U}_i + u_i \tag{11.29}$$

where U_i, $\overline{U}_i$, and u_i are the total velocity, average velocity, and the fluctuated velocity of fluid flow, respectively. The average flow velocity is evaluated by,

$$\overline{U}_i = \int_t^{t+\Delta t} U_i dt \tag{11.30}$$

where t is a time scale that is large relative to the turbulent fluctuation; it should be small relative to the time scale to which the equations are solved.

As a result, RANS can be found by substituting Eq. (11.29) into Eq. (11.26).

$$\frac{\partial(\rho E)}{\partial t} + \frac{\partial}{\partial x_i}(\rho U_i H) = \frac{\partial}{\partial x_i}\left(k\frac{\partial T}{\partial x_i} - \rho\overline{u_i}\overline{H}\right) + \frac{\partial}{\partial x_i}\left[U_j\left(\tau_{ij} - \rho\overline{u_i}\overline{u_j}\right)\right] + q_E \tag{11.31}$$

To deal with the flow velocity and length scale simultaneously in a turbulence model, two-equation models such as the *k-ε* model can be used to achieve a good compromise of numerical effort and computation accuracy. The *k−ε* model uses the gradient diffusion hypothesis to relate the Reynold's stress to the mean velocity gradients and the turbulent viscosity (SolidWorks, 2012). The turbulent viscosity is modeled as the product of a turbulent velocity and turbulent length scale. Because the model has two new parameters k and ε, the momentum equation is modified as,

$$\frac{\partial(\rho U_i)}{\partial t} + \frac{\partial}{\partial x_j}(\rho U_i U_j) = -\frac{\partial p'}{\partial x_i} + \frac{\partial}{\partial x_j}\left[\mu_{eff}\left(\frac{\partial U_j}{\partial x_i} + \frac{\partial U_i}{\partial x_j}\right)\right] + S_M \tag{11.32}$$

where

μ_{eff} is the effective viscosity with turbulence, and
p' is the modified pressure; it is calculated by,

$$\mu_{eff} = \mu + \mu_t = \mu + C_\mu \rho \frac{k^2}{\varepsilon} \tag{11.33}$$

where μ_t is the turbulence viscosity.

The values of k and ε are obtained from the differential transport equations for the turbulence kinematic energy and turbulence dissipation rate,

$$\left. \begin{aligned} \frac{\partial(\rho k)}{\partial t} + \frac{\partial}{\partial x_j}(\rho U_j k) &= \frac{\partial}{\partial x_j}\left[\left(\mu + \frac{\mu_t}{\sigma_{eff}}\right)\frac{\partial k}{\partial x_j}\right] + P_k - \rho\varepsilon + P_{kb} \\ \frac{\partial(\rho\varepsilon)}{\partial t} + \frac{\partial}{\partial x_j}(\rho U_j \varepsilon) &= \frac{\partial}{\partial x_j}\left[\left(\mu + \frac{\mu_t}{\sigma_\varepsilon}\right)\frac{\partial\varepsilon}{\partial x_j}\right] + \frac{\varepsilon}{k}(C_{\varepsilon 1}P_k - C_{\varepsilon 2}\rho\varepsilon + C_{\varepsilon 1}P_{eb}) \end{aligned} \right\} \tag{11.34}$$

where

$C_{\varepsilon 1}$, $C_{\varepsilon 2}$, σ_k, and σ_ε are the constants,
P_{kb} is the turbulence production due to viscous forces,

$$P_k = \mu_t\left(\frac{\partial U_i}{\partial x_j} + \frac{\partial U_j}{\partial x_i}\right)\frac{\partial U_i}{\partial x_j} - \frac{2}{3}\frac{\partial U_k}{\partial x_x}\left(3\mu_t\frac{\partial U_k}{\partial x_k} + \rho k\right) \tag{11.35}$$

P_{kb} and P_{eb} are the influence of the buoyancy forces that are evaluated by

$$\left. \begin{aligned} P_{kb} &= -\frac{\partial\mu_t}{\partial\sigma_\rho}g_i\frac{\partial\rho}{\partial x_i} && \text{when the full buoyancy model is used (Schmidt number } \sigma_\rho = 1) \\ P_{kb} &= \frac{\mu_t}{\rho\sigma_\rho}\rho\beta g_i\frac{\partial T}{\partial x_i} && \text{when the Boussinesq buoyancy model is used (Schmidt number } \sigma_\rho = 0.9) \end{aligned} \right\} \tag{11.36}$$

$$P_{ke} = \max(0, P_{kb})\sin\phi \tag{11.37}$$

where ϕ is the angle between velocity and gravity vectors; it is set as 90° if the directional option is disabled.

The flow simulation in SolidWorks supports the concurrent computational fluid dynamics (CFD) for heat transfer analysis. The temperature changes in both of fluid and solid regions can be evaluated in the simulation. One unique feature of the flow simulation in SolidWorks is to adopt the Cartesian-based meshes for solid–fluid and solid–solid interfaces. It increases the flexibility and reduces computation significantly. In the following section, the flow simulation in SolidWorks is used as a solver to conjugate heat transfer problems.

11.5.2 **Case study—air-quenching process**

In an air-quenching process of aluminum parts, the most important factor to affect material properties of quenched products is cooling curves. In particular, in the temperature range of recrystallization of the given materials. The goal of the numerical simulation is to predict cooling curves in air-quenching processes. The system parameters are geometric dimensions of parts, rack configuration, and the characteristics of airflow. The air-quenching process is a typical conjugate heat transfer process where energy flow from solid to fluid domains.

Fig. 11.8 shows the typical setup of air-quenching work cell. Parts on the rack are quenched in a tunnel-like structure at the right side of the work cell. The forced airflow is formed by the blowers on the left side. The temperature of inlet airflow is controlled as constant to eliminate the uncertainties of air temperature from different weather or season conditions. The blowers have their capacities to generate high-speed air flow with a controlled volume rate. The upper and the lower airflow channels correspond to higher and lower air pressure zones, respectively.

11.5.2.1 Challenges of air-quenching simulation

In air quenching, usually a large number of metal parts are cooled simultaneously in the same furnace. To achieve a high quality of the quenched products, the heat removal rate in the charge should be as uniform as possible (Macchion, 2005). Numerical simulation makes it possible to

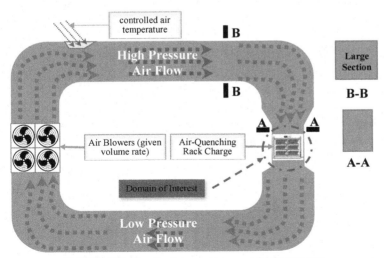

■ **FIGURE 11.8** Schematic of air-quenching process for aluminum parts.

predict the distribution of heat transfer rate in a cost-effective way. However, the simulation for conjugate heat transfers poses two main challenges.

1. **The amount of computation** is the most significant challenge to obtain simulation results in a reasonable time frame. It is in particular true when the model size is large, the time step for iterations is automatically set, or the mesh density is high. Hamman and Berry (2008) introduced their CFD models with the size range of 22—28 GB and 65—85 million elements. High-performance computers were applied to generate meshes, and it took 10 days to run the simulation by an silicon graphics image workstation with 256 processors, 2048 CPUs, 8 cores/node, and 16 GB/node. Moreover, additional challenges included the uncertainties of models, the fidelity of experimental data, and validation of production-type problems.

2. **The characterization of airflow**. The variables for airflow are major inputs for a flow simulation. These variables include properties of air, the magnitude and direction of flow, and turbulence intensity and characteristic lengths. Some parameters, such as the turbulence intensity and length, are hard to be defined; this brings numerous uncertainties of the developed simulation model. Therefore, design variables have been classified in terms of priorities and significances.

11.5.2.2 Air-quenching simulation model

A sample part in Fig. 11.9 is used to illustrate the procedure of FEA modeling for conjugate heat transfer. The parametric study was conducted to evaluate the impact of design variables including part orientation and airflow speed on the heat transfer performance. The heat transfer performance is measured by heat transfer coefficient (HTC) on surface and temperature distribution at a specified time during the quenching process.

Fig. 11.10 shows the continuous domain of the flow simulation model. The domain has a regular volume with six faces, the left and right faces are defined as an inlet and outlet of airflow, respectively. All lateral faces are defined as ideal walls. The boundary conditions are defined based on the operating conditions of a prior test. The airflow is defined as an external flow in the simulation.

Two design variables in the simulation model are the air speed (v ft/min) and the orientation of the part (θ) relative to the air-flow direction. The range of air speed v is determined by the capacity of air pump, which is set as (1157, 1876) ft/min. The orientation of the part varies from 0 to 90 degrees. A conjugate heat transfer is a coupled multiphysics problem, and state variables in the solid domain (temperature T) and flow domain (pressure p and

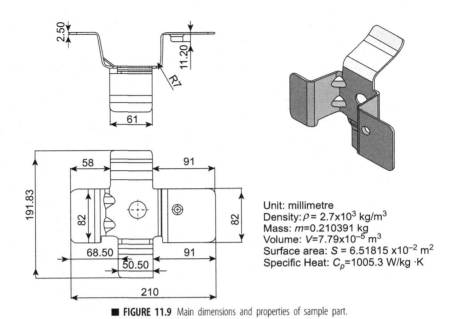

FIGURE 11.9 Main dimensions and properties of sample part.

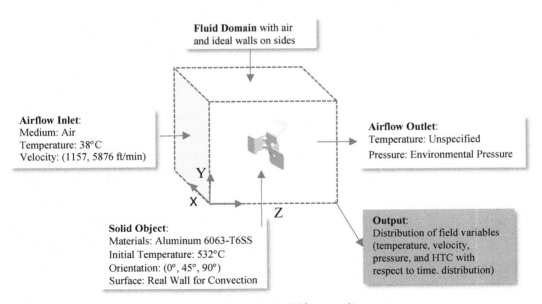

FIGURE 11.10 Flow simulation model for air-quenching process.

velocity v) are solved simultaneously in the flow simulation. These state variables are governed by the equations for mass, momentum and energy conservations. The flow simulation generates the transient data of state variables in both of solid and fluid domain with respect to time.

11.5.2.3 Exemplifying outputs

In air-quenching process, the performance of heat transfer relates greatly on the characteristics of airflow, and these characteristics vary with the conditions of solid—fluid interfaces, in particular, the orientation of part, temperatures on surfaces, and the gradients of temperature on surfaces. Because the conjugate heat transfer is transient, the characteristics of air-flow vary with respect to time. The air-flow is characterized by turbulence intensity and length, density, Prandtl number, and dynamic viscosity, which varies greatly in the regions surrounding solid—fluid interfaces. Fig. 11.11 shows the exemplifying result about the characteristics of air-flow at 6 s for the simulation setup in Fig. 11.11A.

Different from a lumped model with given HTC, HTC over the solid surface varies from one position to another depending on pressure, velocity, and temperature as well as geometric dimensions of part. Fig. 11.12 shows the exemplifying result of HTC distribution at 75 s for the simulation setup in Fig. 11.12A. Fig. 11.12D shows that the maximum temperature difference is around 30°C for the simulated part. Note that due to ununiform HTC over surface, the temperature gradient during quenching may cause residual thermal stress.

11.5.2.4 Exemplifying results from parametric study

FEA can be used for simulation-based optimization. In a parametric study, the simulation is run repeatedly for design variables with different values, the program collects and compares the simulation results to determine the best set of design variables based on the given design criteria.

The first parametric study in this case study is performed to evaluate the impact of the part orientation on heat transfer performance. To this end, the simulation has been run at the fixed flow velocity of 5787 ft/min for a 200-s duration. The part orientation θ is set at five levels (0, 22.5, 45, 67.5, 90 degrees). Fig. 11.13 shows that the results of average temperature and HTC with respect to time, which indicates the impact of the part orientation is ignorable. This result could be explained by the fact that there is no obstacle around the solid domain, so the part orientation does not affect the high-speed forced airflow significantly.

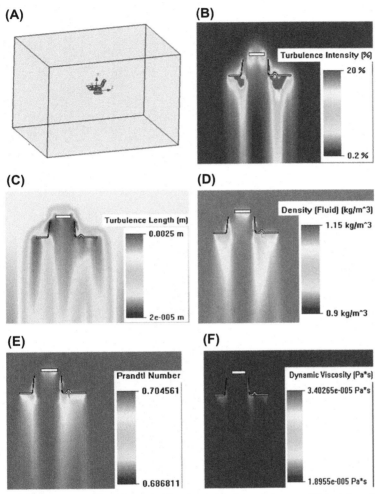

■ FIGURE 11.11 Exemplifying change of airflow characteristics in transient process: (A) simulation setup at 6 s (v = 5787 ft/min; θ = 90 degrees), (B) turbulence Intensity, (C) turbulence length, (D) density, (E) Prandtl number, and (F) dynamic viscosity.

The second parametric study was performed to evaluate the impact of the air-flow velocity on the heat transfer performance. The simulation has been run at the fixed orientation at 45 degrees for a 200-s duration. The airflow velocity from 1157.4 to 5787 ft/min is set at five levels. Fig. 11.14 shows that the results of average temperature and HTC with respect to

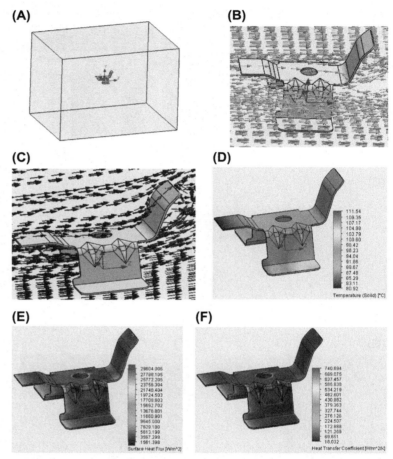

■ **FIGURE 11.12** Exemplifying result of thermal behaviors in transient process: (A) simulation setup at 75 s ($v = 5787$ ft/min; $\theta = 90$ degrees); (B) velocity on H-plane; (C) velocity on V-plane; (D) temperature; (E) surface heat flux; (F) heat transfer coefficient over surface.

time, which indicates the impact of the airflow velocity is significant. This result is aligned well with the results from physical experiments.

Taking a close look of the HTC curve in Fig. 11.14B with a velocity 5787 ft/min, HTC varies significantly with respect to time; this result was verified by the experimental data from the similar operation setting. It is aligned with the results by Shih et al. (2013) as well. Their study showed that HTC measured by the transient techniques differed considerably from the empirical equations in the textbooks, which are based on the assumption of the steady-state condition with isothermal wall.

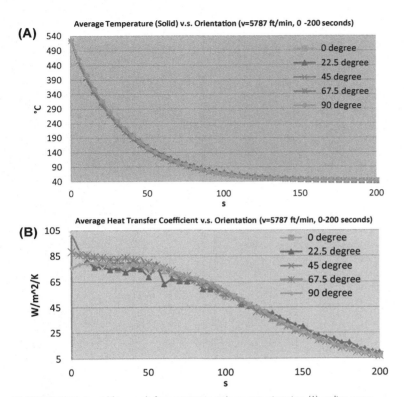

■ FIGURE 11.13 Exemplifying result from parametric study on part orientation: (A) cooling curves; (B) heat transfer coefficients with respect to time.

11.6 FLUID—STRUCTURE INTERACTION

When a fluid flow encounters a structure, flow pressure and drag force apply on the solid object, and this causes the deformation of solid object. In return, the deformation of the solid structure changes the boundary conditions of fluid flow. The fluid—solid interactions are nonlinear multiphysics phenomena that occur to a wide range of applications. The deformation of the structure can be large or small depending on a number of factors such as flow pressure, relative velocity at interface, and the material properties of structure. *Fluid—structure interaction* (FSI) deals with the coupling of structural dynamics and fluid mechanics at interfaces. FSI problems play prominent roles in many scientific and engineering fields. However, a comprehensive study of such problems remains a challenge due to their strong nonlinearity and multidisciplinary nature (Hou et al., 2012).

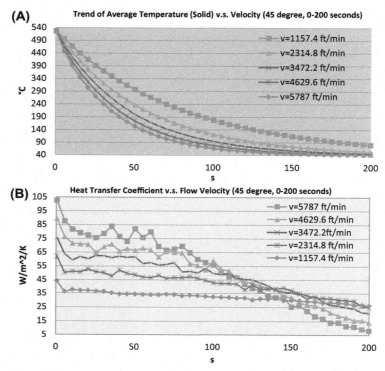

■ **FIGURE 11.14** Exemplifying result from parametric study on airflow velocity: (A) cooling curves; (B) heat transfer coefficients with respect to time.

11.6.1 Description of FSI

An FSI problem is illustrated in Fig. 11.15. The computational domain of an FSI problem is divided into three subdomains: (1) a solid domain (Ω_S), (2) a fluid domain (Ω_F), and (3) an interaction domain (Ω_{F-S}) where the interaction occurs. The solid domain is discretized as a Lagrangian mesh and state variables (u) are nodal displacements. If the structural dynamics must be considered, the derivatives v are calculated as dependent variables of u. The fluid domain is discretized as an Eulerian mesh and state variables are pressures (p) and velocities (v). The state variables in the solid and fluid domains are coupled at the regions of interaction.

Different strategies have been proposed to address FSI problems, and the selection of the most effective approach depends on the characteristics of the problem to be analyzed. Moreover, the mathematical model employed in the representation of fluid behaviors plays a critical role in the selection of the most suited solution procedure. As shown in Fig. 11.1, FSI is a

(A) **(B)**

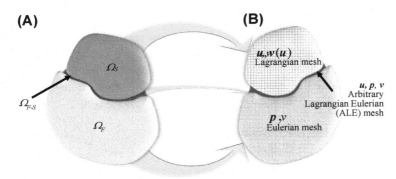

■ **FIGURE 11.15** Fluid—structure interaction problem and discretized domains. (A) Fluid and solid domains; (B) discretized meshes.

multiphysics problem that can be solved by either of a partitioned method or a monolithic method (Gŕetarsson et al., 2010).

In a partitioned method, the structure deformation is quite small and its gradient with respect to time is relatively slow, the dynamic impact of fluid flow on the structure is limited, the fluid—solid interaction can be decoupled by modeling the structure with the resultant pressure and drag force from fluid. For example, if a ship moves at a low or constant speed, the loading conditions of immersed surfaces with water can be simplified as hydraulic water pressure. Pressure and temperature distributions from a computational fluid model are transferred to a structural model as load conditions, although the structural deformation does not affect the flow field. If the impact of solid deformation on fluid flow has to be considered, the solutions of variables involved in the interactions (e.g., force, velocity, and displacement) are passed iteratively from one domain to the other until the convergence is achieved.

In a monolithic method, the structure involves a rigid motion or the structural deformation is relatively large. The gradients of pressure or velocity with respect to time are fast (e.g., more than several hertz), and the structure affects the velocity and pressure of fluid significantly. One needs to treat the FSI in such a case as a coupled multiphysics analysis. The governing equations for the fluid and solid domains are assembled as a holistic system model and solved simultaneously. A transient analysis is defined to handle the coupling of a CFD model and a structural dynamics model. At each time step, loads are transferred from CFD to the structural model and the deformations on the structural model are transferred back for the updating of pressure, velocity, and/or temperature in the next iterative process.

11.6.2 Governing equations of FSI

An FSI problem involves solid and fluid domains where different governing equations are used to represent system behaviors. In addition, the interactions at fluid—solid interfaces impose the constraints of these governing equations. State variables from both of solid and fluid domains must satisfy these constraints.

11.6.2.1 Structural dynamics

If the system dynamics has to be considered in a solid domain, the governing equations can be obtained by including the net force for the acceleration in Eq. (8.4) as,

$$\rho\{\ddot{u}\} - \nabla\cdot[\sigma] - \{f\} = 0 \tag{11.38}$$

where ρ is the density of materials,

$$\{\ddot{u}\} = \begin{Bmatrix} \ddot{u}_x \\ \ddot{u}_y \\ \ddot{u}_z \end{Bmatrix} \text{ is the vector for the accelerations,}$$

$$\nabla = \begin{bmatrix} \dfrac{\partial}{\partial x} & \dfrac{\partial}{\partial y} & \dfrac{\partial}{\partial z} \end{bmatrix} \text{ is the vector for partial derivative operators,}$$

$$\sigma = \begin{bmatrix} \sigma_x & \tau_{xy} & \tau_{xz} \\ \tau_{xy} & \sigma_y & \tau_{yz} \\ \tau_{xz} & \tau_{yz} & \sigma_z \end{bmatrix} \text{ is the Cauchy stress tensor, and}$$

$$\{f\} = \begin{Bmatrix} f_x \\ f_y \\ f_z \end{Bmatrix} \text{ is the vector for the body force.}$$

In the formulation of solid mechanics' problems, local strains $\{\varepsilon\}$ are determined based on state variables $\{u\}$, and the Cauchy stresses $\{\sigma\}$ are determined by their constitutive relations with strains. Therefore, the stress matrix in Eq. (11.38) depends on the state variables $\{u\}$.

The boundary conditions of the fluid domain can be classified into three types as below:

$$\left.\begin{aligned} u_S &= u_{S,BC} & \forall\,(x_S, t) \in \Gamma_{S,1} \times I \\ \sigma_S\cdot n_S &= t_{S,BC} & \forall\,(x_S, t) \in \Gamma_{S,2} \times I \end{aligned}\right\} \tag{11.39}$$

where

$u_{S,BC}$ is the vector of prescribed displacements on the boundary,
$\Gamma_{S,1}, t_{S,BC}$ are the vector of traction forces on the boundary,
$\Gamma_{S,2}, n_S$ are the normal of boundary surface.

The initial condition in the solid domain can be described by,

$$u = u_0, \dot{u} = \dot{u}_0, \quad \forall x \in \Omega_S, t = 0 \tag{11.40}$$

where u_0 and $\dot{u}_0$ are initial displacements and velocities in the solid domain Ω_S.

11.6.2.2 Fluid mechanics

The governing equations for fluid flow have been discussed in Chapter 10. These governing equations include Eq. (10.14) for the mass conservation, Eq. (10.22) for the momentum conservation, Eq. (10.34) for the energy conservation, and the constitutive model of stresses and strains is given in (10.8). The state variables for a fluid mechanics problem are pressure (p) and velocity (v), and these variables are the function of space and time.

The fluid flow consists of the following boundary conditions,

$$\left.\begin{aligned} v_F &= v_{F,BC} & \forall (x_F, t) \in \Gamma_{F,1} \times I \\ \sigma_F \cdot n_F &= t_{F,BC} & \forall (x_F, t) \in \Gamma_{F,2} \times I \end{aligned}\right\} \tag{11.41}$$

where,

$v_{F,BC}$ is the vector of prescribed velocity on the boundary,
$\Gamma_{F,1}$, $t_{F,BC}$ are the vector of traction forces on the boundary, and
$\Gamma_{F,2}$ and n_F are the normal of boundary surface.

The initial condition in the solid domain can be described by,

$$p = p_0, v = v_0, \quad \forall x \in \Omega_F, t = 0 \tag{11.42}$$

where p_0 and v_0 are initial pressures and velocities in the fluid domain Ω_F.

11.6.2.3 Fluid—Structure Interactions

Interactions occur to the fluid—structure interfaces, and this brings the constraints to the state variables in both of solid and fluid domains. The constraints of interactions for velocities and forces can be expressed as,

$$\left.\begin{aligned} \frac{\partial u_s}{\partial t} &= v_F & \forall (x_S, x_F, t) \in \Gamma_{F-S} \times I \\ \sigma_S \cdot n_s + \sigma_F \cdot n_F &= 0 & \forall (x_S, x_F, t) \in \Gamma_{F-S} \times I \\ (v_F - v_R)n_F &= 0 & \forall (x_S, x_F, t) \in \Gamma_{F-S} \times I \end{aligned}\right\} \tag{11.43}$$

where

u_S is the vector of displacements in solid domain,
v_F is the vector of velocities in fluid domain,

$\boldsymbol{\sigma}_S$ and $\boldsymbol{\sigma}_F$ are the vectors of tractions in solid and fluid domains,
$\boldsymbol{n}_S$ and $\boldsymbol{n}_F$ are the normal vectors of solid and fluid boundaries,
$\boldsymbol{v}_R$ is the reference velocity of the fluid domain, and
$\boldsymbol{\Gamma}_{F-S}$ is the fluid—solid boundary.

In both of the partitioned method or monolithic method, the solution to an FSI problem must satisfy the constraints Eq. (11.43) for the coupling of solid and fluid domains. In a portioned method, the constraints of FSI is treated as additional boundary conditions in solid and fluid domains, respectively. The results at each step are exchanged to refine the subsolutions until these constraints are satisfied. In a monolithic method, FEA models for solid and fluid domains are assembled and Eq. (11.43) is treated as a part of boundary conditions to solve state variables in solid and fluid domains simultaneously.

11.6.3 **Mesh adaptation**

FSI often involves in rigid motion or vibration of a structure at the solid—fluid interface; mesh may be changed significantly with respect to time. The conventional Lagrangian method will face the challenge due to a large deformation or the motion of solid object. An arbitrary Lagrangian Eulerian (ALE) mesh can be applied to deal with the overdistortion of the solid domain.

A Lagrangian mesh is used to track nodes of materials. In a Lagrangian mesh, each node is associated with the material particle, and this helps to evaluate the deformation, represent free surface, and define the interfaces of different materials. In addition, it tracks the time dependency of displacements. However, a Lagrangian mesh can be distorted if large displacements occur to structure, the mesh must be modified frequently to adopt the distortion. *An Eulerian mesh* is used to track nodes of volumes. It is widely used in fluid dynamics. An Eulerian mesh is good at handling large distortions for a continuous motion, but the formulation based mesh needs a fine resolution and an intensive computation for the representation of boundary conditions.

Because of the limitation of pure Lagrangian mesh and Eulerian mesh, an ALE mesh is proposed to take advantage of the best features from Lagrangian and Eulerian meshes (Donea et al., 2004). In an ALE mesh, nodal displacements are updated in three ways: (1) be moved with the continuum alike the nodes in a Lagrangian mesh, (2) be held and fixed alike the nodes in a Eulerian mesh, or (3) be moved in an arbitrarily specified way to give a continuous rezoning capability. Due to the flexibility of adjusting the computational mesh during the solving process, an ALE mesh offers the

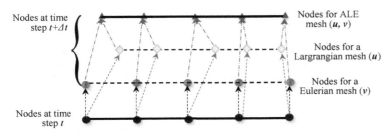

■ **FIGURE 11.16** Arbitrary Lagrangian Eulerian mesh as a trade-off of Largrangian and Eulerian meshes.

better capability of handling larger distortions than a Lagrangian mesh and finer resolutions than a Eulerian approach.

Fig. 11.16 shows the rezoning process of an ALE mesh from time step t to $t + \Delta t$. At any moment, (1) the nodes in a Eulerian mesh are fixed, but the solution to the mesh gives the information of velocities at interactions; (2) the nodes in a Lagrangian mesh are deformed, and the nodal displacements (u) change depending on the level of stress. The displacements can distort the mesh significantly if external forces are large. (3) The nodes at time step $t + \Delta t$ in an ALE mesh are calculated based on the information of v in the Eulerian mesh and u in the Lagrangian mesh.

11.6.4 **Case study—elastohydrodynamic lubrication**

Elastohydrodynamic lubrication (EHL) is a type of fluid-film lubrication where hydrodynamic action can be enhanced greatly by surface elastic deformation; the lubricant viscosity can be increased due to high pressure. It is a typical FSI problem.

As shown in Fig. 11.17, we are interested in the friction characteristics at the fluid—solid interface because the fatigue life of parts at interfaces depends

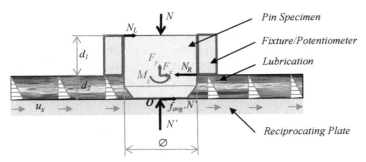

■ **FIGURE 11.17** Free-body diagram of pin specimen in test.

greatly on the magnitude of friction force. The net friction force depends on the magnitude of the normal load and the coefficient of friction (COF). Because COF is indirectly measured by the net friction force. It is necessary to identify any source that can alter the normal load of part significantly. One factor is the fluid flow in the test; the fluid in motion generates the forces and torques on walls. The lifting forces from fluid flow are F_x and F_y along X and Y, and the torque along Z is M_z, respectively. N is the applied load on the pin, u_x is the average sliding velocity of reciprocating plate, N' is reaction force from plate to the pin, and N_L and N_R are the reaction forces from the pin to the fixture.

11.6.4.1 Friction characteristics and major design factors

Friction is the force resisting the relative motion of solid surfaces or fluid layers sliding against each other. NASA (1971) classified friction in term of the differences of lubricant conditions and they provided a guide in formulating design requirements related to friction, wear, and lubrication. Zmitrowicz (2006) gives a comprehensive review on wear caused by friction: depending on the operation conditions, wear patterns were classified into *abrasion, fatigue, ploughing, corrugation, erosion,* and *cavitation.* Hsu et al. (1997) distinguished wear with the rate of damage to the material as "mild wear", "severe wear", and "ultra-severe wear". Different mechanism brings different damage on parts, taking an example of "mild wear", it is mainly caused by tribochemical reaction and plastic deformation.

Generally, friction in tribological components can be characterized by Stribeck curves. A *Stribeck curve* consists of three distinguished regions, including *boundary regime, mixed regime,* and *hydrodynamic regime* (Akbarzadeh and Khonsari, 2010). As shown in Fig. 11.18, a Stribeck curve represents of the change of the coefficient of friction or COF with respect to a Summarfield or Hersey number determined by viscosity, velocity, and pressure (Hironaka, 1984).

Generating a Stribeck curve is an ideal tool to characterize the dependency of COF on various design factors in different applications. Stribeck curves were mostly obtained from experiments. For example, Smith et al. (2001) developed a test setup to obtain the Stribeck curve at the metal-on-metal total hip joints from experiments. In addition, a traditional Stribeck curve uses nondimensional Hersey number (viscosity η) × (speed N)/(Load P). If an additional variable is considered, Stribeck curves could be extended in Stribeck surfaces in 3D or even in higher dimensions. For example, to assess the influence of load and velocity separately, Wang and Wang (2006) defined 3D Stribeck surfaces in journal bearings through numerical

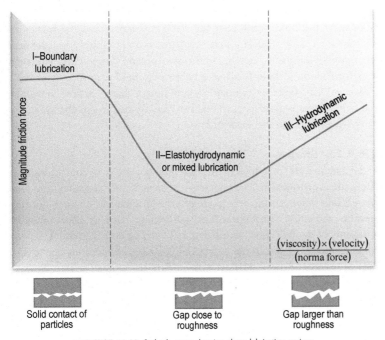

I–Boundary
lubrication

Magnitude friction force

III–Hydrodynamic
lubrication

II–Elastohydrodynamic
or mixed lubrication

$$\frac{(\text{viscosity}) \times (\text{velocity})}{(\text{norma force})}$$

Solid contact of
particles

Gap close to
roughness

Gap larger than
roughness

■ **FIGURE 11.18** Stribeck curve showing three lubrication regions.

simulation. In a 3D Stribeck surface, the x and y coordinates correspond to load and velocity, respectively, and the z-coordinate tells COF when the load and velocity are given.

11.6.4.1.1 Normal loads

In the simplest friction model in solid mechanics, COF is assumed to be constant and the friction is proportional to the normal load. In reality, the normal load is a major factor to determine the friction, and COF is not constant since a varying load changes the deformation and stress distribution of asperity contact. If the load is increased, the gap between two rough surfaces is decreased, more asperities will make contact or change into plastic deformation. The friction is caused by shearing between the boundary layers at the surfaces; a further decrease of the gap will increase the friction until the load is carried solely by asperities.

11.6.4.1.2 Viscosity

Viscosity is a constitutive variable of the Summarfield number in a Stribeck curve. With the given areas and the gap between two surfaces, the resistance from the lubrication depends on viscosity and shear rate. The higher the viscosity of the fluid, the higher the shear resistance, and the higher the COF. However, the viscosity and shear rate are not entirely independent. Accordingly, lubricants can be *Newtonian* or *non-Newtonian*. The shear stress (or friction in fluid flow) of a Newtonian fluid is proportional to the relative velocity between two surfaces, although that of a non-Newtonian fluid is nonlinear with respect to relative velocity due to the change of viscosity with respect to the shear rate.

11.6.4.1.3 Shapes

At the macroscale of contact, the conformance of two contact surfaces affects stress distribution. The contacts can be generally classified into *point contact*, *line contact*, and *area contact*. The macroscale features not only determine the stress distribution for solid contact, but also affect boundary constraints of the fluid flow significantly. To define an EHL model, it poses a great challenge in specifying appropriate boundary conditions due to open inlets or outlets of the fluid flow surrounding the contact points, lines, or areas. Surface textures over a surface, such as dimples and bumps, can be treated as macroscale features in static analysis or flow simulation.

11.6.4.1.4 Surface finish

Surface finish plays a significant role under a boundary or mixed condition. When the film thickness is low in contrast to surface roughness, the film is insufficient to prevent the contacts of asperities; asperity contacts contribute greatly in load sharing. The analysis of lubrication must include the details of surface topography and its characterization, asperity contact pressure, microlubrication pressure, film distribution, and contact temperature (Hua et al., 1997).

11.6.4.1.5 Temperature

Friction generates heat, which might raise temperature in the application environment. Material properties, such as Young's module or viscosity, can vary when the temperature changes. If the range of temperature is significant, its influence on stress distribution, film-thickness, and viscosity must be taken into account. In particular, the reduction of viscosity with respect to a temperature increase is called shear thinning; it reduces the film thickness and changes the load share of asperities and fluid flow consequentially.

11.6.4.2 Mathematic models

The friction at interfaces is affected by lifting and drag forces of lubrication greatly. The interaction at the fluid—solid contact can be formulated as an EHL problem. Here, an EHL model is presented to predict the shearing

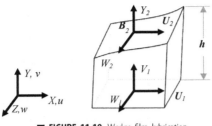

■ **FIGURE 11.19** Wedge-film lubrication.

resistance over surfaces. The shearing resistance relates to many factors such as pressure, part geometry, surface condition, and the attributes of fluid flow.

As shown in Fig. 11.19, let a thin film exist between the two moving surfaces 1 and 2. The former flat and lying in the $X-Z$ plane, the latter curved and inclined. Velocity components U, V, and W occur to X-, Y- and Z-axis, respectively. At any instant moment, two points having the same X and Z coordinates and separated by a distance h will have absolute velocities that give the following boundary conditions,

$$\left.\begin{array}{llll} y = 0, & u = U_1, & v = V_1, & w = W_1, \\ y = h, & u = U_2, & v = V_2, & w = W_2, \end{array}\right\} \tag{11.44}$$

The pressure gradients $\frac{\partial p}{\partial x}$ and $\frac{\partial p}{\partial z}$ are independent of y in a thin film and $\frac{\partial p}{\partial y} = 0$. The mass conservation of an incompressible fluid leads to,

$$\frac{1}{6}\left(\frac{\partial}{\partial x}\left(\frac{h^3}{\mu}\frac{\partial p}{\partial x}\right) + \frac{\partial}{\partial z}\left(\frac{h^3}{\mu}\frac{\partial p}{\partial z}\right)\right) = (U_1 - U_2)\frac{\partial h}{\partial x} - 2(V_1 - V_2)$$
$$+ (W_1 - W_2)\frac{\partial h}{\partial z} + h\frac{\partial}{\partial x}(U_1 + U_2)$$
$$+ h\frac{\partial}{\partial z}(W_1 + W_2) \tag{11.45}$$

Eq. (11.45) was first proposed by Osborne Reynolds in 1886 (Stolarshi, 2000); it is commonly called as Reynold's equation. Because EHL involves fluid—structure interactions, other governing equations including elasticity equations, loading equations, and energy conservation equations should also be included in the system model (Chang, 1995).

11.6.4.3 Case study

The products of interest are sealing parts, which are applied at the interfaces with a reciprocating motion. To improve lubrication performance and

Table 11.3 Give Parameters and Design Variables

Given Parameters

Materials:	Sample:	Turcon polytrafluoroethylene (PTFE) Density: 0.0777–0.0831 lb/in^3 (0.08 lb/in^3) Poisson's ratio: 0.46 $E = 653.2$ (MPa)/94736 (psi) Tensile strength: 15 (MPa)/2176 (psi) Shear strength: 5 (MPa)/725 (psi) Dry friction range: 0.06–0.1
	Plate:	Cast gray iron plate (Brinell = 241, Rc ~ 25) $E =$ **16,000–20,000** ksi (180,000 ksi) Density: **0.258** lb/in^3; Poisson's ratio: 0.211 Tensile strength: **42,500** psi; Shear strength = **57, 000** psi
	Lubrication:	Tonna v68 Density ρ: 880 kg/m^3 (0.03179 lb/in^3); Kinematic viscosity v: 68 mm^2/s (0.1054 in^2/s) Dynamic viscosity: $\mu = \rho v$: 0.00335 lb/in
Operation:	Temperature:	25°

Design Variables

Sample:	Geometries:	Flat and plain bearing surfaces with varying geometric surface profiles
	Textures:	Bumps, flat,...
Plate:	Roughness:	Ra = 0.2–0.4 μm roughness or (7.87 × 10^{-6} –1.57 × 10^{-5} in)
Operation:	Load:	5 to 50 lbf force or (3.33–33.3 lbf/in^2 pressure on 1.5 in^2 area)
	Speed:	5.0–5000.00 mm/min or (0.00328–3.28 in/s)

prolong part lifetime, it is desirable to maintain the full lubrication condition at the interface of two contact surfaces. The lubrication condition relates to many factors, such as part geometry, surface texture and roughness, and optional parameters of reciprocating motions. Table 11.3 has listed the given parameters and variables of the sealing parts.

Three scenarios of friction behaviors at fluid—structure contacts are illustrated in Fig. 11.20. The friction in a fully lubrication condition comes fully from the fluid viscosity, the friction in a partially lubrication condition comes from both of fluid viscosity and particle contact, and the design objective is to predict the operation condition where the partially lubrication condition is transferred into a fully lubrication condition.

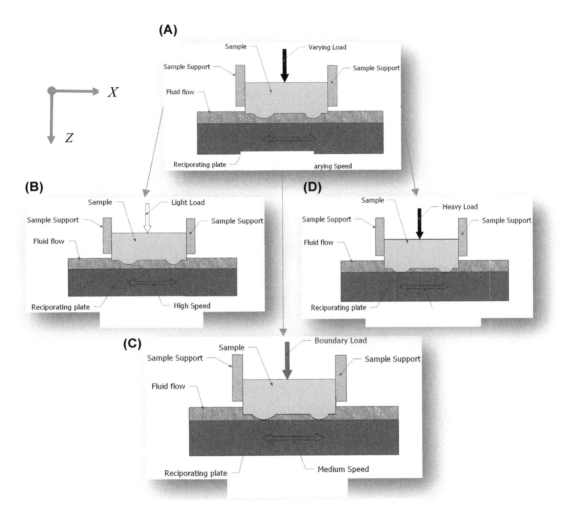

■ **FIGURE 11.20** Test setup and three lubrication conditions: (A) test setup; (B) fully lubrication; (C) boundary lubrication; (D) partially lubrication.

Modeling and simulation on friction are important for a wide variety of engineering disciplines including contact mechanics, system dynamics and controls, air dynamics, fracture and fatigue, and structure dynamics (Berger, 2002). Even though it is a strongly coupling multi-disciplinary problem, there is no integrated model to take into consideration of the contact mechanics and fluid dynamics simultaneously. All of existing solutions adopt an iterative solution; where the constraints in solid mechanics and fluid dynamics were modeled sequentially and iteratively. Because the

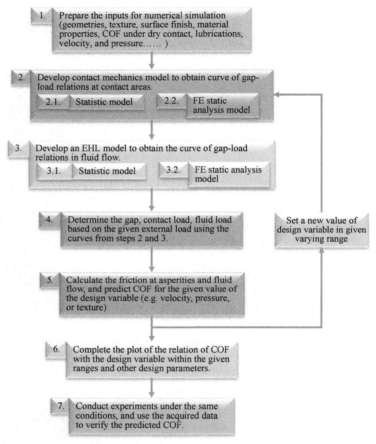

1. Prepare the inputs for numerical simulation (geometries, texture, surface finish, material properties, COF under dry contact, lubrications, velocity, and pressure……)

2. Develop contact mechanics model to obtain curve of gap-load relations at contact areas.
   - 2.1. Statistic model
   - 2.2. FE static analysis model

3. Develop an EHL model to obtain the curve of gap-load relations in fluid flow.
   - 3.1. Statistic model
   - 3.2. FE static analysis model

4. Determine the gap, contact load, fluid load based on the given external load using the curves from steps 2 and 3.

Set a new value of design variable in given varying range

5. Calculate the friction at asperities and fluid flow, and predict COF for the given value of the design variable (e.g. velocity, pressure, or texture)

6. Complete the plot of the relation of COF with the design variable within the given ranges and other design parameters.

7. Conduct experiments under the same conditions, and use the acquired data to verify the predicted COF.

■ **FIGURE 11.21** Proposed finite element analysis modeling procedure for an elastohydrodynamic lubrication problem.

SolidWorks is incapable of modeling EHL problems, the Simulia/Abaqus FEA package is selected to model EHL problems.

The following procedure in Fig. 11.21 is proposed to model and run a simulation for an EHL problem.

Step 1: Prepare inputs for an FEA model and specify a variable to be investigated

To develop a numerical simulation model, it is essential to have sufficient information related to material properties, geometrics, textures, surface finishes, lubrication, boundary conditions, and loading condition. These inputs must be prepared with an appropriate level of accuracy. Design variables,

■ **FIGURE 11.22** Example of simplifying model: (A) complete model; (B) simplified model.

such as load pressure, surface texture, and velocities, have to be investigated individually. It is desirable to specify one design factor for each loop of analysis.

In addition, the contact model under investigation must be simplified as much as possible to minimize the computation in static analysis and fluid flow simulation. As shown in an example of a sample with two bumps on contact surface in Fig. 11.22, the normal load will be shared by two bumps equally, the simulation model might be focused on one bump with half of the external load, and so that the number of elements and nodes can be significantly reduced. Assuming that modeling of one bump takes a million of elements, this simplification requires less than half of memory and computation without affecting the accuracy of prediction.

Step 2: Define a contact model and determine gap-load curve

Asperities will be deformed when a contact surface carries a normal load. The lesser the gap between two nominal surfaces, the higher the external load. With the given characteristics of surface finish, a statistic model or deterministic FEA model can be developed to look into the relation of the external load with the gap between two nominal surfaces. Fig. 11.23

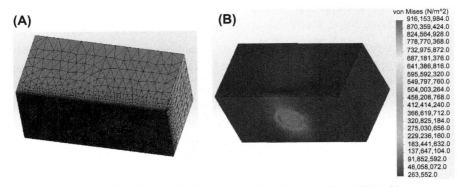

■ **FIGURE 11.23** Example of simplifying model: (A) complete model; (B) simplified model.

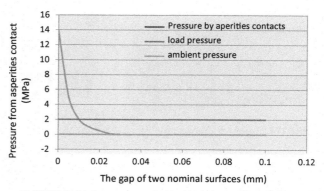

■ **FIGURE 11.24** Average pressure by asperities with respect to surface gap.

shows an example of static analysis where the stress distribution over a bump can be estimated based on the given gap between two nominal surfaces.

If multiple runs under different specified gaps are performed by parametrical design, the relation of mean pressure by asperities and the gap can be obtained. As illustrated in Fig. 11.24, once the relation curve is found; it tells what load the asperities at the contact surface carries under any given gap by an interpolation.

Step 3: Define an EHL model and determine gap-load curve

According to Johnson et al. (1972), the distribution of pressure and film thickness of rough surfaces are very similar to those of two smooth surfaces if the gaps of two rough surfaces or two smooth surfaces are same. Therefore, the tribological behaviors at the contact surfaces are modeled as the thin fluid flow with a given gap between two smooth surfaces. Similarly, to analysis of contact mechanics, either a statistic model or fluid flow simulation can be used to define the film thickness—load curve.

Fig. 11.25 shows an example of flow simulation where the pressure distribution over a bump can be calculate based on the given gap between the two nominal surfaces.

If multiple runs under different specified gap are performed by parametrical design, the relation of mean pressure in EHL model and the gap can be obtained. As illustrated in Fig. 11.26, once the relation curve is found, the load the fluid flow at the contact surface carries under any given gap can be found by interpolation.

(A)

(B)

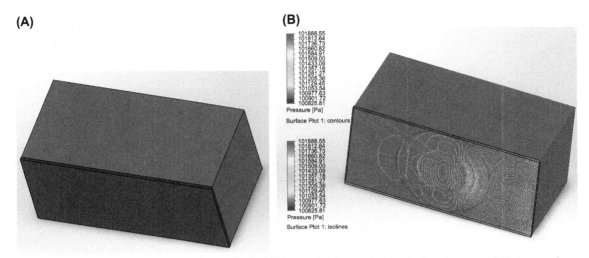

■ **FIGURE 11.25** Example of flow simulation of elastohydrodynamic lubrication: (A) dense meshes of fluid volume; (B) pressure distribution on surface.

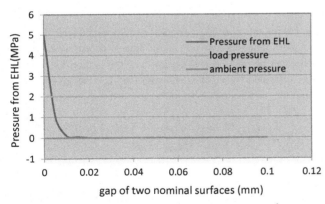

■ **FIGURE 11.26** Average pressure by fluid flow with respect to surface gap.

Step 4: Calculate the gap under the given load based on load-sharing principle

The normal load over contact surface is shared by asperities and fluid flow. The parts of load of both depend directly on the gaps between two nominal surfaces. Using the curves defined in steps 2 and 3, the pivot point can be found based on the force balance over the coupon, i.e., the external load should equal to the sum of the loads by contact and fluid pressure under the given gap. Combining the relations in Figs. 11.24 and 11.26, the sum of pressures must be balanced by external load pressure; it leads to the

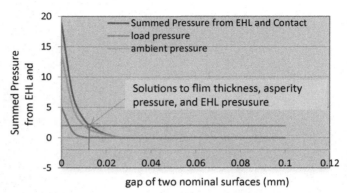

■ **FIGURE 11.27** Determination of film thickness, asperity, and flow pressure based on force balance.

film thickness (the gap corresponding to pressure balance), asperity pressure, and flow pressure under given working condition (in particular, reciprocating speed) as shown in Fig. 11.27.

Step 5: Evaluate the friction from contact and fluid flow

When the gap of the nominal surfaces is determined, the distribution of pressure of both contacts and fluid-flow area is determined, the associate friction force can be calculated accordingly. COF is then found by the total friction force divided by the total normal load for the given value of the design variable.

$$\text{COF} = \frac{\int_{A_{\text{contact}}} \mu(p_{\text{contact}}) \cdot p_{\text{contact}} dA + \int_{A_{\text{EHL}}} \eta\left(\frac{u}{h}\right) dA}{p_{\text{load}} A_{\text{normal}}} \quad (11.46)$$

where

μ, COF for dry sliding condition,
η, viscosity at room temperature,
p_{load}, pressure on pin,
U, reciprocating speed,
h, film thickness,
A_{contact}, asperity contact area,
A_{EHL}, lubricated areas.

Step 6: Change design variable for next iteration

If the plot of COF with respect to a design variable is required, the value of a design variable (e.g., velocity or pressure) is changed in the given range, the steps from step 2 to step 5 are performed repeatedly to define multiple points on the plot as a complete Stribeck curve.

Step 7: Verify results against the measured data from experiments.

The results from a numerical simulation must be verified to ensure the correctness. The most reliable way for the verification is through experiments. It is desirable to conduct a set of experiments under the same conditions specified in the simulation model. The simulation results can be compared with the data collected from experiments to see if the developed numerical simulation is countable.

11.7 **SUMMARY**

Multiphysics is involved in many engineering applications. Even though fully decoupling is not achievable, the principles of complexity decomposition are still applicable in developing concise FEA models for multiphysics problems. The characteristics of various mathematic models are discussed, and some strategies to deal with the coupling of governing equations in different disciplines are introduced. The solutions to multiphysics problems are classified into partitioned methods and monolithic methods. The case studies on structural—thermal problems and conjugate heat transfer problems are used to illustrate the implementation of monolithic methods. A new procedure has been proposed to deal with the strongly coupling in an EHL problem as a demonstration of portioned methods.

REFERENCES

Akbarzadeh, S., Khonsari, M.K., 2010. Effect of surface pattern on Stribeck curve. Tribology Letters 37, 477—486.

Anderson, W.J., 1994. Linear, Static Finite Element Analysis. Automated Analysis Corporation, 2805 South Industrial Highway, Ann Arbor, MI 48104. http://www-personal.umich.edu/~billa/LINEARSTATICFEA.TEXT.pdf.

ANSYS, 2009. ANSYS CFX-Solver Theory Guide. http://www.ansys.com.

ANSYS, 2013. ANSYS Mechanical APDL Element Reference, version 15.0. Southpointe. 275 Technology Drive, Canonsburg, PA. http://148.204.81.206/Ansys/150/ANSYS%20Mechanical%20APDL%20Element%20Reference.pdf.

Berger, E.J., 2002. Friction Modeling for Dynamic System Simulation. http://www.mae.virginia.edu/NewMAE/wp-content/uploads/2012/10/berger_amr.pdf.

Chang, L., 1995. Deterministic modeling and numerical simulation of lubrication between rough surfaces — a review of recent developments. Wear 184, 155—160.

Donea, J., Herta, A., Ponthot, J.-P., Rodriguez-Ferran, A., 2004. Arbitrary Lagrangian-Eulerian methods. In: Encyclopaedia of Computational Mechanics. John Wiley & Sons, Ltd, pp. 413—437.

Felippa, C., 2017. Advanced Finite Element Methods (ASEN 6367). University of Colorado at Boulder. http://www.colorado.edu/engineering/CAS/courses.d/AFEM.d/.

Gasmi, A., Sprague, A., Jonkman, J.M., Jones, W.B., 2013. Numerical stability and accuracy of temporally coupled multi-physics modules in wind-turbine CAE tools.

In: The 51th AIAA Aerospace Sciences Meeting Including the New Horizons Forum and Aerospace Exposition, Grapevine, Texas, 7–10 January 2013. http://www.nrel. gov/docs/fy13osti/57298.pdf.

Gretarsson, J.T., Kwatra, N., Fedkiw, R., 2010. Numerically Stable Fluid-structure Interactions between Compressible Flow and Solid Structures. http://physbam. stanford.edu/~fedkiw/papers/stanford2010-04.pdf.

Hamman, D.K., Berry, R.A., 2008. A CFD M&S process for fast reactor fuel assemblies: experimental and CFD code applications to nuclear Reactor Safety. INL/CON-08–14131.

Hironaka, S., 1984. Boundary lubrication and lubricants. In: Three Bond Technical News. https://www.threebond.co.jp/en/technical/technicalnews/pdf/tech09.pdf.

Hosain, L., Fdhila, R.B., 2015. Literature review of accelerated CFD simulation methods towards online application. Energy Procedia 75 (2015), 3307–3314.

Hou, G., Wang, J., Layton, A., 2012. Numerical methods for fluid-structure interaction — a review. Communications in Computational Physics 12, 337–377.

Hsu, S.M., Shen, C., Ruff, A.W., 1997. Wear prediction for metal. Tribology International 30 (5), 377–383.

Hua, D.Y., Qiu, L., Cheng, H.S., 1997. Modeling of lubrication in micro-contact. Tribology Letters 3, 81–86.

Johnson, K.L., Greenwood, J.A., Poon, Y., 1972. A simple theory of asperity dynamic lubrication contact in elastohydrodynamic lubrication. Wear 19, 91–108.

Laird, G., 2015. Thermal-stress analysis. Predictive Engineering. http://www. predictiveengineering.com/system/files_force/thermal-stress_analysis_theory_and_ practices_-_predictive_engineering_white_paper.pdf?download=1.

Ma, H., 2012. Computational Fluid Dynamics and Heat Transfer Analysis for a Novel Heat Exchanger (MS thesis). Lehigh University.

Macchion, O., 2005. CFD in the Design of Gas Quenching Furnace. Department of Mechanics, Royal Institute of Technology, S-100 44 Stockholm, Sweden.

Malatip, A., Wansophark, N., Dechaumphai, P., 2009. Finite element method for analysis of conjugate heat transfer between solid and unsteady viscous flow. Engineering Journal 13 (2), 43–58.

NASA, 1971. Lubrication, friction, and wear. In: NASA Space Vehicle Design Criteria, NASA SP-8063.

Pawlowski, R., Bartlett, R., Belcourt, N., Hooper, R., Schmidt, 2011. A theory manual for multi-physics code coupling in LIME. In: SAND2011-2195. Sandia National Laboratories, Albuquerque, New Mexico 87185 and Livermore, California 94550.

Rugonyi, S., Bathe, K.J., 2001. On finite element analysis of fluid flows fully coupled structural interactions. CMES 2 (2), 195–212.

Shih, T.I.-P., Ramachandran, S.G., Chyu, M.K., 2013. Time-accurate CFD conjugate analysis of transient measurements of the heat-transfer coefficient in a channel with pin fins. Propulsion and Power Research 2 (1), 10–19.

Smith, S.L., Dowson, D., Goldsmith, A.A., 2001. The lubrication of metal-on-metal total hip joints: a slide down the Stribeck curve. Proceedings of the Institution of Mechanical Engineers, Part J: Journal of Engineering Tribology 215 (5), 483–493.

Sobachkin, A., Dumnov, D., Sobachkin, A., 2014. Numerical basis of CAD-embedded CFD. In: White Paper. NAFEMS World Congress 2013.

SolidWorks, 2012. Enhanced Turbulence Modelling in SolidWorks Flow Simulation.

Stolarshi, T., 2000. Tribology in Machine Design. Elsevier Science. ISBN-13: 9780750636230, ISBN: 0750636238.

Tordini, D., 2014. It Takes Two: Multiphysics Analysis Solutions from Hawk Ridge Systems. http://www.hawkridgesys.com/blog/multiphysics-analysis-solutions-hawk-ridge-systems/.

Wang, Y., Wang, Q.J., October–December 2006. Development of a set of Stribeck curves for conformal contacts of rough surfaces. Tribology Transactions 49 (4), 526–535.

Zienkiewicz, O.C., Taylor, R.L., 2013. The finite element method for fluid dynamics, seventh ed. McGraw-Hill, 225 Wyman Street, Watham, MA 02451, U.S.A. ISBN-10: 1856176355.

Zmitrowicz, A., 2006. Wear patterns and laws of wear — a review. Journal of Theoretical and Applied Mechanics 44 (2), 219–253.

FURTHER READING

ANSYS Mechanical APDL Theory Reference. http://148.204.81.206/Ansys/150/ANSYS%20Mechanical%20APDL%20Theory%20Reference.pdf.

Balaba, G., 2012. A Newton's Method Finite Element Algorithm for Fluid-Structure Integration (MS thesis). Faculty of Mathematics and Natural Sciences University of Oslo.

COMSOL, 2013. Heat Transfer Module User's Guide. Version 4.3.

Jia, X., Zheng, G., Alexander, P.A., Campbell, M.T., Lawlor, O.S., Norris, J., Haselbacher, A., Heath, M.T., 2006. A system integration framework for coupled multi-physics simulations. Engineering with Computers. http://dx.doi.org/10.1007/s00366-006-0034-x.

Petrova, R., Chernev, S., March 28, 2012 (Chapter 16). In: Radostina Petrova (Ed.), Integrated Technology for CAD Modeling and CAE Analysis of a Basic Hydraulic Cylinder, ISBN 978-953-51-0445-2. Under CC BY 3.0 license.

■ PROBLEMS

11.1 A round bar in Fig. 11.28 is with the diameter of $\phi = 1$-in; the bar material is silicon. The upper half surface is coated by ceramic porcelain; the thickness of the coating is 0.001 in. The coating is

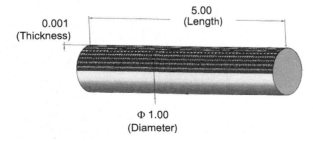

■ **FIGURE 11.28** Coated silicon bar coated by porcelain with temperature change.

applied at room temperature (25°C); the specimen is subjected to the temperature change in (−10°C–60°C), determine the range of maximized thermal stress caused by temperature change.

Properties	Silicon	Ceramic Porcelain	Unit
Elastic modulus	1.63e7	3.2e7	psi
Poisson's ratio	0.28	0.22	
Density	0.0841766	0.0830927	lb/in³
Yield strength	17,404	2.5e4	psi/°C
Thermal expansion coefficient	2.6e-006	6e-6	
Thermal conductivity	0.00165847	2.0e-5	Btu/(in s °C)
Specific heat	–	0.209697	Btu/(lb °C)

11.2 A welded tube frame in Fig. 11.29 is made of case iron, it is fixed at the ground at the room temperature 25°C; analyze the thermal stress and deflection when the surrounding temperature is raised to 100°C. The material properties are elastic modulus: 9,598,321.907 (psi), Poisson's ratio: 0.27.

Density: 0.260116 lb/in³, yield strength: 21,996.13322 psi, thermal expansion coefficient: 6.67e-6/°C, thermal conductivity: 0.000601864 Btu/(in s °C), and specific heat: 0.121811 Btu/(lb °C).

11.3 Fig. 11.30 shows an example of shrink-to-fit assembly. The pin is dipped in liquid nitrogen and allowed to cool to −200°C. The block is heated to 200°C. For ease of assembly, the pin is sized to have a radius 0.025 mm smaller than the hole in the block. Determine if this shrink fit will cause plastic deformation in the block at room temperature 25°C.

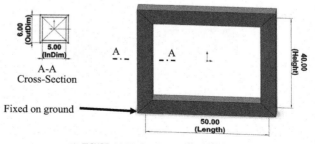

■ **FIGURE 11.29** Dimensions of welded frame.

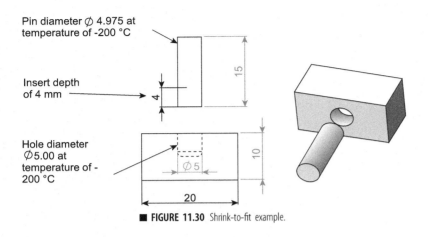

■ **FIGURE 11.30** Shrink-to-fit example.

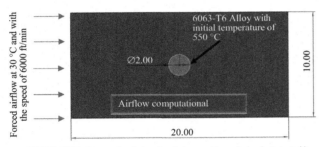

■ **FIGURE 11.31** Computational domain of air-quenching cylinder (unit in inch).

11.4 As shown in Fig. 11.31, a long cylinder with aluminum 6061-T6 is heated to 532°C, it is long and its diameter is 2-in. It is then air quenched by an airflow with flow speed of 5876 ft/min and temperature of 39°C. Create a 2D simulation model to (1) predict the average temperature of the cylinder with respect to time in 500 s, (2) plot the distribution of airflow velocity in your computational domain in 500 s.

Validation and Verification

12.1 INTRODUCTION

Validation and verification (V&V) aims to enhance the reliability of finite element analysis (FEA)-based simulation models. V&V are conducted based on quantified errors from a comparison of simulation and experimental data. V&V must ensure that the errors from computer simulation be acceptable as the solutions to their design problems. However, the FEA modeling procedure involves in many steps, and each step may be error sources, and accumulated errors may lead to unacceptable discrepancy of simulated solution and exact solution.

V&V should be taken at every step to decide if the induced error at that step is within the expected tolerance. If not, addition efforts are required to refine the FEA model and solver. The topic of accuracy is often overlooked in the practice of FEA-based simulation by end users (Pointer, 2004). In this chapter, we will discuss modeling error sources, error quantifications, and some techniques to analyze and minimize errors.

Moorcroft (2012) indicated that the reliability of a numerical simulation depends largely on the quality and accuracy of model. The model quality can be justified by many criteria; for example, what is the credibility of software? What is the solution accuracy of the software? Failing to have appropriate V&V could lead to significant losses. A classic example of simulation failures was the sinking of the Sleipner-A oil platform (see Fig. 12.1). According to Selby et al. (1997) and Collins et al. (2000), the sinking was caused by the inappropriate use of the FEA tool in the detailed design stage. The mesh in an FEA model was too coarse to predict shear stresses of the platform poles accurately. The shear stresses were underestimated around 45%.

FEA solutions are approximated solutions, and the discrepancy of an FEA solution from the exact solution depends on numerous factors such as the assumptions of models, element types, interpolation, or types of solvers. In using FEA methods, such discrepancies should be quantified and

Finite Element Analysis Applications. http://dx.doi.org/10.1016/B978-0-12-809952-0.00012-1

■ **FIGURE 12.1** Sleipner A oil platform. *Adopted from Collins, M.P., Vecchio, F.J., Selby, R.G., Gupta, P.R., March/April 2000. Failure of an offshore platform. Structures 43—48. http://www.civ.utoronto.ca/vector/ journal_publications/jp33.pdf.*

overseen by V&V. V&V are the primary means to assess the creditability of computer models and simulation (Oberkamph and Trucano, 2002).

Verification is a process to determine if a computational model represents the underlying mathematical model and its solution accurately. Verification aims to identify and minimize the errors in a computer model. A closely related concept is validation. *Validation* refers to a process to determine the degree to which a model is an accurate representation of the real world. Validation is conducted from the perspective of the intended use of model. Validation aims to evaluate the results of FEA model against test evidences (ASME, 2006a,b).

The relations of verification and validation are illustrated in Fig. 12.2. V&V is to ensure the correctness of computer models and corresponding simulations. *Verification* is to ensure that a mathematical model can be appropriately solved in computers; although *validation* is to ensure that the simulation result matches the exact solution to original engineering problems. Verification is to check whether or not an FEA user should be going through when developing an FEA model, and validation is about how to

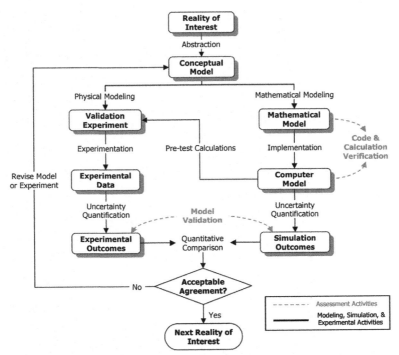

■ **FIGURE 12.2** Validation and verification in numerical simulation. *Adopted from Thacker, B.H., Doebling, S.W., Hemez, F.M., Anderson, M.C., Pepin, J.E., Rodriguez, E.A., 2004. Concepts of Model Verification and Validation, LA-14167-MS, Los Alamos National Laboratory, US Department of Energy. https://www.osti. gov/scitech/servlets/purl/835920/.*

check the FEA model against available test data. The main difference of verification and validation is given in Table 12.1. Validation is supported by verification. The purpose of verification is to assess both the software and the numerical solution it produces (Knupp, 2016).

In a summary, V&V are the processes that collect the evidence of a model's correctness or accuracy for a specific scenario. Reliable results from FEA rely on (1) good understanding of original physical systems and behaviors, (2) good understanding of governing equations of systems, (3) appropriate element types and sizes, (4) adequate V&V in the procedure of modeling, and (5) evaluations of simplifications and assumptions. After all, the users of FEA software tools must understand various error sources. Finally, V&V activities are associated with the specified contents in the simulation. However, note that V&V for one model may not be applicable to other models (Thacker et al., 2004).

Table 12.1 Difference of Verification and Validation

Verification	Validation
<ul><li>Ensure the simulation works to the formulated design problem,</li><li>Define and select correct mathematical models, and solve models correctly, and</li><li>Produce the simulation results for the defined model with the acceptable level of accuracy.</li></ul>	<ul><li>Ensure the simulation works as it is supposed to be,</li><li>Represent a physical system correctly, and</li><li>Check the finite element analysis result with test data or proven results.</li></ul>

12.2 PLANNING OF V&V IN FEA MODELING

To practice V&V, planning is the most critical step to minimize possible errors in FEA modeling. As shown in Fig. 12.3, every step in FEA modeling generates new information about the system to be modeled. The activities in these steps are also error sources. Therefore, along with the information flow, information/data in one format is processed, and converted to the

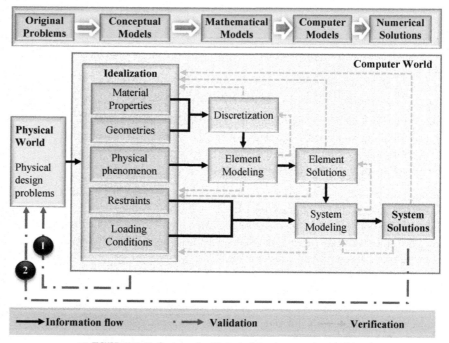

■ **FIGURE 12.3** Verification and validation in finite element analysis modeling.

information/data in another format, corresponding V&V has to be performed to ensure errors are not propagated or accumulated.

V&V relates closely to the objective of FEA simulation. Therefore, the objective of FEA simulation must be clearly stated. This begins with the identification of design variables and system parameters. The design problem must be formulated in a comprehensive way to define a complete FEA model, which includes the source of data, the assumptions from idealization, terminating conditions in solving processes, and the procedures, methodologies, tools, and criteria for V&V. In formulating an FEA problem, the free body diagram (FBD) based approach can be used to identify boundary conditions and loads. In addition, commercial FEA code often provides a taxonomy of element types. This helps users to understand software capabilities and select right element types for given analysis problems.

From the perspective of V&V, it is interesting to look into the role of the idealization in an FEA modeling process. On the one hand, the idealization is based on a real-world design problem, and the simplified model should be a reasonable representation of the original problem. On the other hand, the idealization is directly related to verification. In other words, the idealization must be verified to endure that the conceptual model is converted into a mathematical model adequately.

12.3 **SOURCES OF ERRORS**

As discussed in Section 12.2, an FEA procedure consists of multiple modeling steps and each step brings the possibility of new discrepancy of the computer representation from an ideal one. Understanding error sources is crucial to justify whether or not the obtained results at every step are acceptable; this helps users to select correct analysis types, element types, meshes, and solvers in minimizing errors (Shah, 2002). Identifying possible discrepancies may result in an improvement of the model and the reduction of overall errors of the simulation model (Brinkgreve and Engin, 2013). In this section, the quantification of error is introduced, and the error sources are discussed.

12.3.1 **Error quantification**

Solving an FEA model usually requires an iterative algorithm, in which the terminating condition must be defined to identify an acceptable solution. The terminating conditions are mostly based on quantified errors. For a simplified comparison, an error is mostly quantified as a vector or scalar.

Without losing the generality, the displacements in structural analysis are used as the quantities for the error quantification of state variables.

FEA is applied to find the distribution of state variables such as displacements in structural analysis or associated scalar variables such as natural frequencies in modal analysis. Assume u be a vector of state variables to be determined, the error of an approximated solution can be defined as,

$$\{\delta\} = \{\overline{u}\} - \{\widehat{u}\} \tag{12.1}$$

where

δ is the error of the approximated solution, and
$\{\overline{u}\}$ and $\{\widehat{u}\}$ are the exact and approximated solutions, respectively.

Because a continuous domain has been represented by discretized nodes and elements in a FEA model. The size of vector $\{\widehat{u}\}$ shows how many degrees of freedom are applied to approximate the quantities in a continuous domain. To avoid artificial inaccuracies caused by local features in the quantification, the size of vector $\{\widehat{u}\}$ should be reasonably large enough. For example, a point load is impossible in real life, an idealization on point load will cause an infinitely large displacement and stress at the imposed node; the solution error should be defined for the entire solid domain rather than the local region where the point load is applied.

In the FEA implementation, various norms can be used to convert the vector of errors $\{\delta\}$ in Eq. (12.1) into scalar parameters as summed errors. Taking an example of a system model for structural analysis,

$$[K]\{u\} = \{F\} \tag{12.2}$$

An error of the approximation can be defined in an energy norm as (Shah, 2002),

$$\|\delta\| = \left[\int_{\Omega} (\{\overline{u}\} - \{\widehat{u}\})^{T} \cdot [K] \cdot (\{\overline{u}\} - \{\widehat{u}\}) d\Omega \right]^{1/2} \tag{12.3}$$

where $\|\delta\|$ is a scalar measure from the vector of errors relating to the approximated solution.

In addition, the variation for a relative energy norm error can be defined as,

$$\eta = \frac{\|\delta\|}{\|\overline{u}\|} \times 100\% \tag{12.4}$$

Eq. (12.14) can be extended and applied to quantify the errors for any vector of variables.

12.3.2 **System inputs**

As shown in Fig. 12.3, to represent a real-world engineering problem by a conceptual computer model, many assumptions have to be made in the idealization and these assumptions bring numerous of errors or uncertainties. For example, material properties are essential input for any FEA. Unfortunately, a great deal of uncertainties is raised in defining materials properties. There is a very wide range of materials used for structures with drastically different behaviors. For each material type, it may go through several response regimes, i.e., elastic, plastic, viscoelastic, cracking and localization, and fracture. Although it is impractical and unnecessary to define the accuracy of the constitutive model in modeling, users are responsible to determine how the given materials are used and what assumptions should be made. In defining materials, at least the behaviors of materials in Table 12.2 should be taken into consideration.

When the default settings of materials are applied, a user should be aware of the assumptions underlying these settings. If any of the assumptions is not aligned well with the given application, one has to customize materials in FEA.

Table 12.2 Basic Properties of Materials

Assumption	Explanation
Elasticity	The stress—strain response is reversible, and consequently, the material has a preferred natural state. The natural state is taken at a reference temperature with no load; it is referred as unstressed and undeformed state. When a load is applied or temperature changes, the material develops nonzero stresses and strains, and moves to occupy a deformed configuration.
Linearity	The material behaves in its elastic region. The stress is proportional to the strain at any position. Doubling a stress means to double the corresponding strain, and vice versa.
Isotropy	The material properties are not sensitive to load directions. This is a good assumption for materials such as metals, concrete, and plastics. It is inadequate for the materials with heterogeneous mixtures such as composites and reinforced concrete. Those materials are anisotropic by nature.
Small Strains	A strain is considered small when its magnitude is well within the elastic range of materials. The change of geometry can be neglected as loads are applied on the materials. If a strain exceeds certain level, nonlinear constitutive relation must be defined to model the relation of displacements and strains.

12.3.3 **Errors of idealization**

Idealization is to describe and abstract a physical system as a conceptual model based on the assumptions. A conceptual model is the collection of computer representations for physical objects, processes, or systems (Moorcroft, 2012). Developing a conceptual model involves (1) identifying objects, domains of interest, and the relations of objects in their applied environments, (2) determining the level of agreement between the experiment and simulation outcomes, (3) making assumptions in the representations of physical processes, and (4) specifying the failure modes of interests as well as validation metrics (Thacker et al., 2004). It is the users' responsibility to develop such a conceptual model.

Typical tasks in the idealization: (1) the system of interest has to be isolated from its residential environment to identify its boundary conditions; (2) constitutive models and parameters have to be determined to represent material properties; (3) a virtual model is created to describe the continuous domain of physical object or system; (4) the system physical phenomenon must be clear to identify an appropriate analysis type or mathematic model of elements. Every activity involved in the idealization introduces some simplifications that could cause the discrepancies of a virtual computer model and physical system. The conceptual model should represent the original system adequately, and this can be proven by performing the validations on both of inputs and outputs of FEA models.

FEA treads a continuous domain as a set of discretized elements and nodes, and the state variables in an element are interpolated by shape functions. In addition, shape functions are mostly linear, quadratic, or cubic polynomial equations with a low order. Obviously, the discretization is the source of error because the solution to the continuous domain is approximated by nodal values discretely. Fig. 12.4 gives an example of the strain distribution of a metal plate under an axial load. Fig. 12.4A is for the discontinued strain across elements in a coarse mesh. The results can be improved by using (1) quadratic elements in Fig. 12.4B and (2) more fine elements in Fig. 12.4C.

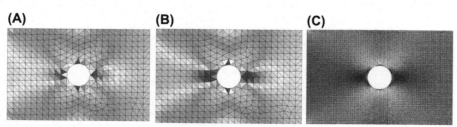

(A)   **(B)**   **(C)**

■ **FIGURE 12.4** Example of strain discontinuity due to discretization: (A) coarse mesh with linear triangle element; (B) coarse mesh with quadratic triangle element; (C) fine mesh with linear triangle element.

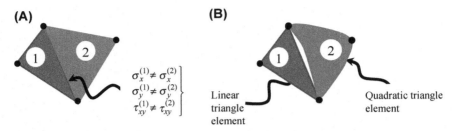

(A) **(B)**

$$\left.\begin{array}{c}\sigma_x^{(1)} \neq \sigma_x^{(2)}\\ \sigma_y^{(1)} \neq \sigma_y^{(2)}\\ \tau_{xy}^{(1)} \neq \tau_{xy}^{(2)}\end{array}\right]$$

Linear triangle element

Quadratic triangle element

■ **FIGURE 12.5** Errors caused by a model with basic elements: (A) stress discontinuity; (B) displacement discontinuity.

However, the errors of discretization exist for whatever sizes and types of elements an FEA model uses.

It can also be seen that a stress crossing a share edge of two elements is discontinued. Fig. 12.5A has shown such a discontinuity. Note that if a share edge relates to nonlinear high-order elements, even the displacement can be discontinued and Fig. 12.5B shows such an example.

From the perspective of FEA modeling, the way of using shape functions to construct mass or stiffness matrices of elements is an approximation of exact mathematic model in a continuous domain. This brings discretization errors. There is no universal rule to warrantee the appropriation of mesh size; however, the convergence study can be performed in the area of interest to test the appropriateness of mesh by a number of iterations.

12.3.4 **Errors of mathematic models**

After the conceptual model is formulated from the idealization, *a mathematical model* has to be defined. Usually, a mathematical model in an FEA model is the representation of governing conditions with specified boundary conditions, initial conditions, and system parameters that are required to describe the corresponding conceptual models. Taking an example of the problems in mechanics, a mathematic model is for the representation of partial differential equations for the conservation of mass, momentum, and energy; the model also specifies the spatial and temporal domain, initial boundary conditions, as well as material properties. On the other hand, a *computational model* refers to the numerical implementation of a mathematic model. A computational model includes a set of discretized nodes and elements, solution algorithms, and terminating criteria. Fig. 12.6 shows an example of the mathematical model for a cantilever beam under the distributed load. The model consists of a differential equation for y-axis displacement (y), boundary conditions for the nodal displacements at $x = 0$, L, and load $w(x)$, and system parameters $(E, I, \text{ and } L)$.

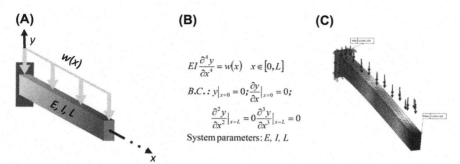

(A) **(B)** **(C)**

$$EI\frac{\partial^4 y}{\partial x^4} = w(x) \quad x \in [0, L]$$

$$B.C.: y\big|_{x=0} = 0; \frac{\partial y}{\partial x}\big|_{x=0} = 0;$$

$$\frac{\partial^2 y}{\partial x^2}\big|_{x=L} = 0\frac{\partial^3 y}{\partial x^3}\big|_{x=L} = 0$$

System parameters: E, I, L

■ **FIGURE 12.6** Example of mathematical model: (A) conceptual model of cantilever beam; (B) mathematical model of cantilever beam subjected to boundary conditions; (C) numerical solution of mathematical model.

A computational model consists of computer programs and the assumptions made in conceptual and mathematic models. To solve a computational model, the inputs of programs are constitutive models and loads, mesh types and density, analysis types, and terminating conditions.

12.3.5 **Errors of model or analysis types**

Element types are selected to represent the most significant features of a physical system or object. To make the complexity of model manageable, not all of the physical features are represented in details. For example, the enclosure of an airplane engine is generally modeled as shell elements, although the wall thickness of some areas is not suitable to be treated as shell elements. If these features are modeled as beam elements, the joining positions should be located at the grid points of shell elements rather than at flanges (Chen, 2001).

It is the primary responsibility of an FEA user to create a computer model. An FEA modeler determines state variables and system parameters to represent real-world objects. This responsibility includes a proper validation of the model and its components. It is also the responsibility of an FEA user to document missed data and the consequences thereof to upper administrator or clients.

12.4 **VERIFICATION**

Verification concerns the mathematic and computational perspectives of an FEA model. Verification is to analyze the introduced errors at every step of FEA to see if these errors are acceptable and within the expected tolerance. As shown in Fig. 12.7, Conover (2008) classified the activities of the verification in terms of modeling stages where errors are introduced, i.e., errors

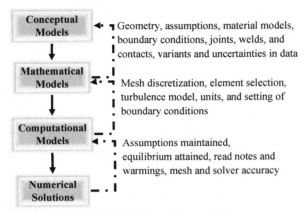

■ **FIGURE 12.7** Verification at different modeling stages.

are introduced from a conceptual model to mathematical model, computational model, and finally to the solution of numerical simulation.

As shown in Fig. 12.8, verification has to be performed at different stages for *geometric correctness*, *element types*, *mesh sizes*, *material properties*, *contact conditions*, *energy balances*, and *abnormal mesh* such as isolated nodes and tangled meshes. To perform these verifications, Thacker et al. (2004) classified the verification in FEA modeling into two basic types. The first

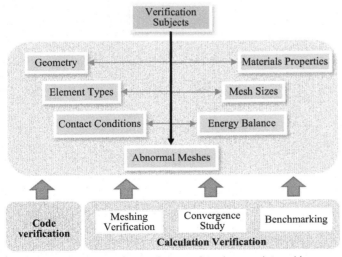

■ **FIGURE 12.8** Verification subjects in a finite element analysis model.

type is *code verification*. Code verification focuses on the identification and removal of errors in the code. The second type is *calculation verification*. Calculation verification focuses on the error quantification introduced during the application of the code to a particular simulation. The most important task in a calculation verification is the study of grid or time convergence to refine the mesh until a satisfactory solution is obtained.

12.4.1 **Code Verification**

Code verification is not used to prove that the mathematic model is a right representation of physical reality. Instead, code verification is to ensure that a numerical algorithm can solve a mathematic model properly. It does not matter if the mathematical model represents the physical system correctly. If the code verification is performed, the errors from the code can be treated separately from the errors of other sources, and this simplifies sequential code validation.

The software developer should be mainly responsible for code verification. As a software product, the software development must follow the standardized procedure to ensure the functionalities and quality of products. Classic waterfall model in Fig. 12.9 can be used as the guidance for the testing and verification of software products (Wall and Kossilov, 1994; IAEA, 1999). The development of a software product experiences several stages from

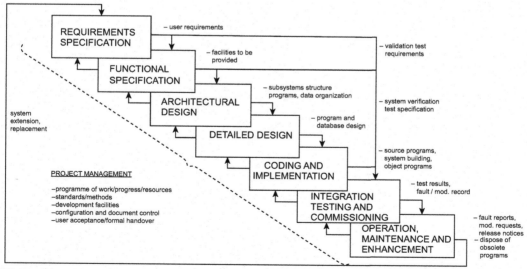

■ **FIGURE 12.9** Code verification in software life cycle. *Adopted from Wall, N., Kossilov, A., 1994. Verification and Validation of Software Related to Nuclear Power Plant Control and Instrumentation. http://www.iaea.org/inis/collection/NCLCollectionStore/_Public/26/045/26045115.pdf.*

conceptual design to final products. At each of these stages, the corresponding verification must be performed to achieve results with expected level of accuracy. When the solution to a mathematic model is programmed, the code must be verified to identify and correct language, grammar, compilation, and assembled errors. After this, the preliminary program check out begins by executing individual program modules to eliminate potential errors. Such tasks have to be accomplished by software developers. To ensure the functionalities and quality of software products, software users should also conduct the independent code verification to prove that the software tool has the expected functionalities for given tasks.

Numerical simulation is to solve differential equations by computers. Differential equations in a continual domain are represented by state variables on discretized nodes to achieve consistency, stability, convergence, and mostly importantly, accuracy of solutions. From this perspective, code verification can be decomposed into two basic tasks: (1) *numerical algorithm property verification*, which is used to determine if the tested algorithm has been implemented in agreement with the proven mathematic properties such as accuracy, stability, consistency, and (2) *numerical algorithm adequacy verification*, which is used to determine the tested algorithm satisfies the accuracy, robustness, and speed requirements of its intended use (Knupp, 2016).

For each verification, the mathematic model and the corresponding program are the inputs for the process of code verification. For the code verification in FEA modeling, a mathematic model includes the detailed description of governing equations, initial and boundary conditions, assumptions, and applicable domains. The mathematic model also includes the inputs and data associated with system parameters and coefficients of model. As a summary, the information about mathematical model must be sufficient to generate a certain solution for the verification purpose.

Example 12.1 (Code Verification)

This example is developed to verify the program for the beam elements in the SolidWorks. The cantilever beam in Fig. 12.10A is fixed at one end and subjected to a concentrated load (200 lbf) on the other end, "plain carbon steel" is used and the cross-section is a rectangle tube with the principal moments of inertia of the area of $I_y = 4.694292$ in^2. The analytical solution for the maximized displacement at the tip is,

$$y_{max} = \frac{PL^3}{3EI} = \frac{(200)(50)^3}{3(3.0458 \times 10^7)(4.6942)} = 0.0583 \text{ (in)}$$

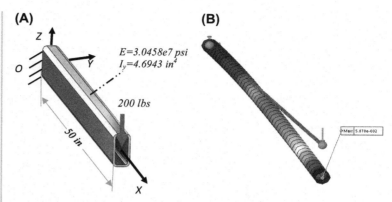

■ FIGURE 12.10 Example of code verification: (A) verification problem; (B) parametric study on the number of beam elements (from 1 to 50 elements).

Solution

As shown in Fig. 12.10B, structural members for the beam are selected in the SolidWorks model, and a design study is conducted by evaluating the impact of the number of elements in FEA on the maximized displacement at the tip. The obtained displacement (~ 0.0588 in) is converged and not affected by the number of elements. The accuracy from the numerical simulation can be verified as,

$$\varepsilon = \frac{\left| z_{max}^{analytical} - z_{max}^{simulation} \right|}{z_{max}^{analytical}} = \left| \frac{0.0583 - 0.0588}{0.0583} \right| = 0.86\%$$

Therefore, the program has been verified to be able to obtain acceptable results for the beam.

12.4.2 Calculation verification

Calculation verification focuses on the removal of errors introduced during the execution of computer programs and the use of software tools. Software users are responsible to perform the calculation verification. In a calculation verification, the simulation result is compared against analytical or validated solutions to determine discretization errors, input data errors, and the overall performance of simulation. Users need to capture various errors in an FEA model such as the errors from *distorted elements*, *disconnected nodes*, *improper material* assignments, inconsistency of *various coordinate systems*, *boundary and interface conditions*, mechanical, thermal, and inertia *loadings*. Calculation verification should ensure that the software correctly yields an acceptable solution instead of accepting the general-purpose software blindly without a valid assessment.

Example 12.2 (Calculation Verification)

Fig. 12.11A shows a simple solid geometry (i.e., a cube with a size of $4 \times 4 \times 4$ in) with the materials of plain carbon, the standard solid mesh is used for the calculation verification. The force equilibrium of the free body diagram is used as the criterion of verification. For the boundary conditions, it is assumed that the base surface is fixed, and the applied loads are $F_x = 2000$ lbf and $F_y = 1000$ lbf.

Solution

As shown in Fig. 12.11B, the cube geometry is modeled in the SolidWorks, the default mesh size and element type are used, and the boundary conditions are applied by fixing the bottom surface and applying two external forces on top surface and the lateral surface on the right side, respectively. After the simulation, the reaction forces from the fixed surface are found as $R_x = -2000$ lbf and $R_y = -999.98$ lbf. The error from the simulation is quantified as,

$$\eta = \frac{\left\| \sqrt{2000^2 + 1000^2} - \sqrt{2000^2 + 999.98^2} \right\|}{\left\| \sqrt{2000^2 + 100^2} \right\|} \times 100\% = 4.0 \times 10^{-4}\%$$

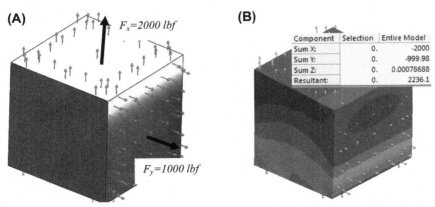

■ **FIGURE 12.11** Calculation verification based on force equilibrium of free body diagram: (A) a cube with two applied loads; (B) list of reaction force at support.

The discrepancy of the simulation result for the force equilibrium of object is ignorable.

The criterion for the force equilibrium is widely used to verify an FEA model. If a system to be modeled is an assembled product. Any component in the assembly can be selected as an object to be verified.

Example 12.3 (Calculation Verification)

A 4-bar mechanism is shown in Fig. 12.12. All of the bars are made of plain carbon steel and with a thickness of 0.5-in. The ground supports are placed at two pivots and the external load 200 lbf is the evenly distributed force on the output bar. The mechanism is assumed at a still position illustrated in Fig. 12.12. Create an FEA model in the SolidWorks and do the calculation verification based on the force equilibrium on the coupler.

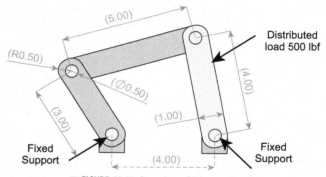

■ **FIGURE 12.12** Dimensions of 4-bar mechanism.

Solution

The model is created in the SolidWorks, the boundary conditions are applied accordingly in the model. The simulation result gives the displacement and stress distributions in Fig. 12.13.

To perform the calculation verification, the coupler is selected as a free body. Fig. 12.14 has shown all of the reaction forces at the contact surfaces, and these forces are completely balanced on the coupler. This passes the calculation verification based on the criterion of the force equilibrium on the coupler.

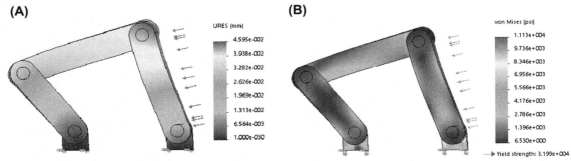

■ **FIGURE 12.13** Simulation result of 4-bar mechanism: (A) max displacement of 1.836×10^{-3} in; (B) max stress of 1.237×10^4 psi.

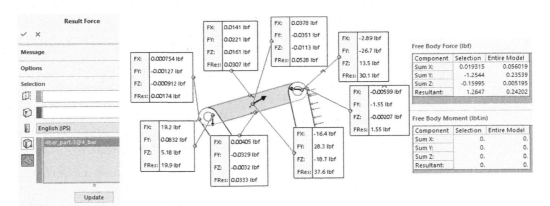

■ **FIGURE 12.14** Example of calculation verification using free-body diagram.

12.4.3 **Meshing verification**

Meshing verification. In an FEA model, a continuous domain is represented by a mesh with discretized nodes and elements. Meshing should be verified to avoid some issues such as large difference of stiffness. For example, it is a common practice in static analysis that the ratio of the maximum and minimum element stiffness coefficients should be less than 1.0×10^8. Otherwise, the rigidity of softer elements will be ignored in a system model, which might lead to unacceptable discrepancy of the simulation solution. Users should verify the appropriation of a mesh from many perspectives,

1. Verify the maximized stress to see if there is an abnormal stress concentration; stress concentration may lead to the stress level beyond yield strength;
2. Verify if all of the displacements are in expected ranges;

3. Verify if the deformed shapes make practical sense;
4. Verify the reaction forces against applied loads to check force equilibrium. For a steady problem, the sum of forces must be balanced by the sum of reaction forces; this can be used to identify misplaced loads, incorrect units, geometric errors or types of input; check reactions at contact pairs;
5. Verify the bonded contacts to see if penetration occurs and generate the plots of stress distribution to verify the reasonableness.

Example 12. 4 (Meshing Verification)

The model in Example 12.1 is used again as an example for meshing verification. Beam elements are used to represent the cantilever beam. Use a parametric study for the meshing verification to prove the convergence of simulation.

Solution

The beam is modeled in the SolidWorks simulation, and a parametric study is defined where the only variable is the number of the beam elements. Fig. 12.15 has shown the result of the parametric study. The max displacement at the tip remains the same for a range of beam members from 1 to 50. This passes the meshing verification.

Due to the limits of computer memory and speed, a user often faces the trade-off between accuracy and computation time. The demand on computing resources can be alleviated by some simplifications such as running an analysis on a partial model for symmetric or antisymmetric

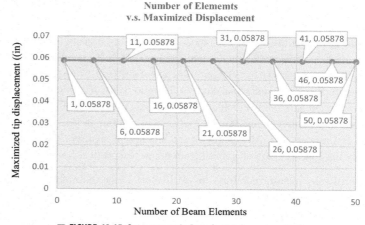

■ **FIGURE 12.15** Parametric study Example 12.1 for mesh verification.

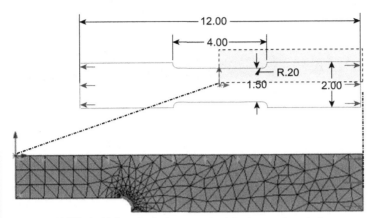

■ **FIGURE 12.16** Example of using symmetric relation in computing reduction.

objects. Fig. 12.16 shows an example of symmetric specimen subjected to a tensile load. Due to symmetric relations, only a quarter of the specimen is modeled in the simulation. However, it is worth to note that for a valid symmetric model, all of the geometries, loads, and boundary conditions must be symmetric.

For a complex object, it is often necessary to reduce the mesh density at uncritical regions. To determine suitable mesh sizes, one should be aware that the accuracy of state variables in a specific region relates to both element sizes and gradients of state variables. Therefore, small-size elements should be used in the areas with a high derivative of state variable; for example, in the regions with estimated high stresses or fluxes. On the other hand, large elements can be used in the regions with low stresses or fluxes. Fig. 12.17 gives an example that the model includes the features of geometric discontinuities. To increase the accuracy of the simulation, the high-density meshing is applied only in local regions with geometric discontinuities. It is also worth to note that the accuracy of one element model is affected by those of surrounding elements; therefore, meshing sizes must be changed smoothly from one region to another.

Automatic mesh generator in a commercial software tool is not visible to users. Users get limited feedbacks or guides when a meshing process fails. To find a resolution for a failure of meshing process, one may try to change the size setting of a mesh to obtain more hints about meshing failures. An assembled model fails more likely in meshing. In particular, the mesher may run out of memory because of tiny features; the decomposition on those features can cause a large number of tiny

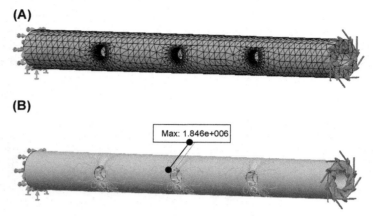

■ **FIGURE 12.17** Estimating gradient distribution to justify mesh control: (A) increasing mesh density in critical regions; (B) high-level of gradient in critical area.

elements. If an assembled model is concerned, users should mesh individual parts before the assembled model is modeled and analyzed.

Major factors in choosing mesh sizes are (1) the balance of computation time and solution accuracy, (2) the avoidance of distorted shapes which lead to near singular stiffness matrices, and (3) the representation of boundary conditions where loads are distributed properly.

To verify the quality of mesh, the concept of aspect ratio in Chapter 3 can be utilized. An aspect ratio is defined as the ratio of the longest and shortest edges. The lower the aspect ratio, the better the shape of the element. Fig. 12.18 has shown that the quality of meshes can be different even with the same number of elements and nodes.

Commercial software tools, such as SolidWorks and ANSYS, provide a collection of test cases to verify and validate the capabilities of the software tools. The simulation can be verified and validated by comparing the results of analytical solutions for numerous classical engineering problems (ANSYS, 2013a,b).

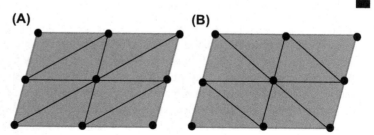

■ **FIGURE 12.18** Impact of aspect ratio on mesh quality: (A) high aspect ratio; (B) low aspect ratio.

12.4.4 **Convergence study**

One primary question in developing an FEA model is how small the elements in a mesh should be so that an acceptable simulation result can be obtained. Note that mesh sizes in the discretization are determined subjectively, and it is hard to justify the errors caused by the discretization. Therefore, one has to ensure that the solving process be converged to correct solutions. A convergence study can be used to evaluate the impact of mesh sizes on the simulation accuracy. The mesh size must be fine enough to ensure that the simulation result will not be changed significantly by further reducing mesh sizes. The requirements of convergence can be met by considering three criteria (Burnett, 1987; Pointer, 2004).

12.4.4.1 *Continuity condition*

The shape functions of elements must ensure that the displacement solutions across elements are continuous. Shape functions are usually required in formulating models of elements. It is the developer's responsibility to ensure the satisfaction of continuity condition.

12.4.4.2 *Completeness condition*

The mesh must be free of singular nodes. The magnitude of a state variable on a singular node tends to be infinitely large; that causes a large gradient of the state variable in the element. The condition of completeness ensures that the nodal values in each element approach the same value when the element size reduces. It is the user's responsibility to verify the condition of completeness.

12.4.4.3 *Convergence of energy*

For an FEA model with the satisfaction of completeness and continuity conditions, the convergence of energy warrantees that the solution to the model is converged. It is the user's responsibility to ensure the convergence of energy.

Mesh sizes relate closely to both computation and accuracy. The finer a mesh is, the better result an FEA model can obtain; however, it demands more computation. A trial and error method can be applied to determine mesh sizes. Once an FEA solution is found, a mesh with a fine size is applied to find a new solution, two solutions are compared to see if the solution is improved by refining the mesh. The iteration will be continued until the difference of two solutions is within the specified percentage of errors.

12.4.5 **Benchmarking**

A comparative reference is essential for code or calculation verification. A *benchmark* is a standard test designed to probe the accuracy or efficiency of

FEA models. A benchmark problem is a well-defined example problem for which a reference solution exists, and *benchmarking* is defined as the process to evaluate the variation in results from different programmers or software codes. If there is no analytical solution to a design problem, numerically derived benchmarking solutions can be applied. The governing equations of an engineering problem are PDEs and benchmark solutions for PDEs can be found in two different methods: (1) a PDE is transformed into an ordinary differential equation (ODE), and the numerical integration is performed on ODE to find the solution; (2) the numerical integration is directly applied to a PDE. In either of two cases, the accuracy of solution must be assessed critically to qualify them for use in code or calculation verification. For a solution from the first method, the benchmarking process has been standardized to assess simulation accuracy. For a solution from the second method, it should be the last resolution for benchmarking and the code used in generating that the solution has been thoroughly verified and documented (ASME, 2006a,b). The credibility of a benchmark solution can be enhanced if it has been obtained by different numerical approaches or software tools. Using a solution agreed by multiple independent sources will mitigate the risk of errors in the verification. Although benchmarking depends on how users translate a benchmark problem into a computer model and how they interpret the results, it is encouraged to benchmark different software packages against each other or against the reference solution.

Most benchmarks are simple practical problems for which no analytical solution exists. Users can use a benchmark to check if they use their own tools, define their own models but still obtain similar solutions to those solutions from the benchmarks. Because the solution is numerical simulation based, a small deviation from the reference solution is acceptable. Even larger deviations may still be acceptable, depending on the type of problem and the level of details that is provided with the benchmark. Some public benchmarks have shown that a large difference can occur; it implies the need for the follow-up validation of numerical models. In summary, benchmarks can serve the following purposes (ANSYS, 2013a,b):

- verify computer software and program modules,
- train inexperienced FEA users, help them becoming familiar with numerical analysis and practice FEA modeling appropriately,
- prove FEA users' competence in solving engineering problems in specific their domains by FEA,
- make users aware of differences of results even for a well-defined problem. This emphasizes the importance of validation of numerical models,

- highlight the importance of providing correct inputs of model, for example, define appropriate constitutive models for the materials, and
- identify the limitations of the state of the art in numerical modeling in practice.

12.5 **VALIDATION**

Validation is to check if (1) a conceptual model represents original physical system appropriately, and (2) an accurate solution has been found for the given analysis problem. Accordingly, validation should be performed at two phases: (1) at the early stage, the inputs to an FEA model have to be validated and (2) at the final stage, the calculated state variables are assessed. Usually, the experimental data should be used as comparative references in the validation.

Even though commercial FEA packages are easy to use, it is challenging to create a good FEA model, which is capable of obtaining an acceptable solution to real-world problem (Brinkgreve and Engin, 2013). An FEA model must be validated to ensure appropriate settings for the assumption of linearity, representation of assembly relations, material properties, constitutive methods, idealization, geometric representation, loading and boundary conditions, element types and shapes, and interpretation of simulation results. Fig. 12.19 shows a few of exemplifying subjects of computer validation. The main methods used for validation purpose include *modal validation*, *uncertainty quantification*, *sensitivity study*, *engineering judgment*, and *benchmarking*.

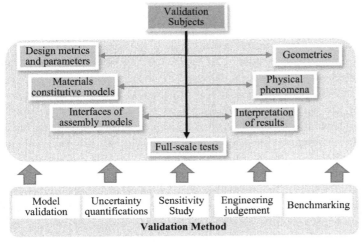

■ **FIGURE 12.19** Exemplifying subjects and methods of validation.

12.5.1 **Modal validation**

Modal validation is performed when a design problem is formulated and a conceptual model is defined. It is also known as *model updating*, *model tuning*, or *parameter calibration*. *Calibration* of an FE model is often undertaken to ensure that modeled features represent realistic mock-ups of system properties. Examples of system properties are the model stiffness in structural analysis and the conductivity in thermal analysis. The model should be adjusted to accommodate the uncertainties of system inputs. Examples of system inputs are the deformation at contacts, damping factors, unmeasurable external excitations, and material variations. For example, if the nonlinearity of a fastener is modeled, the spring rate must be specified and calibrated using the test data. Critical system inputs have to be validated in a certain way, these system inputs include (1) materials and constitutive models, (2) representations of object geometries, (3) contact conditions of parts, (4) element and analysis types, and (5) boundary supports and loading conditions.

12.5.1.1 Materials

Many materials demonstrate heterogeneous and nonlinearity characteristics, which are difficult to be represented by constitutive models. The selection of a constitutive model and the associated system parameters are some important factors to be validated. No matter how complex a constitutive model is, some simplifications and assumptions are made in representing physical behaviors of materials. The impact of these simplifications may be significant enough to lead a wrong answer to the original problem. One has to understand the backgrounds and limitations of constitutive models in defining material properties.

12.5.1.2 Geometric models

Preparing a good computer aided design (CAD) model for FEA never be not a trivial task. A poor solid model can adversely affect the quality, computation, and accuracy of an FEA model (Adams and Askenazi, 1998). Some flaw features, such as tiny lines, interferences, trimming surfaces, and interior voids, most likely lead to failures in a meshing process. Users should make sure that no crack or interference occurs in a model. Most software tools are able to check and repair these problems automatically. However, these functions should not be fully trusted since (1) computer-generated representations might not be optimized ones, and (2) not all of flaws can be detected and fixed automatically. Modeling a physical system possesses more challenging if the contact relations of multiple objects are taken into consideration. It is desirable to examine and mesh individual parts before one creates the assembled one.

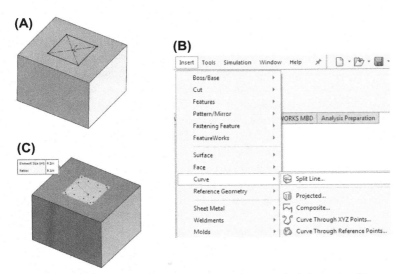

■ **FIGURE 12.20** Example of defining and using split lines: (A) create a sketch for a region of interest; (B) convert a sketch into a split line; (C) region of interest can be specified.

Sometimes, additional sketches are needed to introduce split lines into a part. Split lines can be utilized in many applications such as (1) to be used a reference for mating relations; or (2) to define the regions of boundary conditions such as constraints and loads. As shown in Fig. 12.20, the SolidWorks has the tool called "*split line*" allowing a user to project a sketch on any surface as split lines. Once a split line is projected on a surface, the bounded area can be selected specifically to apply constraints or loads on the split regions.

12.5.1.3 Detailed features

CAD tools are prevailing now. When computer aided engineering (CAE) is integrated with CAD; users tend to utilize the detailed model of a solid object from CAD for engineering analysis. However, the detailed design of a solid object may include many features such as rounds, chambers, which are not significant enough to affect the accuracy of analysis study. These features should be suppressed to facilitate a meshing process. The designer must make a good engineering judgment to determine if a certain feature can be suppressed. Some small features may have a great impact on the accuracy of analysis result. For example, a round feature at discontinuities may greatly reduce the concentrated stress at critical cross-sections. Such a feature can be suppressed only when it locates at a low-stress region of part.

12.5.1.4 Assembly of parts or components

Modern products are mostly assembled. In a virtual assembly model, mates are defined for contact areas of objects in ideal conditions such as frictionless contacts or fully bonding relations. In reality, the geometric mates in a virtual model can be fundamentally different from the mating constraints at actual physical contacts. Contact behaviors of two objects relate to many factors such as displacement, temperature, friction, wear, and surface conditions. In FEA modeling, users may need to set up a number of contact scenarios, run, and compare FEA modes to find the results in most critical cases. For example, if a linear contact model is used to represent physical contacts, an iterative evaluation is required to ensure that a solution satisfies the given constraints at the contacts.

12.5.1.5 Validation of element types

A user should validate if element types in an FEA model have been selected adequately. Element type decides the governing equations of the given problem as well as the shape functions for the interpolation within an element. The higher order of shape functions gives the better interpolation; however, it requires more computation. A commercial FEA package has an element library, which consists of many element types. For example, the element library of ANSYS ADPL has 290 element types (ANSYS, 2013a,b). Each element type is defined based on certain assumptions. Therefore, each type has its theoretical constraints for the specified scope of applications. Selecting ideal element types for an FEA problem will lead to a solution with better accuracy in a shortened time of computation. from this perspective, users need to understand the theory of FEA. In some occasions, the analysis results based on the different choices of element types can be compared with each other for the purpose of verification and validation.

12.5.1.6 Validation of boundary conditions

Idealization of the boundary conditions should proceed with caution. For examples, external loads can be applied on boundaries of objects, they are nonessential conditions (also called as Neumann boundary conditions) for system solutions. However, they are very common in reality and should be represented in an FEA model appropriately. Load conditions should be validated to ensure that the loads are defined to reflect actual situations of concentrated, surface, and body loads of physical objects. In many classic textbook problems, these types of boundary conditions are applied in points or lines. But this should be avoided in an FEA model; because applying point or line loads will cause artificial infinite stresses or heat fluxes on nodes. The simulation may mislead users even if the solving process does not fail. A torque or moment load at a point is also a common simplification in traditional static and dynamic courses, but it does not exist in the real world. As an alternative, a distributed pressure can be defined on small area to represent a couple at its center. To define these loads in an FEA model, they should be modeled as a total net load over a small region with the assistance of split lines. If a body force has to be considered in an FEA model, the user must ensure the correctness of the density and the direction of gravity with respect to the global coordinate system of model. When a load varies within a range, a parametric study should be defined to simulate the system response under different levels of loads. If the load is time dependent, a transient analysis should be defined. If the accumulated damage from dynamic loads is a main concern, the fatigue analysis should be performed on top of static analysis. Similarly, the definition of the dynamic characteristics of loads, such as the order, path, mean, and alterative magnitude of loads should be also validated (Bi and Mueller, 2016).

The boundary conditions applied directly to state variables are called essential boundary conditions. Small changes of supports may lead to significant changes of a simulation result. It is desirable to trail and errors for a number of possible support conditions. If the rigidity of support is not considerably larger than that of objects to be analyzed, it is better to tread supporting components as a part of the FEA model. When a user lacks the confidence of the obtained simulation result or a failure of convergence occurs in the solving process, he or she may change the number of supports or adjust the locations of supports in the critical areas of objects for refinement.

Example 12.5

Fig. 12.21 shows an assemblage of three identical cylindric axial members. Each axial member has the length of $L = 5$-in and a diameter of $d = 1$-in. Two end faces (A and D) are fixed, and the loads are applied in two middle intersections (B and C). Plain carbon steel is used as the materials of cylindrical members. Use the reactional forces as the design criteria for the verification of FEA modeling.

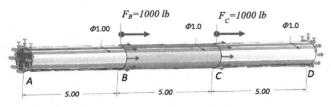

■ **FIGURE 12.21** Example of assembly verification.

Solution

Let the displacements at B and C are δ_B, and δ_C, respectively, the solutions to these displacements can be found from,

$$\frac{EA}{L}\begin{bmatrix} 1 & -1 & 0 & 0 \\ -1 & 2 & -1 & 0 \\ 0 & -1 & 2 & -1 \\ 0 & 0 & -1 & 1 \end{bmatrix}\begin{Bmatrix} 0 \\ \delta_B \\ \delta_C \\ 0 \end{Bmatrix} = \begin{Bmatrix} R_A \\ F_B \\ F_C \\ R_D \end{Bmatrix} \quad (12.5)$$

where

$L = 5\text{-in},$
$A = \pi d^2/4 = 0.25\pi,$
$E = 3.046e7\text{-lb/in}^2,$
$F_B = 1000\text{-lb},$
$F_C = 2000\text{-lb}.$

Solving Eq. (12.5) yield a theoretical solution as $R_A = 1333\text{-lb}$ and $R_D = 1667\text{-lb}$. Fig. 12.22 has shown the result of FEA modeling from the SolidWorks. The same reaction forces at A and D have been found. This verifies the numerical simulation of the given design problem.

For a structure, sufficient constraints are required to eliminate a possible rigid motion of entire structure. Take an example of a rod supported by two rollers in Fig. 12.23, it can be in an equilibrium state in real-life.

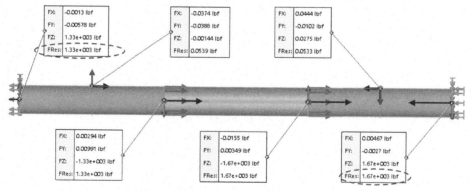

FX:	-0.0013 lbf
FY:	-0.00578 lbf
FZ:	1.33e+003 lbf
FRes:	1.33e+003 lbf

FX:	-0.0374 lbf
FY:	-0.0388 lbf
FZ:	-0.00144 lbf
FRes:	0.0539 lbf

FX:	0.0444 lbf
FY:	-0.0102 lbf
FZ:	0.0275 lbf
FRes:	0.0533 lbf

FX:	0.00294 lbf
FY:	0.00991 lbf
FZ:	-1.33e+003 lbf
FRes:	1.33e+003 lbf

FX:	-0.0155 lbf
FY:	0.00349 lbf
FZ:	-1.67e+003 lbf
FRes:	1.67e+003 lbf

FX:	0.00467 lbf
FY:	-0.0027 lbf
FZ:	1.67e+003 lbf
FRes:	1.67e+003 lbf

■ **FIGURE 12.22** Simulation result of assembly verification.

Additional restraints

■ **FIGURE 12.23** Adding artificial constraints to eliminate a rigid motion.

However, in a conceptualized computer model. One additional restraint has to be included to eliminate the rigid body motion along the axial direction of beam.

12.5.1.7 Symmetry

As shown in Fig. 12.24, symmetric features can be utilized to reduce computation significantly. For example, 2D mesh is widely applied to solve plane stress, plane strain, and axisymmetric design problems. The mesh on a 2D plane can be performed easily and with a high density of elements. To make symmetric relations valid, all of the geometries, loads, and boundary conditions have to be symmetric. In addition, if modal analysis is performed for a structure, symmetric features should be not utilized because the model shapes are unnecessarily symmetric even though the geometry and boundary conditions of model.

If there are some small features affecting symmetry, one has to validate if such interrupts can be ignored.

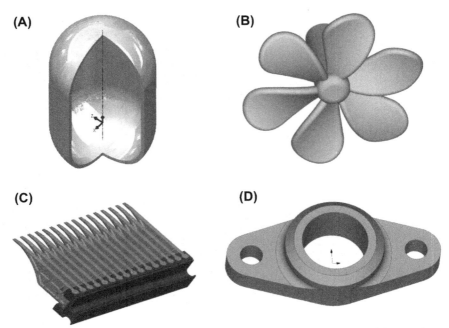

■ **FIGURE 12.24** Types of symmetric relations: (A) axisymmetric; (B) rotational; (C) linear patterns; (D) mirror relation.

12.5.1.8 Singularity of fixed supports

Singularity refers to the location where stress value is unbounded in a finite element model. It is caused by a point or line load or moment, an isolated constraint point where the reaction force acts as a point load, or shape corner. However, there is no stress singularity in a real structure. Point or line loads are artificial terms caused by the simplification of models. In other words, point loads should only be applied for line elements.

When an object experiences the prolongation in one direction, it shrinks in its perpendicular directions. This phenomenon is commonly known as the Poisson effect. However, the fixed geometry on surface will constrain such shrinking in its perpendicular directions. This is also a type of singularity. It may cause the lift-off at the support in real-life. The computer representation of such a phenomenon is illustrated in the example of Fig. 12.25. Taking into consideration the connection of two solid objects with different cross-section areas, the stresses on the joining surface will be continually increased unrealistically due to the singularity.

If the singularities turn into a problem in finding a reasonable solution: the following methods can be applied to alleviate this problem: (1) if the geometry can be refined, add fillets to remove sharp edge; if the load is at a point contact, tap the vertex to create a small load region; (2) try to replace point or line restraints by the boundary conditions on small regions which cover a number of nodes; (3) use plastic materials properties so that the stress will be redistributed locally; and (4) use stress linearization to average out the peak stress.

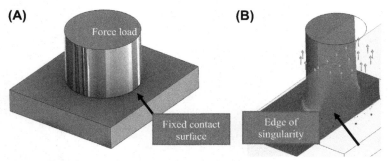

(A) Force load / Fixed contact surface

(B) Edge of singularity

■ **FIGURE 12.25** Example of singularity issue at fixed supports: (A) fixed restraints; (B) singularity in preventing shrinking.

12.5.2 **Uncertainty quantification**

Uncertainty is unavoidable in an engineering system because a physical system entails uncertainties in materials, geometry, and load parameters. The validation for numerical simulation must take into account the uncertainties of experimental data associated with system inputs and simulations. Throughout the modeling process, the uncertainties have to be analyzed and quantified to identify major sources of errors. A simulation model must be able to quantify the effects of the identified uncertainties in simulated results predicted by model.

Uncertainties can be irreducible or reducible. *An irreducible uncertainty* is also called as an *aleatory* uncertainty. An aleatory uncertainty refers to an inherent variation in the physical system to be modeled. An aleatory uncertainty is an intrinsic property of system, which exists all the time with the system itself. The uncertainties in materials and load conditions are aleatory. The variability of system parameters is often characterized by repeating module-level tests that cover the range of operating conditions in applications. If no data of validation testing are available, engineering judgment or prior experience should be used to estimate the variability of system parameters. It is worth to note that the test information is never sufficient to eliminate irreducible uncertainty; however, it can be better quantified by some statistic variables such as mean and fluctuated values and distribution types. To investigate the impact of the uncertainty of system inputs on system outputs, the probabilistic analysis can be used. In a probabilistic analysis, the variability of system parameters can be propagated throughout the simulation, and it leads to an expected variability of the simulation outputs.

A reducible uncertainty is called *an epistemic uncertainty*. An epistemic uncertainty refers to the deficiencies by a lack of knowledge or information. Reducible uncertainties have two main sources: (1) the statistical uncertainty due to the use of limited samples. For example, the mean value based on two or three measurements contains the statistical uncertainty; (2) the model uncertainty associated with the idealization and assumptions of model, for example, an assumption of a constant coefficient in a PDE. In a PDE, system parameters are handled as constants; however, these parameters may be affected by time, position, or temperature in reality. Model-related uncertainties are usually difficult to be quantified (Hasselman, 2001). Uncertainty is traditionally quantified using the probabilistic approaches, which demand a large amount of data for validation (Mulder et al., 2012). When the information is insufficient, probability bounds can be applied instead of probability. The idea of probability bound can be illustrated by using an interval approach to estimate the range of constant π (Muhanna, 2011).

Example 12.6 (Probability Bound)

According to Archimedes (287—212 BC), the constant π was defined as an area of a circle of radius one. Such an area can be approximated by a polygon whose vertices are on the circle. As shown in Table 12.3, an inbound polygon and an outbound polygon specify the probability bound of the circle area.

Table 12.3 Example of Using Interval Approach in Determining the Range of π

N	3	4	5	6	i	
Bounds	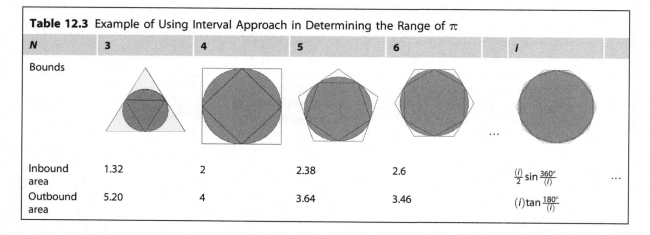					
Inbound area	1.32	2	2.38	2.6	$\frac{(i)}{2}\sin\frac{360°}{(i)}$	...
Outbound area	5.20	4	3.64	3.46	$(i)\tan\frac{180°}{(i)}$	

Table 12.3 tells that the more edges inbound and outbound polygons; the less uncertainty the approximation of the area has. The constant π can be found when the numbers of edges of polygons are sufficiently large.

An interval approach is applied when the information about the uncertainty of parameter is available, i.e., $x = x_0 \pm \delta$ or $x \in (x_0 - \delta, x_0 + \delta)$. It is assumed that uncertainty can be obtained by experimental data, measurements, statistical analysis, or expert knowledge. According to Muhanna (2011), an interval approach has some advantages: (1) it is simple and elegant, (2) it aligns with engineering practices on dimensioning and tolerance, (3) it is capable of representing the uncertainty when the information is insufficient for other probabilistic approaches, and (4) it can serve as computational basis for other uncertainty quantifications such as the fuzzy set, random set, and probability bounds. Indeed, it provides the guaranteed enclosures.

Example 12.7 (Uncertainty Quantification)

Fig. 12.26 shows an example of a simple structure whose geometry has the uncertainty. It is fixed at position *A*, and it is subjected two external loads

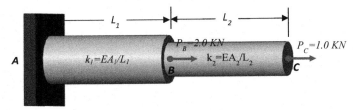

■ **FIGURE 12.26** Example of uncertainty quantification.

$P_B = 2.0$ KN and $P_C = 1.0$ KN. The probability bounds of stiffness coefficients for two segments are $k_1 = 2.00 \pm 0.10$ MN/m, and $k_2 = 1.00 \pm 0.05$ MN/m. Quantify the uncertainty of the displacement at *C* using finite element modeling (Muhanna, 2011).

Solution

The system model of the structure is determined as,

$$-\begin{bmatrix} k_1 & -k_1 & 0 \\ -k_1 & k_1+k_2 & -k_2 \\ 0 & -k_2 & k_2 \end{bmatrix} \begin{Bmatrix} u_1 \\ u_2 \\ u_3 \end{Bmatrix} = \begin{Bmatrix} R_1 \\ P_B \\ P_C \end{Bmatrix}.$$

Taking into consideration of uncertainties of system inputs, the system model can be rewritten as,

$$\begin{bmatrix} (2.85, 3.05) \times 10^3 & (-0.95, -1.05) \times 10^3 \\ (-0.95, -1.05) \times 10^3 & (0.95, 1.05) \times 10^3 \end{bmatrix} \begin{Bmatrix} u_2 \\ u_3 \end{Bmatrix} = \begin{Bmatrix} 2.0 \\ 1.0 \end{Bmatrix}$$

Case	1	2	3	4
k_1 (MN/m)	1.90		2.10	
k_2 (MN/m)	0.95	1.05	0.95	1.05
$\{u\}$ (m)	$\begin{Bmatrix} 1.58 \times 10^{-3} \\ 2.63 \times 10^{-3} \end{Bmatrix}$	$\begin{Bmatrix} 1.58 \times 10^{-3} \\ 2.53 \times 10^{-3} \end{Bmatrix}$	$\begin{Bmatrix} 1.43 \times 10^{-3} \\ 2.48 \times 10^{-3} \end{Bmatrix}$	$\begin{Bmatrix} 1.43 \times 10^{-3} \\ 2.38 \times 10^{-3} \end{Bmatrix}$
Quantified uncertainty of $\{u\}$	$\begin{Bmatrix} u_2 \\ u_3 \end{Bmatrix} = \begin{Bmatrix} 1.43 \sim 1.58 \\ 2.38 \sim 2.63 \end{Bmatrix} \times 10^{-3}$			

Therefore, the probability bound of the max displacement of the structure is (2.38, 2.68) mm.

∎

12.5.3 **Sensitivity study**

Sensitivity analysis aims to discover the effect of uncertain inputs on the results of simulation. It helps to gain the insights of model characteristics. It assists in design of experiments, but sensitivity analysis should be subjected to V&V itself. Sensitivity study is a type of probabilistic analysis to investigate the impact of statistical data of inputs. It requires a number of simulations, which is very time consuming. Positive/negative sensitivity indicates that increasing uncertain variables will increase/decrease the values of simulation results.

12.5.4 **Engineering judgment**

Simulation results can be judged by engineer's experience. The visualization tool in an FEA tool is very useful for a user to justify if the result from simulation makes practical sense. For example, it is difficult to picture system behaviors under the combined load. To validate the capability of an FEA code, loads can be applied in a sequence separately; this allows engineers to check the step results visually and justify the reasonableness of step results. Using their engineering knowledge and experiences, users should be capable of judge whether or not the obtained reaction forces and deflection are reasonable. In addition, users can do manual calculations to check the equilibrium of forces against a free body diagram, or check excessive displacements or unexpected rigid body motions. A special attention has to be paid in the areas with the rapid changes of state variables, which are mostly likely the areas with the omissions or incorrect inputs. If a dynamic load is involved, users have to check the results under changed loads and ensure the consistency of simulation outcomes.

No matter how advanced an FEA software tool is, the success of its application relies mainly on users. As shown in Fig. 12.27, a user should be able to make his or her own engineering judgment to compensate V&V at different phases and aspects of FEA modeling.

12.5.5 **Benchmarking for validation**

Benchmarking for validation is similar to that for verification. Both of them need to have experimental data for comparison. However, benchmarking for

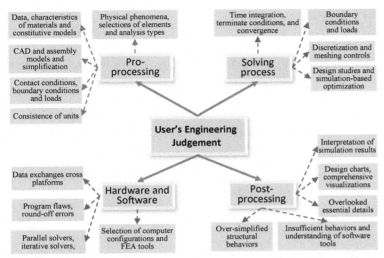

■ **FIGURE 12.27** The user's role for engineering judgment in a finite element analysis modeling.

validation focuses on the fidelity of a simulation result for original physical system. Ivanov et al. (2014) discuss the V&V methodologies for immersed boundary computational fluid dynamics code where different levels of testing are involved. At the most comprehensive level, some applied industrial problems and benchmarks have been provided to demonstrate the application of benchmarking in code and calculation verification. It has been found that for a software tool used by project-oriented engineers, it is impossible to separate the verification and validation procedures because most of the modeling efforts are highly automated in commercial software tools. This means that the activities of code verification and validation are integrated, with the terms being used together when referring to a suite of activities.

12.6 **SUMMARY**

For numerical simulation, V&V is essential to ensure the credibility of simulation results. In this chapter, the roles of V&V in FEA modeling are explained. The errors in numerical simulation are quantified for the purpose of V&V. Main sources of errors of FEA models are classified according to the activities in the modeling process. Some common techniques for V&V, including code verification and calculation validation, have been introduced. Different techniques in code verification and calculation validation are presented. The responsibilities of FEA developers and users in V&V have been discussed. A number of examples have been developed to illustrate V&V procedure.

REFERENCES

Adams, V., Askenazi, A., 1998. Building Better Products with Finite Element Analysis. OnWord Press. ISBN-10: 156690160X.

ANSYS Inc., 2013a. ANSYS Workbench Verification Manual, Release 15. http://www.ansys.com.

ANSYS Inc., 2013b. ANSYS Mechanical APDL Element Reference, Version 15.0. Southpointe, 275 Technology Drive, Canonsburg, PA. http://148.204.81.206/Ansys/150/ANSYS%20Mechanical%20APDL%20Element%20Reference.pdf.

ASME, 2006a. An Overview of the Guide for Verification and Validation in Computational Solid Mechanics. L. E. Schwer, ASME.

ASME, 2006b. Guide for Verification and Validation in Computational Solid Mechanics. http://jacobsschool.ucsd.edu/EEI/academic/courses/06/spring/SE207/pdf/Reference-01.pdf.

Bi, Z.M., Mueller, D., 2016. Finite element analysis for diagnosis of fatigue failure of composite materials in product development. International Journal of Advanced Manufacturing Technology 87 (5), 2245−2257.

Brinkgreve, R.B.J., Engin, E., 2013. Validation of geotechnical finite element analysis. In: Proceedings of the 18th International Conference on Soil Mechanics and Geotechnical Engineering, Paris 2013.

Burnett, D.S., 1987. Finite Element Analysis. Addison-Wesley Publishing Company, Reading.

Chen, G., 2001. FE Model Validation from Structural Dynamics. Imperial College of Science, Technology and Medicine, University of London, South Kensington, London, UK.

Collins, M.P., Vecchio, F.J., Selby, R.G., Gupta, P.R., March/April 2000. Failure of an offshore platform. Structures 43−48. http://www.civ.utoronto.ca/vector/journal_publications/jp33.pdf.

Conover, D., 2008. Verification and validation of FEA simulations. In: Integration of Simulation Technology into the Engineering Curriculum, Cornell University, July 25−26. Cornell University, Ithaca, New York.

Hasselman, T.K., 2001. Quantification of uncertainty in structural dynamics models. ASCE Journal of Aerospace Engineering 14 (4), 158−165.

Holman, J.P., 1997. Heat Transfer, eighth ed. McGraw-Hill, New York. ISBN-10: 0070083002.

International Atomatic Energy Agency (IAEA), 1999. Verification and Validation of Software Related to Nuclear Power Plant Instrumentation and Control. Technical Report No. 384. http://www-pub.iaea.org/mtcd/publications/pdf/trs384_scr.pdf.

Ivanov, A.V., Trebunskikh, T.V., Platonovich, V.V., 2014. Validation Methodology for Modern CAD-embedded CFD Code: From Fundamental Tests to Industrial Benchmarks, White Paper. SolidWorks, Inc. http://www.hawkridgesys.ca/file/solidworks-flow-simulation/validation-methodology-for-modern-cad-embedded-cfd-code.pdf.

Knupp, P., 2016. Code Verification for Phenomenological Modeling and Simulation Software. https://cfwebprod.sandia.gov/cfdocs/CompResearch/docs/cv.pdf.

Moorcroft, D., 2012. Model Verification and Validation Process. In: https://www.niar.wichita.edu/niarfaa/Portals/0/Model%20Validation%20Process2.pdf.

Muhanna, R.L., 2011. Interval based finite elements for uncertainty quantification in engineering mechanics. In: IFIP -Working Conference on Uncertainty Quantification in Scientific Computing, August 1–4, 2011, Boulder, CO, USA.

Mulder, W.D., Moens, D., Vandepitte, D., 2012. Modeling uncertainty in the context of finite element models with distance-based interpolation. In: Proceedings of the 1st International Symposium on Uncertainty Quantification and Stochastic Modeling February 26th to March 2nd, 2012.

Oberkamph, W.L., Trucano, T.G., 2002. Verification and Validation in Computational Fluid Dynamics, SAND2002–0529. Sandia National Laboratories, Albuquerque, New Mexico.

Pointer, J., 2004. Understanding Accuracy and Discretization Error in an FEA Model. http://www.designspace.com/staticassets/ANSYS/staticassets/resourcelibrary/confpaper/2004-Int-ANSYS-Conf-54.PDF.

Selby, R.G., Vecchio, F.J., Collins, M.P., 1997. The failure of an offshore platform. Concrete International 19 (8), 28–35.

Sehnalek, S., Zalesak, M., Vincenec, J., Oplustil, M., Chrobak, P., 2014. Evaluation of Solidworks Flow Simulation by Ground-Coupled Heat Transfer Test Cases. http://www.inase.org/library/2014/santorini/bypaper/SYSTEMS/SYSTEMS2-20.pdf.

Shah, C., 2002. Mesh discretization error and criteria for accuracy of finite element solutions. Cummins, Inc. http://simplorer.com/staticassets/ANSYS/staticassets/resourcelibrary/confpaper/2002-Int-ANSYS-Conf-9.PDF.

Thacker, B.H., Doebling, S.W., Hemez, F.M., Anderson, M.C., Pepin, J.E., Rodriguez, E.A., 2004. Concepts of Model Verification and Validation, LA-14167-MS. Los Alamos National Laboratory, US Department of Energy. https://www.osti.gov/scitech/servlets/purl/835920/.

Wall, N., Kossilov, A., 1994. Verification and Validation of Software Related to Nuclear Power Plant Control and Instrumentation. http://www.iaea.org/inis/collection/NCLCollectionStore/_Public/26/045/26045115.pdf.

FURTHER READING

FEA Best Practices. http://innomet.ttu.ee/martin/mer0070/loengud/fea_best_practices.pdf.

Felippa, C., 2016. Introduction to Aerospace Structures. http://www.colorado.edu/engineering/CAS/courses.d/Structures.d/.

Friswell, M., Coote, J.E., Terrell, M.J., Adhikari, S., Fonseca, J.R., Lieven, N.A., February 2005. Experimental Data for Uncertainty Quantification. 23rd IMAC, Orlando, Florida, USA.

LA-14167-MS, 2004. In: Schaller, Charmian (Ed.), Concepts of Model Verification and Validation. LANL.

Ross JR., H.E., Sicking, D.L., Zimmer, R.R.A., Michie, J.D., 1993. Recommended Procedures for the Safety Performance Evaluation of Highway Features. Report 350. National Cooperative Highway Research Program, Washington, DC. http://onlinepubs.trb.org/onlinepubs/nchrp/nchrp_rpt_350-a.pdf.

Safarian, P., 2015. Finite element modeling and analysis validation. Federal Aviation Administration. http://appliedcax.com/docs/femap/femap-symposium-2015-seattle-area/FEA-Validation-Requiremnents-and-Methods-Final-with-Transcript.pdf.

Stansifer, R., 2016. Some Disasters Attributable to Bad Numerical Computing. http://cs.fit. edu/~ryan/library/Some_disasters_attributable_to_Numerical_Analysis.pdf.

U.S.A. International Atomic Energy Agency (IAEA), 1988. Manual on Quality Assurance for Computer Software Related to the Safety of Nuclear Power Plants. Technical Reports Series No. 282. IAEA, Vienna. http://www.iaea.org/inis/collection/NCLCollectionStore/_Public/19/093/19093322.pdf.

■ PROBLEMS

12.1. (1) Justify if the design problem in Fig. 12.28 can be simplified by an axisymmetric model or a partial model of the whole object, explain the reasons; (2) create and run models with a half and whole object for a comparison of the max displacement and design factor again the static failure (set the mesh size as 0.125 in).

12.2. As shown in Fig. 12.29, Holman (1997) introduced a uniform 2D flows with a laminar boundary layer on a heated flat plate with a length of 0.31-in. The laminar air flow has a Reynold's number of 322. The heat transfer coefficient has been evaluated analytically and illustrated in Fig. 12.30. For the purpose of the calculation validation, create the CAD model, define and run a flow simulation, retrieve the heat transfer coefficient on the boundary surface, and verify if the flow simulation produces the acceptable result.

12.3 Create a heat transfer model as shown in Fig. 12.31 and verify if the conclusion by Sehnalek, S. et al. (2014) about the SolidWorks flow simulation is valid. The required thermal properties are given in Table 12.4.

Hint: the result about the heat flux in the literature seems to be wrong. The correct one should include both of positive and negative heat flux outside and inside of conditioned zone. It should be shown a scenario similar to Fig. 12.32.

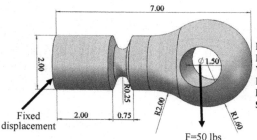

7.00

2.00

2.00   0.75

R0.25

R2.00

R1.60

⌀1.50

F=50 lbs

Fixed
displacement

Materials: Balsa wood
Density: 0.00578 lb/in³
Yield Strength: 2900 psi
Elastic Modulus: 435113 psi
Poisson's Ratio 0.29
Shear Modulus: 43511 psi

■ **FIGURE 12.28** Example of symmetric geometry (unit-inch).

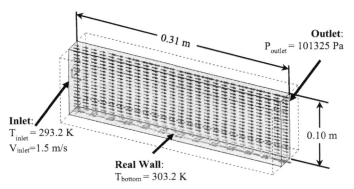

■ **FIGURE 12.29** The conditions for a laminar air flow in Problem 12.2.

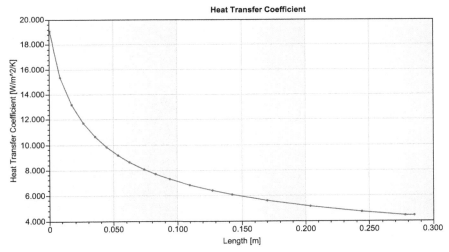

■ **FIGURE 12.30** Theoretical heat transfer coefficient in Problem 12.2.

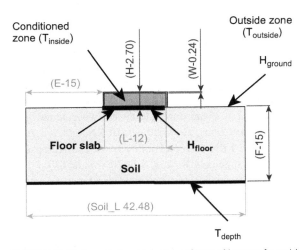

■ **FIGURE 12.31** Exemplified model for the verification of heat transfer model.

Table 12.4 Thermal Properties for Soil, Slab, and Above Grade Construction

	Soil and Slab	Above-Ground Construction
Temperature (°C)	10	30
Surface convection (W/m² K)	100	100
Thermal conductivity	1.9	1.0×10^{-6}
Density (kg/m³)	1490	1.0×10^{-6}
Specific heat (J/kg K)	1800	1.0×10^{-6}

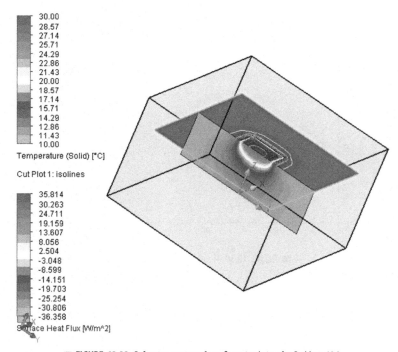

■ **FIGURE 12.32** Reference outcome from flow simulation for Problem 12.3.

12.4. In Example 12.7, if the uncertainties of two external loads should also be considered, i.e., $P_B = 2.0 \pm 0.10$ KN, and $P_C = 1.0 \pm 0.10$ KN. The probability bounds of stiffness coefficients for two segments are $k_1 = 2.00$ MN/m, and $k_2 = 1.00$ MN/m. Quantify the uncertainty of the displacement at C using finite element modeling.

Index